EURASIAN ENVIRONMENTS

PITT SERIES in RUSSIAN and EAST EUROPEAN STUDIES

Jonathan Harris, EDITOR

EURASIAN ENVIRONMENTS

NATURE and ECOLOGY in IMPERIAL RUSSIAN and SOVIET HISTORY

EDITED BY
NICHOLAS B. BREYFOGLE

UNIVERSITY OF PITTSBURGH PRESS

Published by the University of Pittsburgh Press, Pittsburgh, Pa., 15260

Manufactured in the United States of America
Printed on acid-free paper
10 9 8 7 6 5 4 3 2 1

Cataloging-in-Publication data is available from the Library of Congress

ISBN 13: 978-0-8229-6563-3
ISBN 10: 0-8229-6563-1

Cover photo: Nicholas B. Breyfogle
Cover design: Joel W. Coggins

FOR THE FLORA, FAUNA, WATERS, AIRS, AND LANDS OF EURASIA PAST, PRESENT, AND FUTURE

Nature has no such thing as bad weather.

—"A Song about Weather" (Andrei Petrov and Eldar Ryazanov)

CONTENTS

PART III. LAND, ROCKS, SOIL

PART IV. FRUITS of the WATERS

PART V. BODIES and DISEASE, HEALTH and ENVIRONMENT

PREFACE and ACKNOWLEDGMENTS

THIS VOLUME began at the conference "Eurasian Environments: Nature and Ecology in Eurasian History," at The Ohio State University, September 16–17, 2011. The conference, which formed part of the annual meeting of the Central Eurasian Studies Society, brought together an international group of scholars working on issues of tsarist Russian and Soviet environmental history and policy. Papers explored questions of water use and preservation, dam building and irrigation, the environment and public health issues, conservation versus economic development, species extinction and protection, the role of science and scientists in environmental destruction and protection, the sacred aspects of Eurasian environments, and the current realities and policies in the newly independent states of the post-Soviet space. Geographically, the conference focused particularly on the five former Soviet republics of Central Asia; on Siberia, the Russian Far East, and the Arctic regions; and on St. Petersburg, the Crimea and Black Sea regions, and the Eurasian steppe. Throughout, the history of Russia and the Soviet Union was placed in a global, comparative context, and discussions at the conference explored the contemporary implications of tsarist and Soviet environmental history questions. In addition to those whose essays appear in this volume, let me also thank the other participants who made that conference so very productive and exciting: Brian Bonhomme, Megan Dean, Randall Dills, David Hoffmann, Paul Josephson, Scott Levi, Erika Monahan, Chris Otter, Peter Perdue, and Maya Peterson.

It is with pleasure that I thank the many people and organizations that have supported, funded, read chapters, edited, offered moral support, and otherwise helped to bring this book to fruition. For their endlessly helpful editorial contributions, I thank John McNeill, the anonymous readers of the manuscript, Mark Sokolsky, Sarah Cameron, Julia Obertreis, Sarah Siff, and Sabauon Nasseri. I am deeply thankful to Peter Kracht and Andrzej Kaminski for inviting me to present the manuscript of this volume at the conference "Recovering Forgotten History: The Image of East-Central Europe in English-Language Academic Textbooks," in Warsaw, Poland, in June 2015. All the essays benefited from the probing commentary and suggestions of Andrzej Nowak and the other participants. I only wish I had been able to address more fully their appraisal that the volume omits much of the western borderlands (Polish territories, the Baltic region, and much of what is now Belarus and Ukraine)—it was not for lack of trying to solicit essays from scholars working in those geographic areas. It was a tremendous pleasure to work with Peter Kracht and the entire staff at the University of Pittsburgh Press. I thank Bill Nelson for the excellent maps. Generous funding for the conference and the book was provided by the Social Science Research Council and many organizations at The Ohio State University: the Mershon Center, the Center for Slavic and East European Studies, the Middle East Studies Center, the East Asian Studies Center, the Office of International Affairs, the College of Arts and Humanities, and the Department of History. Last but not least, my deepest love and thanks to Jillian, Charlie, and Sam—the hearts and souls of the ecosystem I call home.

Unless otherwise noted, translations in each article are by the author. Two articles are revised versions of essays that have appeared previously and I am thankful for permission to publish revised versions here. An earlier version of Stephen Brain's article appeared as "The Christian Environmental Ethic of the Russian Pomor," *Ecozon@* 2, no. 2 (Autumn 2011): 60–82. Parts of Julia Obertreis's chapter appear in her book *Imperial Desert Dreams: Cotton Growing and Irrigation in Central Asia, 1860—1991* (V&R Unipress, 2017).

ACRONYMS

UNIVERSITIES, INSTITUTES, AND SCIENTIFIC ORGANIZATIONS

AN SSSR — Akademiia nauk SSSR (Academy of Sciences of the USSR)

AOIRS — Arkhangel'skoe obshchestvo izucheniia Russkogo severa (Arkhangel'sk Society for the Study of the Russian North)

DVO RAN — Dal'nevostochnoe otdelenie Rossiiskoi akademii nauk (Far Eastern Branch of the Russian Academy of Sciences)

ICES — International Council for the Exploration of the Sea

IRGO — Imperatorskoe Russkoe geograficheskoe obschestvo (Imperial Russian Geographic Society)

KIVM — Komissiia po izucheniiu vechnoi merzloty (Commission for the Study of Permafrost)

OGAU — Orenburgskii gosudarstvennyi agrarnyi universitet (Orenburg State Agricultural University)

OIAK — Obshchestvo izucheniia Amurskogo kraia (Society for the Study of the Amur Region)

RAN — Rossiiskaia akademiia nauk (Russian Academy of Sciences)

SO RAN — Sibirskoe otdelenie Rossiiskoi akademiia nauk (Siberian Section of the Russian Academy of Sciences)

TONS — Tikhookeanskaia nauchno-promyslovaia stantsiia (Pacific Scientific-Industrial Station)

UrO RAN — Ural'skoe otdelenie Rossiiskoi akademii nauk (Ural Division of the Russian Academy of Sciences)

VASKhNIL — Vsesoiuznaia akademiia sel'skokhoziaistvennykh nauk imeni V. I. Lenina (All-Union Academy of Agricultural Sciences of the Soviet Union)

VNIIGMI-MTsD — Vserossiiskii nauchno-issledovatel'skii institut gidrometeorologicheskoi informatsii—Mirovoi tsentr dannykh (All-Union Scientific Research Institute for Hydrometeorological Information)

VNIICXM — Vserossiiskii nauchno-issledovatel'nyi institut sel'sko-khoziaistvennoi meteorologii (All-Russian Scientific Research Institute for Agro-meteorology)

GEOGRAPHICAL ACRONYMS

ASSR	Avtonomnaia Sovetskaia Sotsialisticheskaia Respublika (Autonomous Soviet Socialist Republic)
DVR	Dal'nevostochnaia Respublika (The Far Eastern Republic)
KKASSR	Karakalpakskaia Avtonomnaia Sovetskaia Sotsialisticheskaia Respublika (Karakalpakstan Autonomous Soviet Socialist Republic)
NAO	North Atlantic Oscillation
RSFSR	Rossiiskaia Sovetskaia Federativnaia Sotsialisticheskaia Respublika (Russian Soviet Federative Socialist Republic)
TASSR	Tadzhikskaia Avtonomnaia Sovetskaia Sotsialisticheskaia Respublika (Tajik Autonomous Soviet Socialist Republic)
USSR/SSSR	Soiuz Sovetskikh Sotsialisticheskikh Respublik (Union of Soviet Socialist Republics)
SSR	Sovetskaia Sotsialisticheskaia Respublika (Soviet Socialist Republic)
Uzbek SSR	Uzbekskaia Sovetskaia Sotsialisticheskaia Respublika (Uzbek Soviet Socialist Republic)
VL	Virgin Lands

MINISTRIES, OTHER GOVERNMENT BODIES, AND INTERNATIONAL ORGANIZATIONS

Dal'ryba	Dal'nevostochnoe upravlenie rybolovstva i okhoty (The Far Eastern Fishing and Hunting Administration)
Gosplan	Gosudarstvennyi planovyi komitet (State Planning Commission)
GUZZ	Gosudarstvennoe upravelnie zemledeliia i zemleustroistva (Department of Agriculture and Land Management)
IWC	International Whaling Commission
Khlopkom	Khlopkovyi komitet (State Cotton Committee)
Komsomol	Kommunisticheskii soyuz molodezhi (Communist Party Youth League)
MGIZ	Ministerstvo gosudarstvennogo imushchestva i zemledeliia (Ministry of State Domains and Agriculture)
Minvodkhoz	Ministerstvo melioratsii i vodnogo khozaiastva (Ministry of Land Reclamation and Water Management of the USSR)
Narkomzdrav	Narodnyi komissariat zdravookhraneniia (People's Commissariat of Health)
Narkomzem	Narodnyi komissariat zemledeliia (People's Commissariat of Agriculture)
NKTP	Narodnyi komissariat tiazheloi promyshlennosti (People's Commissariat of Heavy Industry)
RFWC	Rossiisko-Finliandskaia kitolovnaia kompaniia (Russian-Finnish Whaling Company)
SNK SSSR/ Sovnarkom	Soviet narodnykh kommissarov (Council of People's Commissars)
TsK	Tsentral'nyi komitet Kommunisticheskoi partii (Central Committee of the Communist Party). Between 1925 and 1952, its Russian acronym was TsK VKP(b) (Central Committee of the All-Union Communist Party of the Soviet Union [Bolsheviks]); from 1952 to 1991, its Russian acronym was TsK KPSS (Central Committee of the Communist Party of the Soviet Union)
UVKh Kazakhstana	Upravlenie vodnogo khoziaistva Kazakhstana (Water Management Administration of Kazakhstan)
VLC	Virgin Lands Campaign
VSNKh	Vysshii sovet narodnogo khoziaystva (Supreme Soviet of the National Economy)

ARCHIVE and BIBLIOGRAPHIC ACRONYMS and ABBREVIATIONS

ARCHIVES

Russian Federation

ARAN	Arkhiv Rossiiskoi Akademii Nauk (Archive of the Russian Academy of Sciences)
GAAO	Gosudarstvennyi arkhiv Arkhangel'skoi oblasti (State Archive of Arkhangel'sk oblast)
GAAO-OSPI	Gosudarstvennyi arkhiv Arkhangel'skoi oblasti, Otdel sotsial'no-politicheskoi istorii (State Archive of Arkhangel'sk oblast, Social-Political History Section)
GAMO	Gosudarstvennyi arkhiv Murmanskoi oblasti (State Archive of Murmansk oblast)
GAPK	Gosudarstvennyi arkhiv Primorskogo kraia (State Archive of Primorskii krai)
GARF	Gosudarstvennyi arkhiv Rossiiskoi Federatsii (State Archive of the Russian Federation)
GARO	Gosudarstvennyi arkhiv Rostovskoi oblasti (State Archive of Rostov oblast)
MROKM	Mordovskii respublikanskii ob"edinennyi kraevedcheskii muzei (Unified Local Studies Museum of the Mordvin Republic)
RGAE	Rossiiskii gosudarstvennyi arkhiv ekonomiki (Russian State Archive of the Economy)
RGASPI	Rossiiskii gosudarstvennyi arkhiv sotsial'no-politicheskoi istorii (Russian State Archive of Social and Political History)
RGA VMF	Rossiiskii gosudarstvennyi arkhiv voenno-morskogo flota (Russian State Naval Archive)
RGIA	Rossiiskii gosudarstvennyi istoricheskii arkhiv (Russian State Historical Archive)
RGIA DV	Rossiiskii gosudarstvennyi istoricheskii arkhiv Dal'nego Vostoka (Russian State Historical Archive of the Far East)
RGVIA	Rossiiskii gosudarstvennyi voenno-istoricheskii arkhiv (Russian State Military Historical Archive)

Other Archives

BMDJT	Boygonii Markazii Dawlatii Jumhurii Tojikiston (Central State Archive of the Republic of Tajikistan)
ORMDA	O'zbekiston respublikasi markaziy davlat arxivi (State Archive of the Republic of Uzbekistan). See also TsGARUz
QROMA	Qaraqalpaqstan Respublikasi Orayliq Ma'mleketlik Arxivi (State Archive of the Autonomous Republic of Karakalpakstan)
TsGARK	Tsentral'nyi gosudarstvennyi arkhiv Respubliki Kazakhstan (Central State Archive of the Republic of Kazakhstan)
TsGARUz	Tsentral'nyi gosudarstvennyi arkhiv Respubliki Uzbekistan (Central State Archive of the Republic of Uzbekistan). See also ORMDA.

JOURNALS AND COLLECTED WORKS

AHR	*American Historical Review*
KSG	*Kirgizskaia stepnaia gazeta* (Kazakh Steppe newspaper)
ONU	*Obshchestvennye nauki Uzbekistana* (Social sciences of Uzbekistan)
PSZ	*Polnoe sobranie zakonov Rossiiskoi Imperii* (Complete collection of the laws of the Russian Empire)
SKhT	*Sel'skoe khoziaistvo Turkmenistana* (Agriculture in Turkmenistan)
SKhU	*Sel'skoe khoziaistvo Uzbekistana* (Agriculture in Uzbekistan)
ZV	*Zvezda Vostoka* (Star of the East)

ARCHIVAL AND BIBLIOGRAPHIC ABBREVIATIONS

col.	collection
d.	delo (file)
f.	fond (collection)
Izd-vo	Press (*Izdatel'stvo*)
l., ll.	list, listy (page, pages)
no.	number
ob.	oborot (reverse)
of.	osnovnoi fond
op.	opis' (inventory)
p.	page(s)
pagn.	pagination
t.	tom (volume)
Tip.	Printing house (*Tipografiia*)
vol.	volume

EURASIAN ENVIRONMENTS

1 Toward an Environmental History of Tsarist Russia and the Soviet Union

Nicholas B. Breyfogle

> Nature is not a temple, but a workshop, and man's the workman in it.
>
> —Ivan Turgenev, *Fathers and Sons*

> "Man is king of nature," Andrei prompted.
> "Yes, yes, king. Just reign a bit and you'll be sorry."
>
> —Valentin Rasputin, *Farewell to Matyora*

> "Isn't nature incredible?"
> "To me, nature is . . . I dunno, spiders and bugs. And big fish eating little fish. And plants eating plants and animals eating. . . . It's like an enormous restaurant. That's the way I see it."
>
> —Woody Allen, from the film *Love and Death*

> It's their place, Mac, they have a right to make what they can of it. You can't eat scenery.
>
> —Victor, a Soviet fisherman, from the film *Local Hero*

The essays in this book explore the rich and remarkable environmental history of tsarist Russia and the Soviet Union over the last three hundred years. And, as is often the case in writing the history of Russia, I begin with the insights of literature. The tale of two dams, built about a century and a half apart, offers a window onto the evolving ways in which human political-social-cultural structures were intricately interwoven with ecological, hydrological, and geological systems. In his autobiographical *Family Chronicle*, the famous Slavophile S. T. Aksakov tells a triumphant story of his grandfather, the nobleman Stephan Mikhailovich Bagrov, who set up a family farmstead in Ufa province in the eighteenth century—ousting the previous pastoral Bashkir inhabitants from their land through "savvy" imperialist trickery. In Aksakov's narrative of imperial control, his grandfather was engaged in a great Russian triumph over the natural world and the unleashing of the land's agricultural treasures:

> Dear Land! With Gifts beyond all measure
> By Nature's kindly hand bestown—
> henceforth thy plains' and pastures' treasure
> Not for the herdsman bloom alone![1]

To make the fields bloom and fill the bellies of his family and the hundreds of Russian serfs whom he uprooted and brought with him, Bagrov built a water mill "to imprison the wildly-rushing stream in its appointed basin." For this anthropogenic hydrological transformation, "four mighty oaken posts had been driven firmly into the clayey bed of the Buguruslan [river]," and then more than a hundred men, women, and children threw in bundles of sticks, wood, and straw weighted with stones, dung, clay, and sod. Much was carried away by the river, but more and more caught on the posts and on the detritus dumped in. The work was done feverishly, loudly, and with great excitement and anticipation. "It was no light task to restrain the impetuous stream," according to Aksakov. "But victory was on the side of Man. No longer could the water escape through the stout barrier. The flood paused, as if in doubt; and filling the whole river gorge, rose over the banks, overflowed into the meadows, and by evening lay—a wide and shining lake—unconfined by bank or hedge, and dotted here and there by upstanding clumps of trees."[2] Sometime after the mill was completed, Bagrov took his family to the new calm reservoir for an afternoon fishing party and to inspect the mill's operation with his daughter. As a man (and as a man who "was very expert in all branches of country industry"), Bagrov ruled not just the social and familial roost. He controlled and dictated the structures of nature. As paterfamilias, escorting his subservient daughter around the dam, he was also *pater naturalis mundi*.

More than 150 years later, the celebrated Village Prose author Valentin Rasputin published the evocative, haunting novel *Farewell to Matyora*.[3] Rasputin offers a much more ambivalent view of damming (in his case of the Angara River in eastern Siberia) to create the massive Bratsk hydroelectric complex, which, at the time of its completion in 1967, was the largest such dam in the world. *Farewell to Matyora* is a fictionalized memoir based on Rasputin's own experiences growing up in a village on the Angara that was flooded, displacing him and his family. He explores the slow painful process of removing the people and preparing the ground for the gigantic industrial project. Journalists, poets, and Communist leaders—including no less than Fidel Castro—heralded the Bratsk dam as one of the great triumphs of Soviet socialism. They trumpeted the industrial development of Siberia that would power vast new industries with hydroelectricity, rebuild the USSR after the devastation of

World War II, and propel the people of communism to their utopian future.[4]

Yet the road to the future was paved with the lives and villages of those in the flood zone. Rasputin tells his story from the perspective of the few remaining villagers of Matyora—mostly very old women—who refuse to move from their island home in the middle of the Angara despite the impending rise in water. They lament the coming destruction of their village and the final rupture that the damming will trigger in the centuries-old seasonal lifeways of agricultural planting, care, and harvesting—of the definitive breach in the human-land relationship that had given form and meaning to their lives for generations. Their reluctance to embrace the dam was in some cases made worse by the fact that—in a version of what Mark Carey calls "disaster economics"—the authorities used the hydroelectric flooding to push not only technological and economic but also cultural and social modernization. Officials tried to move the villagers to ostensibly more modern, happier, and productive lives in rationally planned towns and to shift their ways of making a living from agricultural or hunting/fishing ventures to mechanical production or service activities.[5]

Rasputin uses the story of the hydroelectric drowning of the village of Matyora to speak to Soviet humanity's place in larger planetary systems (especially the technologically based transformative projects of the Soviet system). He explores the tensions among large-scale engineering projects, the aspirations of human communities, and the lands and waters they inhabit. In particular, Rasputin challenges the belief, common in the Soviet Union and throughout the world at the time, that humans can and should control and reformulate the natural world so as to better humanity and nature itself. In one of many scenes of generational conflict, the grandmother berates her grandson:

> "Why are you behaving like this? Does this land belong to you alone? We're all here today and gone tomorrow. We're all like migratory birds. This land belongs to everyone—those who were here before us and those who will come after. We're only on it for a tiny time. . . . And we were given Matyora only to take care of. . . . To treat it well and be fed by it. And what have you done with it? Your elders entrusted you with it so that you would spend your life on it and pass it on to the younger ones. And they'll come asking for it. . . ."
>
> "Man is king of nature," Andrei prompted.
>
> "Yes, yes, king. Just reign a bit and you'll be sorry."[6]

Aksakov's and Rasputin's stories of dam building reflect certain continuities in the human relationship with the natural world in the past three hundred years. They underscore the profound, inextricable interconnections of human economy, gender relations, familial networks, imperial structures, power hierarchies, culture, and society with the physical environment and nonhuman living world. They are reminders that human experiences of nature are localized and spatially bound: these are local stories, embedded in local events. Readers can hear the water flowing past and feel the soil under their boots and dirt under their fingernails. Yet, in both cases, the manipulation of rivers was directly tied to larger human and planetary processes: empire-building, forcible efforts to "modernize" existing lifeways, building communism, and technological transformations and electrification, among others. The stories of river control resulted from decisions made by leaders hundreds of kilometers away and the imposition of Russian power on ethno-religious communities across Europe and Asia. Both stories are saturated with Russian nationalism and imperial conquest. Yet, the indigenous peoples—Bashkirs and Buriats—remain silent and voiceless while the waters and lands around them are altered.

The two stories also highlight change over time. Aksakov's dam built of wood, mud, and dung, designed to power a mill for grinding grain, is but a quaint forebear of the massive concrete Bratsk dam, with its churning turbines and surging electricity. The change in scale, scope, and technology was accompanied by an evolution in the human response to the dams—shifting from the unmitigated enthusiasm of Aksakov and his deeply held views of patriarchal progress to Rasputin's much more critical understanding of the Bratsk dam's impact on humans and the natural world. Although most Soviet observers held views even more triumphalist than those of Aksakov, by Rasputin's time decades of "reigning" over nature had raised fundamental doubts about the wisdom of conquest for a growing group of concerned Soviet citizens.

If Aksakov and Rasputin provide two literary entrées into the fertile world of Russian environmental history, the essays in this volume offer an in-depth overview of the human-nature relationship in the tsarist/Soviet past and showcase the wave of field-changing research currently being written on tsarist/Soviet environmental history.[7] Through a series of interconnected and reinforcing chapters, the international group of contributors ask us to rethink our understandings of tsarist/Soviet history by placing the human experience within the larger environmental context of flora, fauna, geology, and climate. A pressing contemporary crisis in Eurasia is the Soviet Union's legacy of environmental degradation.[8] Yet we are only beginning to understand how the peoples of the tsarist and Soviet empires viewed or utilized the "natural" world historically, or how the experience of recent pollution and physical manipulation fits into long-term patterns. By exploring the intersections among humans and the ecologies, landscapes, and waters of Imperial Russia and the Soviet Union, the authors in this book (1) contextualize the ecological traumas of the past two-plus centuries, (2) analyze the relationships found at the nexus of the region's diverse peoples and environments, and (3) discuss the changing understandings of "nature" and the development of conservation efforts.

Geographically, the authors focus on the lands and waters of the former tsarist and Soviet empires, including the five former Soviet republics of Central Asia; the steppe lands of what are now Ukraine, southern Russia, and Kazakhstan; the Black Sea region; Siberia, the North Pacific, and the Russian Far East; and the Russian North and Arctic regions. They examine a wide range of natural phenomena and natural "resources:" from rocks, soil, and sand to fish and whales, trees and plants, rivers and waters, pathogens and disease vectors, ice, snow, and permafrost and from forests to steppe grasslands, deserts, oceans, and seas. Throughout, they place the history of Eurasia in a trans-chronological, comparative context, seamlessly linking the local and the global, rooting the chapters in the ecological and geological specificities of place and community while unveiling the cross-planetary patterns of human-nature relationships.

Although historians of Russia have for generations highlighted the importance of environment, climate, and geography to the patterns of Russian history, it is only in recent years that insights and aspirations of the field of environmental history have begun to influence broader understandings of Russian and Soviet scholarship.[9] Environmental history has had precursors and "deep intellectual" roots (such as the Annales School), but it developed earliest and most fully in the historiography of the modern United States. From its roots in the 1970s, it has rapidly come to influence scholarship on other regions and eras of human and natural history, developing into a rich, self-conscious, diverse, and inclusive field.[10] Scholars of environmental history explore the intersections of the biological, ecological, geological, hydrological, seismological, climatic, and astrophysical, with all areas of human life, uniting the material and the cultural. They recognize that humans are deeply embedded in the natural world, transformed by the natural systems in which they live, just as humans simultaneously alter the environment through everything from breathing, growing food, and expelling bodily waste to building nuclear power plants, driving cars, and releasing chemicals and plastics into the world's waters.

Indeed, it is a foundational insight of environmental history that the human past is only rarely the history simply of humans. The human experience can only be understood when embedded in its larger contextual webs with other organisms; climate and weather; rocks, mountains, soil; water, lakes and rivers; and earthquakes and volcanoes.[11] This approach to the nexus of environmental and human history contrasts with earlier scholarly approaches that saw climate, geography, and environment as static (an unchanging stage on which humans acted) and also deterministic (at times crudely so).

The chapters in this volume reflect the inclusiveness and rich diversity of approaches found in the field of environmental history. They explore economic activities and productive capacities, from grain agriculture, nomadic pastoralism, hunting and fishing, forestry, and industry and industrialization. They delve into the place of water in human life, including drought, floods, and the hopes and perils of water engineering projects (from irrigation to hydroelectric dams). The authors analyze the reinforcing interrelations of bodily health, disease, medicine, and environment; the cultural and religious understandings of nature; and the growth of scientific-technological knowledge and the expert role of physicians and engineers in environmental transformations. They also interrogate the history of disasters ("natural" and human); waste, pollution, and toxicity; conceptions and effects of climate change; behaviors of risk and resilience; human-animal relations; the built environment; and nature protection, preservation, and conservation.[12]

All the chapters are engaged with questions of the intersection of nature with human political power, geopolitics, and colonialism and the ways in which different political systems lead to differentiated environmental outcomes. They interrogate the work of Karl Wittfogel on the link between hydrology, irrigation, and autocratic/despotic systems; James Scott on modernity, technology, and interventionist states; and Donald Worster and others on the relationship of capitalism and the environment.[13] The authors explore the marshaling of environmental knowledge to ensure more efficient or wealth-producing exploitation of natural resources as well as to facilitate conservation and protection efforts. In the process, they highlight the importance of "native" or "local" knowledge in the ever-evolving relationship of humans to the natural world.[14] The authors engage with a variety of different methodologies, from actor-network-theory, the new materialist history, evolutionary history, and niche theory. They thoughtfully examine the ways in which the natural world possesses various forms of "agency" and influence in the ideas and actions of humans.[15]

In their articles the contributors to this volume both challenge us to reconsider some of the prevailing understandings of Russian/Soviet/Eurasian history and to make a contribution to the larger framework of global environmental history. In particular, several authors offer new temporal markers of tsarist/Soviet history, underscoring the continuities across time and political regime. Here the October or Stalin revolutions, for instance, serve less as moments of foundational change than as readjustments in larger patterns of human behavior in the Eurasian ecological context.[16] The contributors also speak to the characteristics of the Soviet political system, underscoring how environmental factors ensured that improvisation and adaptation overshadowed plan and ideology as the favored ways to mitigate risk and to ensure socioeconomic resilience. Throughout their chapters, the authors are consciously and assertively comparative, tending to "de-exceptionalize" the tsarist/Soviet case. They find that the patterns and processes at the core of tsarist/Soviet environmental history match very closely other parts of the globe during these periods.[17]

THE IMPORTANCE OF RUSSIA/SOVIET UNION TO THE FIELD OF ENVIRONMENTAL HISTORY

The tendency of scholarship on the tsarist and especially Soviet environment has been to focus on "ecocide," "prometheanism," "gigantomania," pollution, and the self-defeating destruction of nature as the primary—essentialist—characteristics of the human-nature relationship. What, then, can we learn from Russia, other than a cautionary tale? The answer is "a great deal"—and in seven areas in particular: (1) geographic, ecological, and cultural diversity; (2) the generation of environmentally derived, locally specific scientific knowledge; (3) explorations of the political-environmental relationship; (4) agriculture and food production; (5) the linkages of industrialization, technology, and environment; (6) the ecologies of empire-building; and (7) comparative discussions of the variety of conservation and preservation practices. To make these points is not to deny the wide-ranging and life-damaging environmental contamination unleashed by the tsarist and Soviet systems. As Stephen Brain has noted, "The historical record shows that the Soviet Union pursued simultaneously, from the outset, astonishingly progressive and cataclysmically destructive policies, and continued to do so until its fall in 1991." Soviet citizens were exposed to unhealthy levels of pollution, radiation, and toxins through water, air, soil, food, and their built environments. There were multiple nuclear disasters, of which Chernobyl is only the most well-known. Industrial, agricultural, and domestic wastes were rarely disposed of in sustainable healthy ways, leaving behind a poisoned landscape, at times barely habitable. But, as the chapters here illuminate, there is more to the Eurasian environmental story than a focus on "ecocide" alone tells.[18]

The first point of importance is Eurasia's geographic, climatic, and ecological heterogeneity. Stretching across as much as one-sixth of the world's surface, the lands that made up the tsarist and Soviet empires offer exceptional laboratories for studying the many varieties of human-nature relationship. These territories include examples of a wide range of biomes: ice sheets and polar desert, tundra, the vast boreal forests of the taiga, temperate broadleaf forests, forest steppe, arid steppe, and semiarid to arid desert, to name the most prominent (see map 1.1). And within these biomes come five of the longest seventeen rivers in the world (and three of the top ten), as well as the planet's oldest lake, the largest by volume of water and depth. This is not to forget the waters and coasts of the Baltic, Black, Caspian, Barents, and White Seas and the Arctic and Pacific Oceans, with their teeming life and their influence on both local and global weather. There are large swaths of permafrost, the mountains of the Caucasus, Urals, Altai, Verkhoyansk, and Yablonovyi ranges, and the magnificent volcanoes of Kamchatka. Throughout, the lands, waters, and airs of Eur-

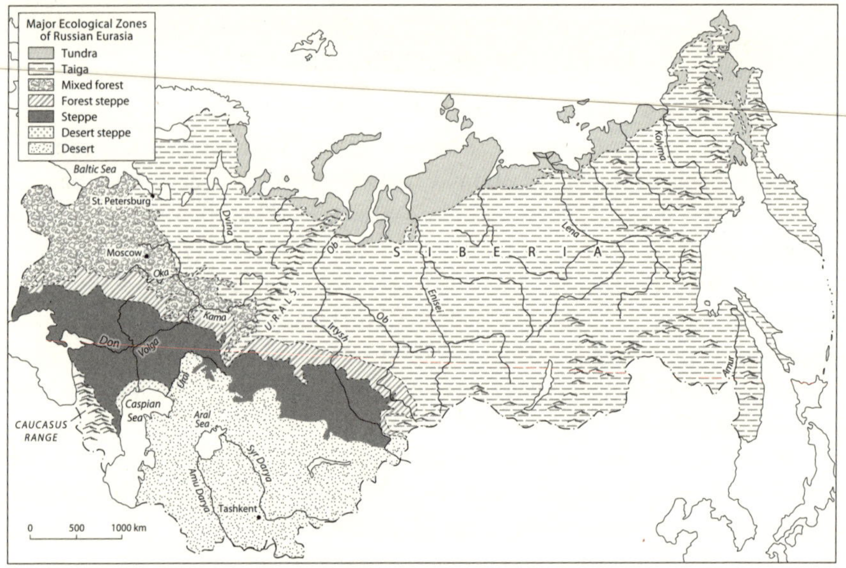

Map 1.1. Major Ecological Zones of Russian Eurasia.

asia offer habitat to some of the most diverse and fascinating flora and fauna on the planet, with hundreds of endemic species.[19] This ecological multiplicity was (and is) matched by a human diversity in terms of religion, culture, language, lifeways, familial structures, economic practices, and social patterns. The ecological complexities and characteristics of Eurasia offered both opportunities and constraints for its myriad peoples.

Taken together, the sheer size and unparalleled diversity of environments of the former tsarist/Soviet empires—and especially the wide swaths of cold and frozen climates—offer new perspectives on global environmental history.[20] The unification of the massive Eurasian region within the tsarist and Soviet states created an integrated ecological-political-economic zone in which the demands of certain regions transformed the ecologies of others. Animals, plants, materials, and humans were extracted from one location and transported, traded, consumed, and then discarded in others. The desire for furs by tsars and elites led ultimately to the massive killing of sable and sea otters and to the extinction of Steller's Sea Cow. The demand for fossil fuels, various minerals (nickel, uranium, iron, manganese, asbestos, among so many others), and cash crops like cotton fundamentally remade local ecologies in such diverse locales as Noril'sk, Khanty-Mansiisk, the Urals, the Kola Peninsula, Central Asia, and the Arctic region.[21] At the same time, the integration of these lands under tsarist/Soviet control allowed for the movement of peoples, commodities, flora and fauna, pathogens and disease vectors, and humans across large landmasses and waters, binding Europe, Asia, and North America. Pathogens, in particular, found their way through the interconnected Eurasian landmass, from smallpox decimating native Siberians to cholera moving from South Asia through the Caspian Sea and the Volga and on to other parts of Europe (not to mention plague, HIV, and tuberculosis).[22]

The second point of importance is that a dynamic, world-changing process of scientific discovery and technological enhancement was an outcome of this environmental diversity. Geographer David Livingstone persuasively argues that the specificities of place matter in the creation of science; that "issues of space—location, place, site, migration, region—are at the heart of scientific endeavor."[23] Indeed, the essays in this volume underscore that we cannot fully understand the development of scientific, technological, and medical/health ideas and techniques outside of the local Eurasian environmental context in which humans produced them.[24] Pey-Yi Chu describes how the study of permafrost began in tsarist Russia, was institutionalized as a scientific discipline in the early Soviet period, and then spread from the Soviet Union to other parts of the Northern world. Important concepts of sustainability and perceptions of climate change also developed in the Russian and Soviet contexts, as Marc Elie highlights in his chapter. George Lywood shows the ways in which the specificities of Crimea's local environment led physicians to develop a wide range of distinctive medical cures and therapies for those visiting health resorts. Moreover, David Moon's chapter highlights the staying power of locally produced environmental ideas: how cultural understandings about forests grew in one ecological location and then were transported and applied as German and Russian settlers made their way into the steppe. Mieka Erley and Julia Lajus both underscore that Western ideas of soil, risk, and fishing were reformulated in consequential ways as they came into the tsarist environmental context and into Russian scientific and intellectual circles. In this way, the study of Eurasian environmental history reminds us of the importance of Russian and Soviet contributions to global science—contributions that were all too often ignored, downplayed, or summarily dismissed during the Cold War years. Certain scientists—such as I. P. Pavlov, D. I. Mendeleev, V. V. Dokuchaev, and V. I. Vernadsky, to take a small list—have become household names for their scientific discoveries, but many others remain obscure in the global scientific canon despite their tremendous influences on scientific research and conceptual understanding.[25]

Third, the study of Eurasia from an environmental perspective allows us to consider more fully how different political-economic structures—whether monarchical, liberal democratic, theocratic, familial, authoritarian, socialist/communist, or capitalist—affect the human-nature nexus. Despite the dramatic shifts in political structures over the past three hundred years, the authors in this book generally underscore the continuities and similarities in human-environmental relations across modern political and economic systems regardless of location or governing structure. Moreover, they offer important insight into the links between environmental consciousness and ecological movements on one hand and the development of an active public sphere, civil society, and professional networks of expertise on the other hand, as Julia Obertreis, Lajus, and Erley discuss.[26]

Fourth, the tsarist and Soviet experiences expand our understandings of agroecological history, which has been at the foundation of the environmental history field since its inception. As well as providing essential calories to sustain life, the human-nature relationship embedded in agriculture has been at the core of state formation processes, power structures, and social hierarchy in the Holocene.[27] The story of tsarist/Soviet agriculture sheds light on the socioenvironmental changes associated with the (often forced) shift in land-use patterns and social systems from nomadic pastoralism to settled, grain agriculture, as Sarah Cameron, Ian Campbell, Elie, and Moon emphasize in this volume. Furthermore, Christian Teichmann and Obertreis speak to the ethno-hydrological complexities of irrigated agriculture in the twentieth century—a topic that has important resonance in irrigated lands from the American West to Australia and the Middle East.[28]

In addition, the tsarist/Soviet cases are characterized by repeated episodes of famine that were often catastrophic for the people involved and served as turning points in their relationship with land, soil, and agricultural practice. These included the famines of 1891, civil war, collectivization, and the Virgin Lands experiments, to name but some of the most notorious. The study of tsarist and Soviet famines speaks to the question of causation—was it climate variations or human land use patterns (or some mutually influential combination) that led to insufficient yields and famines? Elie (this volume) explores this question in depth.[29] Moreover, agrarian reform in late imperial Russia and the Soviet Union was characterized by efforts to make agriculture more efficient and productive—to "modernize" food production so as to ensure sufficient amounts of food in the right places and at the right times to keep the population (especially the urban population) fed. Such motives informed the transition away from serfdom and bonded agriculture, the Stolypin reforms, and Soviet collectivization. Soviet approaches to agriculture became a model of human land usage for other modernizing communist states in the twentieth century—both to emulate and to avoid.[30] The tsarist/Soviet cases also speak to the history of silviculture and fishing, as highlighted in the articles here by Moon, Mark Sokolsky, Stephen Brain, Lajus, and Ryan Jones.

Fifth, the experience of Russia and the Soviet Union with industrialization offers a lens on the diverse ways in which the rapid socio-technological changes of the "great transformation" affected different communities. Russia was not among the initial select group of northwestern European states to go through the great industrial technological shift to factory manufacturing and fossil-fuel energy sources. However, after playing catch-up, it quickly joined the ranks of these polities as one of the planet's earliest adopters. In the twentieth century, the USSR became an alternative model of development to the capitalist West.[31] How did culture and politics affect how humans experienced this shift ecologically? In what ways did innovative technologies, energy sources, engineering systems and ideologies, and structures of human-animal-technological ecologies transmute environments, climates, hydrologies, and geologies?[32] As the articles by Andy Bruno, Chu, Teichmann, Obertreis, Sokolsky, Lajus, and Jones show, industrialization brought new understandings and uses of nature—from rocks and geology, to hydrological and irrigation systems, the ethno-technological transformation of fishing, and the heroism of the whale hunt. The embrace of industry, Bruno and Obertreis remind us, introduced unprecedented pollutants and toxins into the environment and into human and animal bodies.[33] Similarly, Lisa Walker illustrates the linkages among the spread of malaria, disease control efforts, and Soviet modernization efforts in Tajikistan. Moreover, the Eurasian region is important to understanding the global transformations of human energy production and use, from wood and dung to fossil fuels, nuclear power, and hydroelectric dams.[34]

Sixth, the story of Eurasian environmental history also affects how we should understand the mutual relations of empire and environment—explored in all of the articles here. The environmental aspects of imperialism have been a tremendously rich vein of environmental history writing: from Alfred Crosby's *Columbian Exchange* to William Cronon's *Changes in the Land*, Richard Grove's *Green Imperialism*, and J. R. McNeill's *Mosquito Empires*, to name just a few. As one of the planet's great imperial entities in the nineteenth and twentieth centuries, the tsarist/Soviet cases offers important insights into the nexus of empire and ecology—especially as an empire that was simultaneously contiguous and overseas and that controlled vast and diverse landscapes, waterscapes, and peoples. The movement of Russians across Siberia and the Pacific generated a mass slaughter of fur-bearing mammals and tremendous, rapid, and uncomfortably visible change to animal and fish species. This ecological cleansing led, in turn, to fundamentally new scientific and cultural understandings of the human-nature relationship—a theme explored here by Jones, Sokolsky, Lajus, and Brain. Moreover, in the Russian/Soviet case, the story of imperial and ecological expansion spotlights the multiple ways in which empires exacerbated the ecological vulnerability of the native peoples for the benefit of the metropolitan leaders—increasing their risk and reducing their resiliency. It also illuminates how colonial settlers brought with them a portmanteau of ecological ideas and practices (as Moon, Campbell, Cameron, Teichmann, Obertreis, Sokolsky, Jones, Lajus, Brain, and Walker all make clear). The essays here make visible the linkages between empire-building and disease. For Walker, Soviet efforts to control Tajikistan were mediated through the Soviet encounter with malaria. Lywood, by contrast, illustrates how the colonized lands and waters of Crimea became a "coast of health" to cure metropolitan visitors. At the same time, the tsarist and Soviet states' multi-confessional character permits comparison of the ways in which Christians, Muslims, Jews, Buddhists, Shamanists, and other spiritual communities understood and explained the human-nature relationship. Brain, for example, demonstrates how the practices and beliefs of the Pomor Christian faith affected (and were in part determined by) their approaches to fishing.

Finally, the tsarist/Soviet case is especially important to comparative environmental history because of its distinctive conservation and preservation practices. Russia's *zapovednik* (nature preserve) structure developed in the tsarist era and then expanded rapidly and extensively in the Soviet years and has since been applied elsewhere in the world. In comparison to the multi-use national park system of the United States (usually held up as the archetype for nature protection globally), the zapovednik model represents a distinct Russian/Soviet contribution to global conservation efforts. These segregated, closed lands were to be used for both nature protection and scientific research: by separating out examples of a range of different ecologies, scientists could examine and understand the development of these natural communities absent human influence.[35] In addition to the celebrated zapovedniki, the authors here underscore that tsarist and Soviet ideas of conservation and preservation also developed from Christianity (Brain), geopolitical competition (Sokolsky and Jones), disappearing fish stocks (Lajus), imperial racism (Sokolsky), industrial development and pollution (Bruno and Obertreis), and tourism (Lywood).

THE IMPORTANCE OF ENVIRONMENTAL HISTORY TO RUSSIAN/SOVIET STUDIES

Through environmental history we are reminded that "human" history is inextricably linked to natural agents of change, both organic and inorganic. We explore the interrelationships of human and nonhuman worlds and discover that there is no meaningful division between humans and nature; that humanity and the nonhuman world exist in an ongoing web of mutual transformations and influences.[36] The insights of global environmental history have important lessons for Russian/Soviet historiography in four areas in particular: (1) historical causation, the relationship of structure and agency, and change over time; (2) chronology, timescales, and periodization; (3) space, place, and the relationship between local and global; and (4) the types and uses of sources and methodologies. The field is also productively comparative across time and place, allowing historians to situate their stories in a global human context (as all the authors here do explicitly). Moreover, environmental history promotes new categories of analysis to explore existing historical questions. Indeed, environmental historical approaches offer fresh ways to approach long-standing core questions of Russian/Soviet history: for instance, political power and the governmental characteristics of tsarist autocracy and Soviet communism, the long-standing and thorny question of civil society and public sphere, the practices and meanings of law and policing, questions of imperial expansion and interethnic contact, and the parameters and restrictions of agricultural and industrial development.

Environmental history advances a new understanding of causation—affording nature an active role in the making of history, as Bruno does (this volume). These new approaches to structure and agency should challenge historians of Eurasia to rethink core questions of the field. For example, John McNeill's recent study of the role of malaria and yellow fever in the Atlantic world reminds us that understanding just who won in the imperialist struggles in the Caribbean—and why—is less directly connected with military and naval strategy than has traditionally been understood. Instead, the fate of the region was also sealed by the ways humans created and fostered, however unwittingly, new breeding grounds and homes for mosquitoes and therefore malaria and yellow fever, which differentially affected European settlers and soldiers. These disease vectors transformed the political facts and military engagements of the Caribbean just as surely as any military leader or political decision. In her chapter on malaria in Tajikistan, Walker echoes these approaches. Similarly, the field of environmental history has also underscored the ways in which human technological-behavioral changes in the recent past have produced all sorts of new environments. For example, while refrigeration made possible the movement of food across vast distances without spoilage, it also opened up brand new environments for cold-loving "novel foodborne pathogens," especially E. coli O157 H7 and listeriosis, which have threatened human health. The environmental context in which humans exist is regularly changing—and, in no small measure, this is because of human action.[37]

Environmental history requires scholars to explore chronology and timescale from a different vantage point. "Big history" and deep climate history remind us that the modern moment, which so often consumes the attention of the majority of historians, is but one small (if consequential) blip in a much longer history of the *Homo sapiens* species and the earth. We must fully understand the patterns deep in the planet's history as we strive to understand what is new and what is old in the Holocene—and in the Anthropocene (the much discussed theory of a new geologic epoch defined by human transformations of the earth system) and the "great acceleration" (the rapid increase in human production and consumption after 1945, accompanied by massive population growth and environmental change).[38] At the same time, we are reminded that changes in the earth's geological, climate, and biological systems (let alone the larger structures of the universe) often take centuries and millennia. The human timescale remains limited and all too often focused on singular moments of human life rather than on the broader patterns of geological or climate change. Scholars need to be able to weave back and forth between the longue durée of planetary life and the short, contained events of a human life span. While Soviet history remains the most popular era to study in Eurasian history, environmental historians remind us that the human experience—and the webs and linkages of human-nature relations—goes ever so much farther back in time.[39]

In addition, environmental history helps historians to rethink periodization and to pinpoint the defining moments of human and planetary change. There are, for instance, many candidates for turning points in the history of the modern world—the French or American Revolutions, World War I, the Bolshevik Revolution and the rise of Communism globally, colonization and decolonization, among so many others. And yet, in environmental history we are asked to consider the role of other, nonpolitical events in defining the chronology of the Anthropocene: especially, the Neolithic shift to settled agriculture, the Little Ice Age, the industrial revolutions of the eighteenth and nineteenth centuries, and then more recently, the atomic detonation and the nuclear era and the "great acceleration." Here, environmental historians underscore the importance of avoiding a fixed focus on grand political events as the primary markers of human change.[40]

Eurasian history also benefits from the insights of environmental historians for what they tell us about the structures of place and space.[41] Much history written in recent years has been told through the lens of states and nation-states and often restricted to the boundaries and borders of such imagined geographic delineations. The "area studies" model in American universities, which divides the world into geopolitical-cultural blocks of study, institutionally reinforces this tendency. Environmental history offers fresh understandings of spatial and geographic interrelationships—and of the meaning of place to human communities. For instance, historians of water often utilize a watershed approach to mapping events and relations—organizing space through hydrological linkages to encompass interconnected water and terrestrial systems. In the mid-nineteenth century, the famed one-armed explorer of the American West, John Wesley Powell, submitted an extensive report to the US Congress arguing unsuccessfully that the West needed to be divided administratively around river systems and watersheds. He aspired to make one state out of all the lands and waters connected with the (say) Colorado or Columbia River and not simply drop straight lines through the dry western terrain. Politicians at the time ignored this catchment boundary-making approach, and the hydrological, political, and economic outcomes for the American West were dramatic. Indeed, the actual creation of political boundaries globally has generally ignored catchment connectivity in charting and dividing land and water. Rather than mapped as a connected space, bodies of water have been parceled out to different political administrations, thus creating geopolitical tensions around transboundary rivers.[42]

Last but not least, in the field of environmental history the historian discovers innovative professional tools and different types of sources, theoretical models, and methodologies. The study of the human-environment nexus merges the methodologies and approaches of the sciences with those of history. Here, for instance, biological understandings of the human species (from epigenetics to microbiomes and evolutionary and co-evolutionary patterns), ecological systems of flora and fauna, knowledge of disease spread and infection patterns, geological and seismological research, and explorations of climate shifts, all offer new methodologies and categories of analysis for understanding the human past. The study of the human-environment relationship has underscored the ways in which humans, animals, and the natural environment help to coproduce each other, and the work on evolutionary history has been especially fruitful here. Humans are intricately and inextricably embedded in the larger web of environmental relations. As humans struggle to exert some modicum of control over their surroundings, so too do the species that exist in relationship with them. Moreover, our increasing understanding of microbiomes underscores how human bodies are in fact multispecies bodies—where "nature" is both within and a part of humans—and humans are shaped physically and emotionally every day by these interspecies interactions. In addition, new sources such as sediment varves, ice cores, pollen samples, tree rings, and genomic analysis of long-buried bodies and bones are equally as important to reconceptualizing the past.[43] Finally, in environmental history new methodologies are brought to historical study, especially active fieldwork, historian-nature interactions, and historian-scientist interdisciplinary projects. Here, historians supplement the essential work in archives and libraries with additional research in situ, getting their boots muddy and their skin sunburned in the places they study. They interact with the ecologists, geologists, seismologists, biologists, and other scientists who work in these areas (both past and present) and examine their environment of study through the eyes of the people who live(d) in it.[44]

EURASIA AND CONTEMPORARY ENVIRONMENTAL CONCERNS

While this is a historical book, the collection speaks to a series of important contemporary strategic and climate issues. First, these case studies offer theoretical insight into the relationships among economic development, resource exploitation, and environmental degradation and management. Given the central importance of natural resources to the economic future of Russia and Central Asia and to their political clout on the world stage (such as Russia's wealth in and strategic use of oil and natural gas, and the importance of fossil fuels to Kazakhstan's economic growth), it is crucial to understand the historical patterns of environmental exploitation in Eurasia. In addition, the study of this region—covered as it is with tremendous amounts of forest and permafrost—is significant for our understandings of global climate change and the atmospheric balance of carbon dioxide and methane. Exploring the historical causes and outcomes of environmental pollution and toxicity helps us to determine the health threat to the population and the economic implications of cleanup today. Moreover, Russia has long had a presence in the Arctic regions, both via lands within the Arctic Circle and an Exclusive Economic Zone in the waters of the Arctic Ocean through the 1982 UN Law of the Sea. In the current "race for the Arctic," set off by melting Arctic ice cover, Russia is playing a lead role (with Canada close behind).[45] In addition, many of the chapters in this collection deal with questions of water. Water resources and the deterioration of water supplies will likely represent one of the most contentious problems of the twenty-first century. If the predictions are at all accurate, Russia (with more than 20 percent of all the surface freshwater on the planet) will perform a pivotal role in the future of the world's water. As global fish stocks dwindle, understanding past patterns of Russian/Soviet fishing will be important to scholars, policy makers, environmentalists, and fishing workers alike.

Second, the geopolitical relations of the former Soviet republics, both with each other and with their other neighbors, are in many respects determined by environmental factors such as struggles over resource usage and pollution. Here, too, a thorough examination of the environmental history of the region is instructive, as the articles by Sokolsky and Jones clearly reveal. Tajikistan and Uzbekistan's competition over scarce water resources for hydroelectric power, agriculture, and drinking sources have been intensifying for several years. China's seemingly unquenchable thirst for Russian natural resources—from timber to water, minerals, furs, oil, and gas—lies at the heart of the current relationship between the two countries. And the potential riches from these resources have led to high levels of illegal resource extraction and the rise of organized crime groups working across the Russo-Chinese border. One feature of post-Soviet Russian-Ukrainian tensions and violence was energy supplies and energy delivery structures. And the growing exploitation of Arctic resources has embroiled Russia in diplomatic struggles with Canada, the United States, Denmark, and Norway, which are now playing their way through the legal structures of the United Nations.

PART I

Steppe Environments

2 Planting Trees in Unsuitable Places

STEPPE FORESTRY IN THE RUSSIAN EMPIRE, 1696–1850

David Moon

Why, in the wake of the Russian annexation of the western Eurasian steppe, did the Russian imperial authorities and the settlers they sponsored insist on and persist in planting trees in the region in spite of the unsuitable environment? The Eurasian steppe was an immense, largely flat, semi-arid, windy grassland with alkaline subsoils, where few trees grew naturally outside river valleys and ravines. In seeking to explain the insistence on tree planting in such conditions, I will consider both cultural factors (trees and forests are central to Russian identity) and practical reasons—Russians and other Slavic peoples have used wood and timber for many everyday needs. Further, the settlers included Germanic peoples, for whom trees and forests have also been important. I contend that, when people from largely forested environments—the Russian heartland, today's northern Ukraine and German lands—moved onto the treeless steppe in large numbers from the turn of the eighteenth century onward, they took with them a preference for wooded landscapes and also their practices of using forest products to meet many practical needs. They were further influenced by contemporary arguments that forests had wider environmental benefits.

The focus of this essay is on the period from the end of the seventeenth to

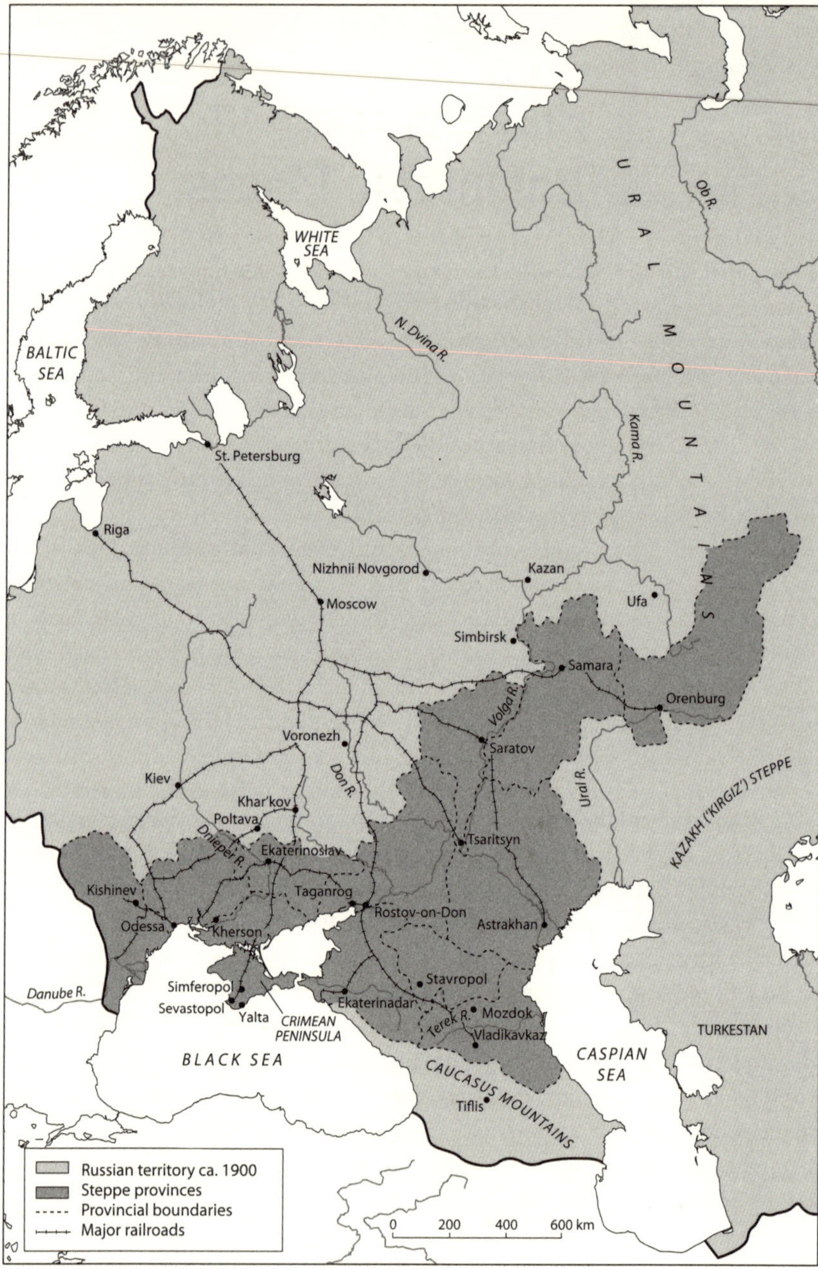

Map 2.1. The Steppe Region of European Russia, c. 1900.

the mid-nineteenth centuries, during which many attempts to cultivate trees on the steppe failed. In 1696, Tsar Peter the Great (r.1682–1725) reputedly ordered trees planted near Azov, which he had recently taken from the Ottoman Empire. It was only by the mid-nineteenth century, however, that forestry specialists had worked out how to grow trees on the steppe with some chance of success. The region under consideration is the open steppe across the south of present-day Ukraine and the Russian Federation to the west of the Ural Mountains and the Ural River that was annexed to the Russian Empire from the successor states of the Mongol Empire and the Ottoman Empire between the mid-sixteenth and late eighteenth centuries (see map 2.1). This was part of the partition of the vast Eurasian steppe, following the demise of the Mongol Empire, between the empires on its periphery: the Chinese in the east, the Persian and Ottoman in the south, the Habsburg in the west, and the Russian in the northwest. These empires employed various strategies to incorporate portions of steppe and bring them and their inhabitants under their control. These included ways of managing and exploiting the environment of the steppe and creating identities that unified the empires' domains.[1]

Many of the Slavic and Germanic settlers who moved onto the western steppe, and their descendants, seemed almost obsessed by the lack of trees. They saw it as a hindrance to their way of life and a deficiency that needed to be corrected. Shortly before his death, Viktor Skarzhinskii (1787–1861), a noble landowner in the largely treeless steppe province of Kherson in today's southern Ukraine, recalled a dream he had had a few decades earlier:

> I dreamed I was standing on the bank of great river, under the canopy of tall trees; after walking across a bridge . . . I found myself in the bowl of a dense forest, amid pleasant coolness; . . . birds fluttered among the . . . trees. . . . A few minutes earlier I had been breathing the sultry hot air of the steppe, and it was so pleasant to be in the cool of the forest, that I involuntarily said: if on this earth there is such a paradise, I could not hope for a better one. With these words I saw [my recently deceased uncle and brother], who assured me that indeed I was in paradise.

He exclaimed: "I, an inhabitant of the steppe and of a small orchard planted by my father, was in ecstasy from these wonderful places and involuntarily cried out: . . . what if I could cultivate such a forest on my estate in the steppe!"[2]

Skarzhinskii's affection for forested landscapes is notable, since he was a native of the steppe and spent much of his life there. There were a few forested interludes in Skarzhinskii's life. As a child he studied for two years in far-off St. Petersburg (which was surrounded by forests), but he suffered ill health and returned home to continue his studies. He went back to the northerly capital in 1805 for six years' government service. Skarzhinskii attributed his dream to the very beautiful gardens and parks he had seen abroad in 1813. From October 1812 to May 1813, as a patriotic subject of the Russian Empire, he saw service against Napoleon's Grande Armée with a cavalry squadron he had raised himself. In 1813 he took part in the siege of Toruń, in present-day Poland.[3] It is possible that he visited the local botanical garden. It had been founded in 1797 by Dr. Johann Gottlieb Schultz, who had created a beautiful garden in the romantic style with water features, bridges, arbors, an aviary, and trees and plants from all over Europe.[4] Schultz's park, therefore, resembled Skarzhinskii's dream.

The contrast between romantic botanical gardens in Polish lands and the forests of Russia's heartland, on the one hand, and the treeless steppe of Kherson province could hardly be starker. The reasons that it is so difficult for trees, whether seeded naturally or artificially, to thrive on the steppe were the subject of much discussion among naturalists, scientists, foresters, and landowners in the Russian Empire from the eighteenth century onward. Their discussions were motivated by the desire to overcome the difficulties so that trees could be grown successfully. Over time, as repeated attempts met with little success, they concluded that the climate and soils, which were ideal for grasses and to a lesser extent cultivated grain, hindered the growth of many species of trees. They also speculated that the steppe had once had much more extensive forest cover but that the previous inhabitants—nomadic pastoralists—had destroyed the forests with fire and by grazing their herds of livestock. The settlers compounded the shortage of woodland by felling the existing trees to provide timber and firewood and for other purposes.[5]

Ivan Palimpsestov (1818–1901) was a passionate believer that the steppes had once been forested, an opponent of continued felling, and a staunch advocate of planting trees. He was an Orthodox clergyman who left the church to concentrate on teaching agronomy in Odessa on the Black Sea coast of Kherson province. From 1853 Palimpsestov served as secretary of the Imperial Society for Agriculture in Southern Russia, which promoted agriculture and forestry on the steppe. Like Skarzhinskii, Palimpsestov was a native of the steppe, but this did not inhibit their passion for forests, which they seem to have inherited from their forebears who had moved to the steppe from the forest lands to the north.[6]

The Russian conquest of the steppe began in the mid-sixteenth century with the defeats of two of the successor states to the Mongol Empire: the Khanates of Kazan and Astrakhan on the middle and lower Volga River, which flowed south to the Caspian Sea. Russia's conquest of the western steppe was completed by Empress Catherine the Great (r. 1762–1796), who defeated the Ottoman Empire and annexed its vassal, the Khanate of Crimea, thus extending Russian power to the northern coast of the Black Sea. She named her new southern lands Novorossiia (New Russia), and this was later divided into Kherson, Ekaterinoslav, and Tauride provinces. Russian conquest was followed by waves of settlers, mostly farmers in search of more fertile lands. They moved from forested lands in central Russia, the north of today's Ukraine, and central and eastern Europe. The settlers included many Germans and also Mennonites (whose forebears originated in the Netherlands) from the Danzig (today's Gdańsk) region. The "foreign colonists" had responded to invitations to settle on advantageous terms by the Russian authorities, which hoped they would become model farmers and tree-planters.[7]

For millennia before Russian conquest and the influx of European agricultural settlers, the steppe had been the domain of successive nomadic empires, most recently the Mongol Empire, whose populations lived largely by herding livestock but also engaged in some crop cultivation. Their mostly pastoral way of life, both mobile and settled, was well adapted to the semi-arid grasslands.[8] During this period, for the Slavic peoples eking out a living in the forests to the north, the steppe had been largely an alien environment with a hostile population. The Mongols had conquered much of the Russian lands in the mid-thirteenth century; the Crimean Tatars rode north to burn Moscow for the last time in 1571; nomad raids deep into Russian territory to seize property and slaves continued into the eighteenth century. The Russians protected themselves by building fortified lines right across their steppe frontier and hiring Cossacks, who lived beyond the lines, as guards.[9]

DIFFERING CULTURAL PERSPECTIVES ON FOREST AND STEPPE

The steppe is abutted by different environments, including arid deserts to the south and southeast and humid forests to the north and northwest. Since ancient times, travelers have visited the steppe or crossed it to reach the regions beyond. Some travelers left descriptions of the landscapes they encountered, and most compared them implicitly or explicitly with the familiar landscapes of their homelands.[10] These accounts by people from various cultures and landscapes offer differing perspectives on the landscapes of the Eurasian steppe and the Russian forests. And they provide a wider context for Skarzhinskii's dream and the compulsion felt by Slavic and Germanic peoples to view the steppe as "treeless" and to plant it with trees.[11] In contrast, travelers from the south and east favored open landscapes and disliked the dense forests to the north. Thus, landscape preferences may be cultural, favoring the familiar.

In medieval times, Muslim geographers and historians described the lands to the north of the Caucasus Mountains and the Caspian and Black Seas, including the exotic forests of the Russian lands, which they contrasted with the landscapes of their homelands. In 921, Ahmad ibn Fadlan, an envoy of the Abbasid caliph Muqtadir, traveled from Baghdad, around the Caspian Sea, and thence north across the deserts and steppe to the Bulgar capital on the mid-Volga (near present-day Kazan). Among the people he encountered there were the Rus' (Slavs). En route, Ibn Fadlan described the cold, the "flat, desert-like steppe," and the nomadic peoples and their customs. Further north, he noted the short summer nights and early sunrise, which caused alarm among the Muslims lest they did not wake in time for dawn prayers. He also noted unfamiliar types of trees and "great forests filled with immense trees."[12] At around the same time, Ibn Rusta, a native of Isfahan in present-day Iran, wrote about the Khazars, whose empire encompassed much of the steppe. He was interested in their religion (Judaism), the Muslim minorities, and their annual raids on their Pecheneg neighbors. He wrote little about their "extensive" lands; perhaps the steppe was not worthy of attention as he was already familiar with flat, arid, treeless lands from his homeland. When he described the lands to the north, however, he was on unfamiliar territory. The "Saqaliba [Slavs]" lived "ten days' march from the lands of the Pechenegs, and to reach them one crosses steppe and trackless wilderness, with springs and thick forest. There are no vineyards or cultivated fields." He described their customs but especially the "extreme cold," which was "so harsh that the inhabitants are forced to construct underground dwellings, roofed with wood." For Muslim travelers from the south, the steppe was a familiar landscape, but the strange lands to the north had short nights in the summer and long nights during the cold winters and were covered with thick forests of enormous trees.[13]

The forests of the Russian lands also attracted comment by travelers from China. In 1712 the Chinese emperor sent an embassy, led by Tulisen, to the Kalmyks (the "Tourgouth Tatars"). The Kalmyks were a nomadic Mongol Buddhist people who lived on the steppe along the lower Volga in the Russian Empire. To reach them, the embassy traveled northeast across Mongolia to Siberia, through Siberia and eastern Russia to the Volga, thence downstream to the Kalmyk steppe. At the start of his journey Tulisen was not impressed by the Mongolian desert, but as they approached the Russian border, to the south of Udinskii (present-day Ulan-Ude), Tulisen noted "well-wooded vallies," "thickly wooded hills," "thickly wooded river banks," as well as timber buildings. Thereafter, his account is full of descriptions of the trees and forests. The route from Eniseisk "lay through thick forests." Along the Irtysh River, "the woods here are more close and thick than in other places." The banks of the Tobol' River were "so overgrown with wood that there was no tracking path." The Urals were "covered with trees," and on the other side, the "country was still covered with woods." This is just a selection of Tulisen's many descriptions of forests, which contain more than a hint that he found them both tiresome and oppressive. As they made their way south along the Viatka River toward Kazan, however, Tulisen exclaimed with evident relief: "The woods and deserts, we now saw no longer, except at a distance." The route south from Kazan toward Simbirsk and the steppe was: "over open and extensive plains. Here there are no trees, but a considerable portion of the surface is divided into fields and cultivated." The Chinese ambassador used more lyrical tones when the Kalmyk khan asked him to describe his own homeland. He spoke of high mountains, great rivers, sweet and excellent water, woods and other "extremely luxuriant" vegetation, an abundance of beasts and birds, and fields sown with five types of grain.[14]

Thus, travelers from the south and southeast of the steppe who ventured to the lands to the north described the features that were most striking to them—in particular, Russia's forests—but they were unimpressed by them and preferred open spaces. When a Russian naturalist from St. Petersburg, Ivan Lepekhin, traveled a similar route to Tulisen, out of the forest and on to the steppe, later in the eighteenth century, his impressions were a mirror image of the Chinese ambassador's. On his way from Moscow to Simbirsk in 1768, Lepekhin complained about the steppe landscape and the lack of trees. He noted the poverty of the peasants, who had fertile land but no forests and had to buy their firewood and timber. He found fault with the site of the city of Simbirsk: it was built on the steppe without trees and so timber had to be shipped down the Volga. He attributed the shortage of trees to careless destruction of forests.[15] Elsewhere, Lepekhin remarked that there was little worthy of comment as they had reached the place where "the vast steppes and little-forested areas begin."[16] He was oppressed by the sight of fields sown with grain but with no trees. He was happier when they came across a few "pleasant" places with some trees.[17] The nomadic Kalmyks did not meet with Lepekhin's approval either. They resisted attempts by the Russian authorities to make them settle, take up grain cultivation (which, as pastoralists, they considered humiliating), and were reluctant to observe the rites of the Orthodox religion. Worst of all for Lepekhin, he observed them felling trees without any care that they were creating shortages for the local Russian population.[18] Lepekhin was a participant in the Russian Academy of Sciences expeditions that crossed the steppe in 1768–1774. Most of the naturalists were German or, like Lepekhin, educated at German universities. To a man (and they were all men), they deplored the wasteful destruction of woodland and foresaw shortages of timber and firewood for the local population, echoing contemporary German concerns for forest protection.[19]

In the early 1840s the German baron August von Haxthausen's extensive travels in the Russian Empire included journeys down the Volga as far as Saratov and, in today's Ukraine, from Khar'kov in the forest steppe to Ekaterinoslav on the steppe proper. Haxthausen was from Paderborn in present-day North Rhine–Westphalia, in the forested German lands. Haxthausen made particular note of moving from the forest, which was familiar, to the steppe, which was not: "To the traveler coming from the north the Steppe becomes gradually perceptible by the forests appearing more and more in isolated patches, and the grass plains growing larger in extent. All at once the wood ceases, not a bush is anywhere to be seen, and the Steppe stretches out in its immensity before us."[20] Elsewhere he wrote: "[We] found ourselves in a genuine Steppe country, extending as far as the horizon on every side, and for many hours nothing but Steppe—at that time of year [late July] anything but beautiful! The soil was dry . . . the grass parched. . . . Of trees or forest no trace was to be seen, here and there in the hollows were tall green reeds and willow-bushes."[21]

Haxthausen's views mirrored those of a prominent Mennonite settler, Johann Cornies, who in 1827 traveled from the steppe of today's southern Ukraine to Saxony in the German lands to purchase merino sheep to improve the Mennonites' stock. Cornies noted in his travel diary that Dresden "lies in a very pleasant valley, surrounded by many beautiful villages with forest and fruit trees." In a letter to an official of the Guardianship Committee that administered the foreign colonists, he wrote with evident approval that Saxony's "beautiful, well-built towns, villages, and manorial estates are set into a continuous dark green, forest-like orchard." Describing his return journey, however, he was struck by the lack of trees when he got home: "We wandered happily through the Polish forests, across beautiful, flat Ukraine, and over forestless New Russia."[22] For Slavic and Germanic peoples (and many more accounts could be cited), therefore, the steppe was a landscape without trees and forests, which they saw as a fault that needed correcting.

SLAVIC, GERMANIC, AND EUROPEAN PREDILECTIONS FOR TREES

In spite of the Russian conquest and settlement of the largely treeless steppe, there is a strong association between the forest and Russian—and Slavic, more generally—life and culture. As an American environmental historian of Russia, Stephen Brain, put it: "Russia is unimaginable without its forests."[23] Forested lands were the homeland of the Russian people, and until the nineteenth century most Russians spent most of their lives in the forest heartland. They used wood to build their houses and churches and to make their farming implements, carts, sledges, boats, and many household utensils. They heated their houses over the long winters with firewood, wore footwear made from bast (inner bark), and painted their religious icons on wood.[24] A list of the ways in which Russians have utilized forest products would be longer even than Tulisen's list of the interminable forests his embassy encountered in Russia.

A consequence of the practical importance of the forest and its products in Russian life is the large role that forests and trees have played in Russian culture and identity, and the importance of the forest persisted, although Russians incorporated other landscapes, including the steppe, into their notions of "national" landscapes over the nineteenth century.[25] The forest environment was celebrated in Russian literature and landscape painting, the latter most notably by Ivan Shishkin (1832–1898) in his paintings of Russian forests.[26] Palimpsestov, a Russian inhabitant of the steppe, took the argument for the cultural and national importance of forests further than most. In the 1880s he asserted that forests were the "most precious representative of the plant world" and that "for us, Russians, inhabitants of a vast and also harsh continent, the forest question is a national [*narodnyi*] and state question; it is connected with the life and fate of Russia."[27] Thus, for Palimpsestov, the benefits of planting trees seems to have included asserting Russian ownership of the steppe, much of which had been annexed only in the late eighteenth century. Or, as Palimpsestov wrote, "returned to us" by Catherine the Great.[28]

In her book on forests in nineteenth-century Russian culture and identity, evocatively entitled *Heart-Pine Russia*, Jane Costlow drew on some of Russia's greatest writers, artists, and forestry specialists to take her readers into the realm of myths and legends, folkloric and sacred, that grew up over the centuries Russians lived in a wooded environment. The forests, real and imagined, presented by Costlow represent a mythic past, an old Orthodox Rus' of legends and folklore, saints and wooden churches, a refuge from persecution and modernity, mostly north of the upper Volga—a past that retained its importance even when many Russians no longer lived in the forest.[29]

The continued cultural affinity for forested landscapes among Russians and Ukrainians was shown by a survey by two Russian conceptual artists in the mid-1990s. Vitalii Komar and Alex Melamid investigated "aesthetic preferences and taste in painting" in several different countries. Russians preferred outdoor scenes depicting forests and water. Ukrainians were slightly less enthusiastic about forests but still rated them highly. Komar and Melamid produced landscape scenes based on the aesthetic preferences of Russians and Ukrainians.[30] Both bear striking similarities to Skarzhinskii's dream a century and a half earlier. Skarzhinskii's dream of a forested paradise is reminiscent, moreover, of portrayals of an arboreal heaven on Russian Orthodox icons of the sixteenth and seventeenth centuries.[31]

Russians and Ukrainians shared the settlement of the steppe with colonists from German lands. Forests have also been prominent in the life, history, and culture of Germanic people. German writers from the time of the Reformation looked back to Tacitus's account of Teutonic fighters from their densely forested Germania defeating Roman legions. Like the Russians, Germans used forest products for many day-to-day needs. The German ethnographer Wilhelm Heinrich Riehl in 1852 asserted: "We must preserve the forests, not simply so that the oven is not cold in winter, but also in order that the pulse of German folk life continues to beat warmly and cheerfully, in order that Germany remains German."[32]

In addition to such generalized and apparently unchanging cultural relationships with forests among Slavic and Germanic peoples, over the eighteenth and nineteenth centuries a different set of positive attitudes toward forests developed across Europe, which emphasized the wider benefits endowed by trees and forests on what we today term the "environment." The German naturalist and explorer Alexander von Humboldt (1769–1859) and his many followers believed that among other environmental advantages forests moderated the climate. This was an argument to protect existing woodland and to plant more trees, especially in regions where trees were few in number and the climate harsh. In the Russian Empire in the 1830s, such arguments were taken up by the new Society for the Encouragement of Forestry, the Imperial Academy of Sciences, and the Ministry of State Domains, all of which organized expeditions and further research.[33] The drought, crop failure, and famine of 1833 hit the steppe region particularly badly and prompted serious discussion of the causes and how to avert future such disasters. In 1834 the Society for the Encouragement of Forestry published an article advocating planting trees on the steppe for this purpose. It pointed to the mutual relationship between the absence of vegetation, especially treelessness, and droughts, and argued that tree planting would have the reverse effect, i.e. preventing droughts.[34] Other environmental benefits were believed to ensue from planting trees on the steppe, such as sheltering fields and villages from the wind and protecting against soil erosion, dust storms, and drifting sands.[35] Such was the enthusiasm for forests, that in 1835, F. L. Breitenbach, the German-born director of the Forestry Institute in St. Petersburg, asserted that "forests can even partly help prevent that most frightening and calamitous phenomenon on the Earth—earthquakes."[36]

Such attitudes to forests were common also among European colonial powers in their overseas empires. Colonial administrations sponsored tree planting in attempts to arrest climate change and soil erosion brought about by careless exploitation. The British in India and Germans in Southwest Africa pursued "scientific" forestry in their colonies, planting trees for practical and environmental reasons and protecting remaining woodland.[37] There were also underlying cultural motives behind tree planting in Europe's colonies. Over the nineteenth century in the lands of British settlement, Thomas Dunlap has argued, settlers "destroyed existing ecosystems and made new ones on a scale and with a thoroughness unprecedented in human history." They imported vast numbers of plants, animals, and birds in order to make their new lands familiar.[38] This was part of a larger story, related by Alfred Crosby, of the extraordinary success of European peoples with their plants, animals, and diseases in transforming temperate parts of the globe into "neo-Europes." European colonists, to varying degrees, displaced indigenous populations and reproduced familiar ways of life, which they transplanted from their homelands. Crosby's "neo-Europes" did not, however, include the Eurasian steppe, and trees did not feature prominently in his account of "ecological imperialism."[39] Indeed, when the Russian authorities and the settlers they encouraged to move to the steppe tried to re-create familiar, forested landscapes, they encountered serious difficulties as the steppe environment was not amenable to afforestation.

THEIR IMPERIAL MAJESTIES ENCOUNTER DIFFICULTIES IN PLANTING TREES ON THE STEPPE

Russian attempts to plant trees on the steppe date back to the end of the seventeenth century and involved the rulers. Tsar, later Emperor, Peter the Great, a native of the forested heartland around Moscow, was reputed to have been the first Russian to try, when, allegedly, "with his majestic right hand," he sowed acorns in a gulley near his new port of Taganrog on the Sea of Azov in 1696. The acorns, so the story goes, grew into the wood of Dubki.[40] According to a more measured official account, in around 1707 Peter ordered the governor of Azov to plant groves of oak and other trees near Taganrog by transplanting saplings from the Don River valley. Peter also ordered Governor Volynskii of Astrakhan to plant acorns on the steppe, "in the manner forests were planted in Europe."[41] Peter's main interest was in growing timber to build ships for his navy. To this end he ordered existing woodland to be surveyed and forests with naval timber protected.[42] In 1724 the publicist Ivan Pososhkov advocated tree planting on the steppes. His rationale was less majestic than Peter's. He simply noted that the steppes were "bare" and forests "far away."[43]

Later in the eighteenth century, Empress Catherine the Great (who was born in Stettin on the Baltic coast of Germany) took an interest in forestry. Promoting forestry was one of the aims of the Free Economic Society she founded in 1765, which published articles on steppe forestry in its journal.[44] In 1774 Catherine sent Grigorii Potemkin as viceroy to New Russia. After the foundation of the city of Kherson in 1778, Potemkin ordered trees to be planted to make the damp air healthier. He later imported young oaks trees from the Russian heartland to plant on the steppe.[45] The settlers did not find it easy to grow trees. In 1781–1782, Vasilii Zuev, the leader of an Academy of Sciences expedition to Kherson, wrote: "In many places, they are trying to satisfy their needs for timber by planting trees, but make progress with difficulty."[46] Catherine was not deterred. On her tour of her new southern lands in 1787 she ordered trees to be planted near Kherson. She allocated ten thousand rubles as prizes for success.[47] In spite of Catherine's and Potemkin's orders, however, there was not much progress in cultivating trees on the southern steppe. In 1795 an inhabitant noted that no one had even thought about conserving and planting trees in Ekaterinoslav province.[48] The decrees of autocrats, therefore, made little headway in overcoming the unsuitability of the steppe environment for trees.

Foreign Colonists Achieve Some Success in Planting Trees on the Steppe

The foreign colonists—in particular people from the German lands—also took a keen interest in planting trees on the steppe. Indeed, they were required to do so by the Russian authorities in return for their grants of land and privileges. The regulations for colonists in New Russia of 1801 stated that they were to plant trees along the streets to provide shelter against storms and the wind, especially mulberry trees that could be used for silkworms, and fast-growing trees suitable to the soil in well-plowed land. Moreover, colonists were enjoined to conserve existing woodland and to divide it into woodlots to be felled and allowed to regenerate in turn.[49] Revised regulations two years later required every householder to plant twenty mulberry trees, fruit trees, and fast-growing trees to provide timber each year.[50]

The colonists were given some assistance. Samuel Cortenius, the head of the Guardianship Committee, established an experimental plantation near Ekaterinoslav and distributed seeds and saplings.[51] The forestry specialist Georgii Vysotksii (1865–1940) later described the painstaking way the colonists planted trees. They plowed the soil several times to depth of one *arshin* (twenty-eight inches). They then dug large holes, where they planted saplings around two *arshiny* high. The colonists watered the trees regularly and hoed the land constantly to destroy weeds. But in spite of all this preparation and care, the majority of the trees dried up and died after two or three years.[52]

One group of colonists, the Mennonites who settled along the Molochna River in Tauride province, achieved some success, but it was very hard work.[53] They took the obligation to plant trees seriously, and had experience of planting trees to reinforce river banks in their previous home on the Vistula delta near Danzig. By 1825 they had planted over 200,000 trees on their new lands. From 1830 the Mennonite forestry society under their elder Johann Cornies gave out seeds, saplings, and advice. Cornies set up a model plantation on his own estate, where he conducted experiments with different types of trees and implements. He stated proudly that the trees he planted on his estate survived the drought of 1833–1834 when those on other local plantations dried up and died. Cornies, moreover, enforced the colonists' obligation to plant trees.[54] He reported the results to the Russian authorities. In 1843 he wrote that in the Molochna colony there were 434,993 trees of various broad-leaved and coniferous species in nurseries and in addition 126,325 young mulberry trees. In total in the Molochna colony, there were thirty-nine plantations, covering 162 *desiatiny* in the Molochna colony (which covered 26,000 desiatiny). Most of the trees had been planted on low-lying land and on slopes and had grown successfully.[55] By 1848, Cornies reported, the total number of trees planted had increased to 636,579, with a further 752,248 in nurseries.[56]

Visitors described the success of the Mennonites' tree planting. Haxthausen, who visited in 1843, wrote: "I have been informed that it is only the Germans, indeed only the Mennonites, who in recent times have planted wood."[57] The Mennonites had "begun . . . to plant wood of all kinds, one of the colonists, Herr Kornies [Cornies], having done more in this way than any other person in Russia: he possesses . . . nurseries from which an extensive forest might be planted out." He continued, that a forest "of about a hundred and twenty square miles" might be planted by "five thousand soldiers" under the "direction of Herr Kornies."[58] Haxthausen's optimism ignored the fact that the Mennonites' relative success, which entailed an enormous investment of labor, was not typical across the steppe.

THE ENVIRONMENTAL CONSTRAINTS REVEALED IN ODESSA

The environmental constraints on cultivating trees in the region were shown by efforts in Odessa. The city's botanical garden, founded in 1820, struggled to establish its arboreal collection.[59] An early director, Karl Dessemet, imported trees from all over Europe and North America. Some grew successfully—or at least seemed to for a few years.[60] The Free Economic Society awarded him a gold medal for his work in 1833.[61] His successor, however, had to deal with the problems as the young trees grew. Aleksandr Davidovich Nordman, who was born in Finland, was director of the botanical garden in the 1830s and 1840s. He believed that for the most part conditions on the steppe, in contrast to his densely wooded homeland, were not suitable for trees. The problems were caused, he argued, by the hot summer sun, southerly winds, and constant droughts. He also believed that the subsoil was fatal for trees.[62] Nordman believed that only certain areas in the region, such as river valleys, were suitable for trees. The German geographer Johann Georg Kohl, who visited Odessa in the late 1830s and met "the celebrated Professor Nordman," wrote about the "almost insurmountable difficulties" experienced by "lovers of trees and gardens": "As long as the trees are young and the roots continue in the upper soil, if the seasons are not too unfavorable, they usually flourish, but as soon as their roots reach the bed of chalk, they begin to grow sickly, their growth ceases, and most of them soon die. . . . [The] crippled and miserable state [of trees in the region] plainly indicates the unsuitableness of the soil, and the great forests which occupy the valleys of the Dnieper, the Dniester, &c. have never been able to gain a foot of land on the high steppe." Kohl described the enormous labor required to nurture trees in land from which the "injurious stratum of chalk" had been dug out and replaced with "good soil." He concluded: "From this may be judged what foundation there is for the hopes expressed by some travellers 'that the steppe will one day be covered with luxuriant forests.'"[63]

Kohl did not succeed in ending such hopes among all travelers. Haxthausen visited Odessa for a week in October 1843. He does not seem to have included the botanical garden in his itinerary nor to have numbered Professor Nordman among his acquaintances. He did note, however, that the previous governor of Odessa, Aleksei Levshin, had planted trees on infertile and sandy soil along roadsides, where they were thriving. For Haxthausen, this was evidence that people could create forests on the unfavorable mixture of soil in the area. In general, Haxthausen did not consider either the soil or the climate of the steppe to be unfavorable for trees, and wrote that he thought there were few parts of the steppe where some sorts of trees could grow well. The baron firmly believed in the benefits of afforestation. He wrote that the lack of forests was the greatest hindrance to the wider cultivation of the steppe region. He asserted that if there was more woodland in the steppe, then there would be a more humid climate, there would be more springs and lakes, and the soil would be moister. These changes would make the entire region more amenable to settlement and cultivation.[64] Haxthausen was convinced that the German colonists, among whom he included the Mennonites, could overcome this obstacle.[65]

Haxthausen doubted that Russians would plant trees on the steppe, as it was "opposed to the habits and ideas of the majority," who saw "no sufficient immediate advantage," and in any case they were "too little educated to understand the scientific planting of trees." In the same tone, he continued: "There may perhaps be some landowners who have imitated the example [of the Mennonites], but more for amusement and vanity than with much success."[66] Haxthausen urged the Russian government to take more action to encourage planting woods on the New Russian steppe. He noted that a decree of September 14, 1828, had given incentives and rewards to state peasants in southern Russia for planting trees, but that it had had "scarcely any perceptible effect."[67]

STRUGGLES TO PLANT TREES ON DON COSSACK AND KALMYK LAND

Haxthausen was too quick to praise the achievements of those he considered his compatriots and too dismissive of attempts to cultivate trees by other inhabitants of the steppe region, of whom he knew little. Above all, he underestimated the environmental constraints that took no account of the ethnicity of the tree planters. The constraints were underlined by unsuccessful attempts to plant trees on the Don Cossack and Kalmyk steppes in the 1840s and 1850s. The attempts also demonstrated the motives of the authorities who were behind them: to provide timber and firewood for the population (rather than draw on the practices of indigenous peoples who used other materials) and to promote a settled agricultural way of life (rather than mobile pastoralism). In other words, planting trees was part of wider efforts to transplant the way of life from the forested regions to the steppe.

The Don Cossacks were required to plant trees by law in 1835. The aim was to provide them with timber and other forest products.[68] In the early 1840s a scheme was launched to plant trees on arid land south of the Don, where no trees grew naturally. It proved very difficult, as saplings were killed off by droughts and winds. Investigations revealed that insufficient research had been carried out into the soil and climate of the locations selected for planting. Nevertheless, tree planting continued, with similar, mostly unsuccessful results, until 1863, when the authorities in St. Petersburg finally put a stop to it.[69]

At around the same time, there was an official plan to plant trees on the Kalmyk steppe in Astrakhan province. The plan originated in a scheme to plant shrubs to reinforce sandy soil with a view to future afforestation. But, under the leadership of the Imperial Corps of Foresters of the Ministry of State Domains and the official guardian of the Kalmyks, it evolved into a more grandiose plan to plant forests on the steppe to link up with those being planted on adjoining Don Cossack land. The plan had practical objectives: to grow trees to provide firewood and shelter for the Kalmyks and their livestock. It had wider environmental aims to conserve moisture in the soil and atmosphere and thus improve the climate. It was also part of an official policy of encouraging the Kalmyks to abandon nomadic pastoralism and take up settled farming and forestry. According to the plan, which began in 1846, in the first two years the Kalmyks were to plant one *desiatina* with trees for every twenty tents (*kibitki*). There were a total of 17,545 tents of Kalmyks, whose land covered between 11 and 14 million *desiatiny*. Thus, in the first two years, they were to plant over 700 *desiatiny*. The planned area increased considerably after two years, and over forty years, the ambitious target was to plant 17.5 thousand *desiatiny*.

The results fell far short. The Kalmyks were not provided with sufficient seed, so they planted only 4.25 *desiatiny* in the first year instead of 350 *desiatiny* but managed 107 *desiatiny* in the second year. Even then, their efforts showed again just how difficult it was to grow trees on the steppe. In some areas the seeds did not germinate; in others they did, but the shoots dried up and died in a few months. Some areas were planted with cuttings, which were damaged by drought, frost, and by snowmelt water running through ravines. Further damage was caused by hares. After two years, 90 percent of all plantings had been lost. The failure had other causes as well. The Kalmyks, who were accustomed to life on the treeless steppe, were unused to planting trees, and unwilling. They provided poor-quality laborers—children, the elderly, and the infirm—as well as women, who sometimes abandoned the work. There were too few foresters to supervise the Kalmyks, who did some of the work at the wrong times of year. By 1860 only 743 *desiatiny* of land had been planted, less than 5 percent of the target. On two-thirds of this area, the plantings perished. Of the remaining third, trees on only 100 *desiatiny* showed any promise of survival. The rest were on land with salty soil, which was totally unsuitable for trees, or on good soil but in areas that needed artificial irrigation, which was not available. The plan was halted by the authorities in 1861.[70] The evolution of the original plan into a larger-scale one, without adequate preparation and with almost inevitable failure and waste of resources, was a precursor to similar such grandiose—and failed—plans of the Soviet era.

RUSSIAN ESTATE OWNERS PLANT TREES ON THE STEPPE

More successful attempts to plant trees on the steppe were made by some Russian estate owners, such as Skarzhinskii, whose dream introduced this essay and who will return at the end. Estate owners were motivated by practical and aesthetic concerns and, crucially, had the resources to accumulate experience in how to overcome the obstacles. Estate owners employed specialists to assist them. Among them was Franz Meyer. He traveled from his native Hannover as a young man in 1801 to accompany a collection of exotic plants bought by Count A. K. Razumovskii. Meyer stayed on and spent the rest of his life as an estate manager in Novosil' district in southern Tula province in the forest steppe region. His first employer was Prince Golitsyn, and from 1821 until Meyer's death in 1860 he worked on the estate of Mokhovoe, which belonged to the Shatilov family. He planted some trees on Golitsyn's estate, but it was on the Shatilov estate that his work really took root. His initial motive was purely aesthetic as he had grown up in forested Germany and wanted to liven up the bare black-earth steppe. Indeed, his early plantings were largely ornamental. Over time, his motives developed into more practical ones: supplying firewood, timber for construction, and other forest products for making household items.

Meyer experimented with different methods of cultivating trees and discovered that planting saplings in plowed furrows worked well. He worked hard to cultivate various coniferous and broad-leaved species on black earth. He planted trees on land that was unsuitable for grain cultivation, in particular on the slopes of ravines, in gullies, and along the sides of roads. The first two were suitable for trees, as there was more moisture than on the high steppe. Plantings in all these places protected the soil against erosion and sheltered adjoining fields. By the time of his death Meyer had planted almost 130 *desiatiny* with trees. He was very influential. Many foresters and estate owners, including the writer Lev Tolstoy, visited Mokhovoe. Meyer was a member of scientific societies (including the Free Economic Society), wrote articles on steppe forestry, and advised the Imperial Society for Agriculture in Southern Russia and Ministry of State Domains.[71]

Another estate owner who pioneered tree planting in the steppe region was Vasilii Iakovlevich Lomikovskii of Mirgorod district, Poltava province, in present-day Ukraine. Lomikovskii started his activities in 1809 and published an account of his work in 1837. He described the local environment, which was in the forest-steppe region, and the conditions—in particular the types of soil and available moisture—that were most suitable for growing trees. He learned from experience which species grew best in particular conditions, the value of growing seedlings first in a nursery before transplanting them, how to prepare the ground for planting saplings (he spent a long time explaining how best to dig holes for them), the need to protect saplings against livestock. He also described the uses that could be made of different sorts of wood for fuel, craftwork, fencing, construction, and so on. Like Meyer, he also realized the benefits of rows of trees to shelter fields of crops from the influence of the climate.[72]

Lomikovskii was the model for the improving estate owner Konstantin Kostanzhoglo in Nikolai Gogol's satirical novel *Dead Souls*, first published in 1842. In real life Gogol was also a native of Poltava province and a near neighbor of Lomikovskii. In his novel the main character, Pavel Chichikov, visits Kostanzhoglo's estate in the company of Kostanzhoglo's brother-in-law Platonov. "'Look,' said Platonov 'that's where his land begins. It looks different, doesn't it?' And indeed, trees had been planted across a whole field and the trees were all of the same height and as straight as arrows; behind it there was another tree plantation, a little taller and also recently planted; behind them was an old wood, and each wood was taller than the last. Then there was another stretch of plain covered with thick woods." Chichikov asks Platonov how he did it:

> You'd better ask him. He's such an expert agriculturalist that he uses everything up. He knows not only the property of the soil, but also which crops to rotate with which and what trees to grow or plant next to what cereals. Everything with him has to serve three or four purposes. His woods, for instance, are grown not only for timber, but also for providing more moisture for the fields, for helping manure the soil with leaf-mold, and for giving a certain amount of shade. When there's a drought in this district, his land suffers from no drought.[73]

Gogol may have been satirizing attitudes to hardworking estate owners striving to improve their estates and the very notion of planting trees in such an unsuitable environment. Such attitudes were found elsewhere in the region. Palimpsestov—the Orthodox priest-cum-agronomist who championed forests on the steppe—recalled receiving a "strange glance" and a "mocking smile" for asserting at a meeting in Odessa in 1853 that no region was more favorable for afforestation than southern Russia.[74] On the other hand, Gogol, whose forebears had lived on the steppe for several generations, was an early example of a Russian—or, rather, Ukrainian—who took a pride in and valued the landscape of the steppe of his Ukrainian homeland. In contrast to many Slavic inhabitants of the grassland, he was not so inclined to view the lack of trees as a fault that needed correcting.[75]

Nevertheless, the Russian authorities and some settlers on the steppe persisted in trying to cultivate forests in the unsuitable environment. The mixed results showed just how difficult it was, and how much research would be necessary if the environmental barriers were to be overcome. The start of "scientific" research into steppe forestry is usually given as 1843, when the Ministry of State Domains founded a forestry plantation at Velikii Anadol' in eastern Ekaterinslav province. The first head was, perhaps characteristically, a Russian-born German forester named Viktor von Graff. Two years later the ministry appointed the Mennonite Johann Cornies to head another plantation in Berdiansk, in Tauride province. It took a great deal of research and effort to make some progress in overcoming barriers such as the alkaline subsoil and periodic droughts. Foresters at these and later steppe forestry plantations worked out which species of trees were best suited to the environment, which combinations of species to plant, how to prepare the soil, and so on. By the late nineteenth and early twentieth centuries, however, they had given up on large-scale afforestation schemes and focused instead on planting trees in limited quantities for specific purposes, such as combating soil erosion and in shelterbelts to protect land from the wind.[76] Steppe forestry reached its culmination in 1948 when the Soviet authorities launched the Stalin Plan for the Transformation of Nature, which envisaged wholesale planting of shelterbelts across the steppe to combat the recurring droughts that damaged agriculture. The plan had mixed results, on account of both environmental constraints and poor planning.[77] Some specialists, such as Vysotskii, were never convinced that planting trees in large numbers on the steppe was viable or worthwhile.[78]

CONCLUSIONS

From 1696 to 1850, Russian authorities and settlers—mostly Slavic and Germanic—persisted in planting trees in the unsuitable environment of the steppe because they needed timber and forest products for many everyday needs, because they had a cultural preference for forested landscapes, and because many believed forests bestowed wider environmental benefits. Their persistence paid off to a certain extent, and by the mid-nineteenth century they had accumulated some experience of how to grow trees, with some success, on the steppe. The Russian experience of trying to plant trees on the steppe was part of a wider story of Europeans in their colonies displacing indigenous populations and seeking to re-create ways of life and landscapes they were familiar with in their homelands. For much the same reasons as on the steppe, in some European colonies, the authorities planted trees where none had been before, took action to protect native forests and to replant woodland that earlier settlers had destroyed. In all cases, tree planting was part of larger desires to make unfamiliar environments familiar and to make them more suitable for European settlement and farming.[79] Thus, the Russian strategies to incorporate parts of the Eurasian steppe by finding ways to manage and exploit their resources and transforming the environment by introducing plants—in this case, trees—from their homelands were part of the larger story of European colonialism around the world as well as the partition of the Eurasian steppe by the empires that surrounded it.

The travelers from Muslim and Chinese lands to the east and southeast of the steppe who ventured north to the Russian lands preferred open landscapes to forested landscapes, which some found oppressive. The nomadic pastoralists who lived on the Eurasian steppe for millennia prior to Russian conquest and settlement also preferred open landscapes. Their way of life was well adapted to the steppe environment, and in turn, they adapted the steppe to suit their needs. Pastoral nomads periodically set fire to the steppe to promote the growth of fresh grass as fodder for their livestock. The periodic burning and incessant grazing contributed to the formation and endurance of the grassland and prevented the regrowth of woody vegetation. Thus, the nomads—like the Slavs, Germans and others who largely displaced them from the steppe—adapted and changed the environment to suit their way of life.[80] Indeed, Palimpsestov had long believed that it was the nomads who had destroyed the forests (which, he insisted, had once covered the steppe) and that "replanting" them would reassert not just Russian control but a Russian identity.[81]

Viktor Skarzhinskii realized his ambition to cultivate a forest on the steppe. After he returned home from the Napoleonic Wars, he planted trees in the fields and along the banks of the River Arbuzinka on his estate of Trikraty in Kherson province. Like other pioneers of steppe forestry, he learned from experience how to deal with the environmental constraints. In the winter of 1834–1835, for example, some of his tulip trees perished from the cold. His plantings were not just ornamental. Skarzhinskii supplied saplings to other parts of the region, including the Odessa botanical garden, the military settlements in Kherson province, and the city of Piatigorsk in the north Caucasus.[82] In 1872 a marble bust was erected to his memory outside the Imperial Society for Agriculture in Southern Russia in Odessa. The inscription read: "To Viktor Petrovich Skarzhinskii, who cultivated on the sultry steppes of Kherson province 400 *desiatiny* of forest of coniferous and broad-leaved species."[83] The

Trikratskii forest survives. It is now an attraction where visitors can find shelter from the hot air of the steppe under the canopies of the trees—including oak trees over 120 years old—by the banks of the river. The photographs of the forest on the websites of tour companies advertising trips bear a striking resemblance to Skarzhinskii's dream.[84]

3 "People Arrive but the Land Does Not Move"

Nomads, Settlers, and the Ecology of the Kazakh Steppe, 1870–1916

Sarah Cameron

When the Bolsheviks seized power in October 1917 they did not begin with a blank slate. Rather, they contended with a range of political, social, economic, and ecological changes put in motion by the fallen Russian Empire as well as by the catastrophe of total war. The role that these legacies played in shaping the formation of the new Soviet state has been explored in numerous studies with a focus on political, social, and economic factors.[1] By contrast, the ecological legacies of Russian imperial rule and the ways in which they influenced the development of the Soviet state have remained relatively unexplored. This omission is particularly surprising given the outsized role that agriculture was to play in Soviet history. The Soviets adopted a particular form of modernization, one built upon the backs of the peasantry. The fate of industry was closely intertwined with that of agriculture and vice versa. The first decades of Soviet rule were scarred by food shortages and famine, most notably the 1917–1920 famine in Russian Turkestan, the 1920–1921 famine in the Volga and Ural River regions, and the collectivization famines of 1930–1933, which hit Ukraine, Kazakhstan, the Volga Basin, and the Don and Kuban regions of the North Caucasus with particular intensity. Millions perished in these famines, and the resultant upheaval would profoundly influence the trajectory of the new Soviet state.

To illustrate the promises that an ecological approach to the 1917 divide holds for Russian and Soviet history, in this essay I focus on one collectivization famine, the Kazakh famine of 1930–1933.[2] Although this famine is little known to most scholars in the West, its effects were catastrophic. More than 1.5 million people perished, one-quarter of Soviet Kazakhstan's total population. Over a million refugees fled abroad or to neighboring Soviet republics, many resettling there permanently. Prior to the famine, most Kazakhs practiced pastoral nomadism, a way of life characterized by seasonal migrations along predefined routes to pasture animals, including horses, sheep and camels. But during the crisis, some 90 percent of the republic's livestock herds perished. Due to death of their herds (the basis of their nomadic existence), Kazakhs were forced to abandon their nomadic way of life for a settled one, a process that sparked a painful and far-reaching reorientation of Kazakh identity. The famine transformed Kazakhs from a pastoral nomadic society into one that was largely sedentary.

Like the better known famines that afflicted Ukraine, the Volga Basin, and the Don and Kuban regions of the North Caucasus during the early 1930s, the most important cause of the Kazakh famine of 1930–1933 was forced collectivization. Under the auspices of Stalin's first Five-Year Plan, a far-reaching program to transform the economy and society of the new Soviet state, activists shunted nomads into collective farms and collected onerous grain and meat procurements, programs that quickly sparked hunger. But the legacies of Russian imperial rule—principally, changes induced by massive peasant colonization of the Kazakh Steppe during the late nineteenth and early twentieth centuries—intensified the effects of the Soviet regime's brutal collectivization policies.[3] These changes, which included shifts in the diet and migration patterns of Kazakh nomads and ecological changes in the steppe itself, affected Kazakhs' ability to command food (what the famed theorist of famine Amartya Sen would refer to as "entitlements"), making Kazakhs more susceptible to hunger.[4] Although Moscow's assault on the countryside was the primary cause of the Kazakh famine, the legacies of Russian imperial rule must be considered an important contributing factor.

To date, most existing studies of the Soviet collectivization famines adopt a political, state-centered approach, focusing closely on the period following Moscow's launch of forced collectivization in the winter of 1929–1930. Their authors analyze Moscow's actions immediately prior to the outbreak of famine and during the crisis, with an eye to unearthing how these decisions, such as the levying of additional grain procurements or the distribution of food aid, might have contributed to or mitigated hunger.[5] The politicized nature of the Soviet collectivization famines has played a major role in the adoption of this particular approach. Both Kazakhs and Ukrainians suffered disproportionately because of famine, outcomes that raise the issue as to whether Stalin sought to use famine to punish particular ethnic groups. Buoyed in part by an active Ukrainian expatriate community, the question of whether the Ukrainian famine should be considered a genocide has been the subject of polemical long-running debate in the West.[6]

Close scrutiny of Moscow's decision making is essential to understanding the nature of these crises, and it is doubtful that famine would have broken out anywhere in the Soviet Union from 1930 to 1933 without the Soviet regime's violent assault on the countryside.[7] But to fully understand the terrifying toll of the Soviet collectivization famines, we also need to add a broad array of environmental issues, from ecological change to alterations in land use patterns and food distribution networks, to the list of the legacies of Russian imperial rule. These questions help us assess the vulnerability or resilience of particular groups to hunger, thereby revealing the role that Russian imperial rule played in shaping the character and intensity of the Soviet famines.[8]

This approach would bring the study of the Soviet collectivization famines into conversation with a broader literature on comparative famines, which has shown how both abrupt change and slower moving structural processes can combine to produce a famine.[9] The adoption of an ecological approach that includes Russian imperial rule also offers scholars a chance to examine similarities and differences between the Russian Empire and the Soviet state. When examined through this lens, some of the Soviet regime's interventions—such as the tightening of border controls, which reduced the pasturage available to pastoral nomads—appear more an amplification or intensification of Russian imperial techniques rather than something "uniquely" Soviet.

Adopting an environmental methodology, I illustrate how the Soviet Kazakh famine of 1930–1933 was profoundly shaped by the legacies of Russian imperial rule, specifically the period 1896–1916, when more than 1.5 million peasants from European Russia settled the Kazakh Steppe.[10] In a historic shift this territory, a place long synonymous with pastoralism, became a mixed economic region, one populated by large numbers of settled, agrarian peoples in addition to pastoralists. Although the available data does not allow a full investigation of ecological change in this period (the systematic collection of temperature and precipitation data in the Kazakh Steppe began only in the late nineteenth century), other materials, including archival sources and ethnographic accounts, illustrate important shifts in the relationship among humans, animals, climate, and environment.[11] I find that these changes would make Kazakh nomads particularly susceptible to famine.

KAZAKH NOMADS AND THE KAZAKH STEPPE

The practice of pastoral nomadism has a long history in the steppe zone of Central Eurasia, dating back at least four millennia.[12] For many, the history of the steppe is synonymous with nomadism, and it conjures up images of mounted, raiding warriors who wandered free from the trappings of settled life. But as researchers have shown, the history of pastoral nomadism in Central Eurasia is far more complex than an image of wandering, raiding peoples might suggest. Throughout the centuries, nomadism did tend to be the predominant economic activity in the steppe zone, with sedentary populations concentrated in oases or irrigated river valleys. But the environmental constraints on sedentary agriculture in this region were neither precise nor immutable.[13] Nor was the practice of pastoralism itself either timeless or unchanging. Pastoral nomads regularly adapted their practices according to opportunities and risks. Environmental changes such as shifts in temperature or precipitation could cause some pastoral nomads to migrate year-round rather than seasonally. Social and political changes such as the intrusion of new peoples or shifts in political structures could prompt pastoral nomads to increase or decrease their reliance on animal herding in favor of other economic strategies such as agriculture or hunting.[14] Pastoralists regularly sought to alter the landscape of the steppe to suit their needs by burning areas of the steppe to prevent the spread of shrubs and trees or by constructing wells to tap groundwater reserves.[15] In this region, humans, animals, environment, and climate were connected in an intricate and mutually transformative relationship.[16]

The origins of the Kazakhs, a Muslim Turkic-speaking group, date to the late fifteenth century when Janibek and Kirey, sons of Barak Khan of the White Horde of the Mongol Empire, broke away from Abu'l-Khayr, khan of the Uzbeks, forming what came to be known as the Kazakh khanate. According to popular tradition, in the late sixteenth century the Kazakh khanate split into three supra-tribal confederations, each of which was known as a horde (*zhüs*). Over time, the Elder (*ulu*) Horde came to control Semirech'e, the Middle (*orta*) Horde the central steppe and southwestern Siberia, and the Little (*kishi*) Horde the west of the steppe.[17] By the early eighteenth century, the Kazakh khanate began to fragment further, as it came under sustained attack from several nomadic confederations. In the eighteenth century Russia completed the construction of a defensive perimeter manned by Cossacks across the northern reaches of the Kazakh Steppe. Then, in the nineteenth century, as the disorder in the Kazakh hordes increased, Russia began to tighten its hold over the Kazakh Steppe. By the late nineteenth century the Russian Empire's incorporation of the lands of the three Kazakh hordes was complete.

By this time, the region where Kazakh nomads could be found comprised several ecological zones, including steppe, desert steppe (sometimes referred to as the "semi-desert"), desert, and mountains.[18] The steppe zone lay north of the Aral Sea and south of the forests of western Siberia. It formed part of a larger steppe belt of grasslands stretching across the Eurasian continent. The desert steppe and desert zones that Kazakh nomads frequented lay south of the steppe zone. They encompassed several major arid areas, including the Betpak-Dala, the Üst-Yurt plateau, the Qara-qum at Aral Sands and part of the Qïzïl-qum desert. The Betpak-Dala (the ill-fated steppe), the largest of these territories, was an immense plateau located between the Sary-Su River in the west and Lake Balkhash in the east. Today, this territory, which is also known by a Russian name Golodnaia step' (the hungry steppe), lies at the heart of contemporary Kazakhstan, just south of the city of Karaganda.[19] The Üst-Yurt, a smaller plateau, lay between the Caspian and Aral Seas. The other two major arid areas where Kazakh nomads could be found, the Qara-qum at Aral Sands and the Qïzïl-qum, were classical deserts, with unsecured shifting sands.[20] Finally, there were two mountain ranges, the Tian Shan Mountains in the southeast and the Altai Mountains in the east, at the edges of the desert and the desert steppe zone. The area around these mountain ranges generally received more precipitation than the surrounding zones, and the Tian Shan overlooked a particularly fertile region, known as Zheti-su or Semirech'e (seven rivers). With the exception of the mountain ranges, the distinctions among these zones were not precise: the climates of steppe, desert steppe, and desert were closely related to one another, and the borders of these ecological zones could also shift over time because of environmental change.[21]

All of these zones had a sharply continental climate, with hot summers and very cold winters. Given the region's continentality (or distance from oceans), its climate was much harsher than similar zones further west in European Russia. Winter temperatures could dip below –40 degrees Fahrenheit, while summer temperatures could reach well over 100 degrees Fahrenheit. Only the southern reaches of the steppe, such as the city of Vernyi (present-day Almaty), enjoyed average January temperatures above freezing.[22] The region's precipitation was shaped by winter westerlies from the Atlantic. Shifts of the North Atlantic Oscillation (NAO) determined the distribution of this rainfall over the centuries. A negative NAO mode might bring ample rainfall to the region, while a positive NAO mode brought little rainfall. During the Little Ice Age (roughly 1500–1850 CE), the steppe appears to have enjoyed plentiful rainfall. By the late nineteenth century, however, the climate of the steppe was entering a warm dry phase.[23] Aridity had become one of the Kazakh Steppe's defining features.

Not only was the amount of rainfall low but rainfall patterns in the Kazakh Steppe were more unstable than in the steppes in the European part of Russia. The amount of rain might fluctuate dramatically from year to year and from season to season. For the desert and desert steppe zones, the period of maximum precipitation was generally in the spring, with the hot dry summers bringing little or no rainfall. By contrast, the wettest period for the steppe zone was usually the summer. Thus, seasonal vegetation such as grasses appeared first in the desert and steppe desert zones and only later, in June and July, in the steppe zones.[24] The overall amount of yearly rainfall decreased markedly the farther south one traveled, from steppe to desert steppe to desert. The steppe might receive between ten to twenty inches of rain, the desert steppe six to ten inches, and the desert less than four inches. In each zone, the most important indicator of climate was not the amount of rainfall but, rather, the ratio of precipitation to evaporation, which was generally quite low. Because of long, sunny cloudless days and high temperatures, the rain that fell in these regions usually evaporated quickly.[25]

In the north, the region had *chernozem*, or "black earth" soils. These contained a high percentage of humus and could produce good agricultural yields. Chestnut soils could be found further south, in the desert steppe zone. Like the black earth soils, chestnut soils could be quite fertile, yet they had far less humus than black earth soils, and repeated plowing could quickly exhaust their fertility. Two other soils, *solonchak* and *solonetz*, were sprinkled throughout, occurring with greater frequency in the desert and desert steppe regions. Solonchak soils were formed as groundwater rose to the surface. As this water evaporated, it left white salty patches across the surface of the soil. Solonetz soils were solonchak soils that had been leached through irrigation or increased rainfall. As water passed through the soil, sodium carbonate formed, and the soil became alkaline. Neither of these soils was fertile, but their distribution was uneven—chestnut soils could be interspersed with solonetz and solonchak soils—and changeable, as weather or irrigation patterns shifted.[26]

To utilize this landscape Kazakh nomads migrated along predetermined routes to pasture their animal herds. These migrations were not carried out by individual households but, rather, were conducted in small groups, nomadic encampments known as *auly*. Each aul was generally made up of two to eight households, and together all the members of the aul assumed responsibility for the care of their herd. The number of animals associated with a given aul varied widely (sheep, for instance, generally took less effort to tend than horses), and these numbers could also fluctuate, depending upon seasonal or environmental variations. New auls could form when animal numbers grew too large to be supported by the original aul's pastures.[27]

Most auls pastured mixed herds, rather than relying on a single type of animal. Given the steppe's climate, animals that could travel great distances and easily paw at fodder trapped under the snow, such as sheep and horses, were of particular importance. Sheep were raised for their mutton and wool, and horses were prized for their meat and fermented mare's milk (*qïmïz*; Russian, *kumis*), in addition to serving as a source of transport for people and goods. The camel, an animal ill-suited to uncovering fodder trapped under snow, was much less common than sheep or horses in the Kazakh Steppe. Camels, however, could be found in the west of the steppe. In this arid region, the camel was the sole animal capable of sustaining long journeys and, by extension, was the only means of transporting goods.

Kazakhs' migrations were seasonal, and they followed the appearance of vegetation in the steppe, assuming a circular south-to-north-to-south pattern. Each aul generally had separate spring (*kökteu*), summer (*zhaylau*), fall (*küseu*), and winter (*qïstau*) pastures. The summer pastures were particularly important, as they provided the thick grasses that were necessary for fattening the nomads' livestock. In the winter months it was critical to find a sheltered pasture, one where the snow was not too deep and the animals might find some relief from the wind and the cold. Both the spring and the fall pastures were usually located close to the winter pasture. Nomads would bring their herds to the spring pasture when the snow had melted but before the summer pasture's grasses had begun to bloom. When the vegetation in the summer pasture began to fade, they would move their herds to the fall pasture, then migrating to the winter pasture with the first snowfall.[28] To supplement their practice of pastoralism, many Kazakhs regularly incorporated other activities, including trade, hunting, and seasonal agriculture.

Kinship, Kazakhs' allegiances to particular clans, was an important source of identity, in addition to governing crucial economic aspects of pastoral nomadic life. In a practice known as *ata qonïs*, individual clans claimed grazing rights over particular pastures, and these privileges were then passed down along genealogical lines. When nomads' migration routes were disrupted or pressure on pastures increased, such as during the Zunghar invasion of the steppe in the eighteenth century, the leaders of various clans would meet to reallocate the usage of pastures.[29] Pastoral nomads also relied upon practices of mutual aid as protection against the steppe's regular occurrences of drought and *zhŭt*, the periodic frost of ground cover that could result in massive herd losses.[30] In *saün berü*, a Kazakh with large numbers of cattle might loan a poorer kin-member a milk cow for the winter. This poorer kin-member would be responsible for pasturing and feeding the cow, yet he would gain the benefits of its milk and, potentially, its offspring. Such mutual aid practices constituted a necessary survival strategy in a landscape where distinctions between "rich" and "poor" could shift rapidly over the course of a single season.

THE ARRIVAL OF "THOSE WHO CAME WITHOUT INVITATION"

The settlement of the Kazakh Steppe by Slavic peasants was part of a broader migration of Slavic peoples to Siberia, the Russian Far East, and Central Asia during the late nineteenth century.[31] In the aftermath of the emancipation of the serfs in 1861 many peasants—known as *samovol'tsy*, or "self-settlers"—came illegally, seeking fertile lands to farm and relief from the poverty and land hunger that characterized peasant life in parts of European Russia. By 1889 St. Petersburg, seeking to regulate this flow of migrants and convinced of the civilizing role that Slavic settlers could play in these regions, issued the Resettlement Act. This act marked the first central government effort to coordinate this migration. It codified settlement as official state policy, setting up settlement programs in European Russia, Western Siberia, and the *oblasti* (provinces) of Akmolinsk, Semipalatinsk and Semirech'e. In 1893 construction on the Trans-Siberian Railroad began, and one of the last remaining obstacles to large-scale peasant colonization of these regions, the arduous trip across European Russia by oxen and cart, was erased (see map 3.1).

There was a close relationship between peasant settlement and the ecological conditions of the Kazakh Steppe. By 1898 the area where many Kazakhs migrated had been split into several different administrative regions: The Steppe Governor-Generalship, which roughly corresponded to the northeast of the Kazakh Steppe (the oblasti of Akmolinsk and Semipalatinsk); the Governor-Generalship of Turkestan, which included, among other parts of southern Central Asia, the oblasti of Semirech'e, Syr Darya, and Transcaspia, all of which had large numbers of Kazakh nomads; and Turgai and Ural'sk oblasti, which were not part of any larger administrative unit.[32] In the steppe provinces, peasant settlement came to resemble a belt, which corresponded to the amount of annual rainfall that the land received. With its fertile soils Akmolinsk oblast quickly became the favored destination for peasant settlers. Over time the northern reaches of Turgai, Semipalatinsk, and Ural'sk oblasti also became important sites of settlement. But because of the lack of rainfall, few settlers could be found further south within these oblasti, and the eight-inch annual precipitation line roughly marked the southern extent of peasant settlement in these provinces.[33]

Further south in Russian Turkestan, the pattern of settlement in Kazakh lands was a bit different, but it was also closely linked to the region's environmental conditions. Fearing the potentially disruptive effects of large-scale peasant settlement in this region, administrators had closed Turkestan to settlement in 1896, adding Semirech'e to this list when it was returned to the auspices of the Governor-Generalship of Turkestan in 1898. But *samovol'tsy*

MAP 3.1. The Russian Empire, c. 1914.

continued to arrive and the pressure for new, fertile lands grew. In 1890–1891 parts of European Russia experienced a devastating famine, and impoverished, starving peasants fled as far south as Turkestan in search of better lands. In 1910, as part of the far-reaching agrarian reforms spearheaded by Prime Minister Petr Stolypin, St. Petersburg permanently lifted the ban on settlement in Turkestan. Settlers concentrated heavily in Semirech'e, with its abundant rainfall and good soils, and to a far lesser extent in the river valleys of Syr Darya. Its aridity, extreme heat, and poor soils meant that few settlers colonized Transcaspia, and this oblast was largely untouched by the phenomenon of peasant settlement.[34]

The peasants who settled in Turkestan or the steppe provinces came primarily from the middle Volga and the left bank of Ukraine.[35] But within these peasant communities, there were differences. Those who had settled the steppe at the beginning of this migration tended to be known as the *starozhily* (old-timers) while more recent arrivals were often referred to *novosely* (new settlers).[36] In Semipalatinsk oblast, a group of peasants, including many Old Believers, migrated to the Altai Mountains. They became known as the "Altai peasants" or the *kamenshchiki* (stone people), and over time they developed traditions distinct from other peasants in the region.[37] Kazakhs referred to all these peasant settlers by a number of nicknames, including *qarashekpendĭler* (black coats, a reference to the peasants' traditional dress) and, more snidely, *kelsĭmsekter* (those who came without invitation).[38]

Peasants found the land they occupied in various different ways. Some contracted with land scouts (*khodoki*), who would arrange for the rental of suitable land from Kazakh or Cossack communities prior to their arrival. After the legalization of this practice, scouts could enroll migrants directly in government-surveyed areas.[39] As part of an effort to solidify the Kazakh Steppe's new border with China, officials offered special subsidies to those who settled in the eastern parts of Semirech'e and Semipalatinsk oblasti.[40] The most destitute arrivals simply poured into existing peasant settlements. Once there, they often appropriated "unused" pasturelands from nomads' seasonal migration routes, and frequent conflicts over land, water, and livestock rights broke out. These tensions, as well as fears about the peasants themselves, particularly their poverty and supposedly uncivilized nature, caused regional officials to plead, albeit unsuccessfully, with the Internal Affairs Ministry to halt the flow of settlers.[41] The Kazakh saying "khalïq osedï, zher ospeidï [people arrive, but the land does not move]" hints at the overcrowding that occurred on particularly fertile lands.

Upon arrival settlers struggled to cope with severe problems such as droughts, frosts, and plagues of locusts. In 1891, after several years of poor harvests, the Governor-Generalship of the Steppe temporarily closed the steppe to colonization.[42] In 1907 settlers in Semirech'e suffered from droughts and destitution so severe that officials in the region began to organize food aid.[43] That same year a detachment of the Russian Society of the Red Cross, after surveying the steppe region, observed that there were people in every village suffering from scurvy because of a poor diet.[44] In 1911 part of the Kazakh Steppe experienced a severe drought. Officials in Omsk *uezd* (district) estimated that the harvest of wheat and other grains was 80 percent less than the previous year. These harvest failures meant that peasants were not able to collect enough seeds for the following year's sowing or sufficient hay to feed their livestock. During the drought the grass on the steppe perished, and peasants' livestock sickened from eating it.[45]

These environmental challenges, as well as other factors such as conflict with local officials and Kazakh nomads, caused large numbers of settlers—perhaps more than 20 percent—ultimately to return to European Russia prior to 1917.[46] Many of those who remained sought to diversify their activities, moving away from an exclusive focus on grain farming as a way to manage hazards such as drought. Some peasant settlements in Turgai and Semipalatinsk oblasti were primarily or exclusively focused on livestock production rather than grain farming.[47] With the construction of the Trans-Siberian Railroad and the introduction of ferry service along the Irtysh River in the 1880s, settlers began to import new technologies to the Kazakh Steppe, including mechanized means of clearing hay (*senokosilka*) and heavy iron plows that could break up the sod more effectively than wooden plows.[48] Settlers' continuing ties with regions in European Russia, including the travel back and forth of family members, brought to the steppe new seeds (especially hardier varieties of wheat, the settlers' favored crop) as well as animal breeds, most notably the black sheep prized by Russians and Cossacks.[49]

As peasant settlement continued, Kazakhs began to adapt their practice of pastoral nomadism to the changing environmental, political, and social circumstances of life on the steppe. The demarcation of borders and the physical presence of areas of Slavic settlement meant that Kazakh nomads could not use the land as they had previously. In Akmolinsk and Semipalatinsk oblasti, where many Middle Horde Kazakhs migrated, nomads began to cultivate hay as a source of winter fodder, to compensate for the loss of some of their pasturelands. To protect these hay fields and their winter pastures from encroachment by peasants or other nomads, some Kazakhs began to spend nine months of the year on their winter pastures, migrating just a short distance to their summer pastures in the remaining three months.[50] Some Little Horde Kazakhs, by contrast, increased the length of their seasonal migrations, as peasant settlement forced them to travel farther in search of reliable water sources and good pasturelands.[51]

Many pastoral nomads changed the composition of their herds in part because of this shift in land use practices. Those Kazakhs who had shortened their migrations began to include animals suited to minimal migrations, such as cattle, followed by sheep and then horses. Prior to the late eighteenth century, cattle had not formed a major part of Kazakhs' herds. They were difficult animals to herd across long distances, and compared to sheep or horses their grazing habits could be fickle; cows tended to subsist entirely on the top layer of grass rather than grazing more intensively.[52] But in a significant shift, cattle now constituted an important part of many Kazakh nomads' herds.[53] By contrast, the numbers of camels in the steppe began to decline, as fewer nomads were able to carry out the long migrations needed to pasture them.[54] Although the steppe's overall animal population was growing rapidly, the number of animals that a particular aul pastured was going down, as few auls were able to sustain the long migrations needed to pasture large herds.[55]

Russian imperial officials anticipated that peasant settlement would encourage Kazakhs to abandon their nomadic way of life for a settled one, a goal that St. Petersburg had pursued to varying degrees since the reign of Catherine the Great. The spread of agriculture, it was believed, would "civilize" native peoples and make the lands in these regions "more productive."[56] The ideas of evolutionary theorists, who held that disappearance of nomadic life was part of the inevitable outgrowth of modernity—as a lower civilizational form, nomadism, gave way to a higher one, settled life—underpinned and buttressed such assumptions. But Kazakhs' responses to peasant settlement did not always take the form that St. Petersburg might have hoped. In 1906, at a meeting of the Turgai oblast administration on the question of land apportionment, officials noted that many Kazakh households in Kustanai and Aktiubinsk uezdy had begun to plow up their pastures for grain, an outcome that seemed to confirm the predictions of those who had heralded the end of nomadism. On closer inspection, however, they discovered that the Kazakhs in these districts had secretly rented these lands to Russian settlers, who were farming them. They concluded: "In spite of the growth of agriculture, the Kirgiz [Kazakhs] in these districts are a long way from giving up nomadism."[57] Although most nomads began to reduce their mobility, they also adopted a number of other strategies, such as the rental of their pasture lands, to maintain their nomadic way of life and adapt to the changing social, political, and environmental circumstances of life on the steppe.[58]

In addition to the rental of their pasturelands, Kazakhs increased their trade with Russia. This steppe trade was distinct from the caravan trade between Russia and Central Asia that passed through the Kazakh Steppe, and it had grown in importance as the Russian empire advanced into the region.[59] Kazakhs sold live animals as well as animal products such as hides, butter, and wool, and they purchased items such as grain, tea, kerosene, matches, pottery, and Russian manufactured goods from traders, mostly Muslim merchants. This trade was conducted near areas of peasant or Cossack settlement—at seasonal trade fairs or deep within the steppe itself at so-called mobile markets, where traders would travel with nomadic encampments. The biggest markets tended to take place at the end of the spring, just as the livestock was beginning to gain weight, and in the fall, so that Kazakhs did not have to pasture these animals over the winter.[60] By the late nineteenth century, Kazakhs became increasingly focused on raising animals for Russian markets. They sold them directly to Slavic peasants or to traders who, with the construction of the Trans-Siberian Railroad, shipped them to consumers in European Russia. The railheads of Omsk and Petropavlovsk became centers for this trade. In 1908 the steppe provinces and parts of Russian Turkestan exported 400,000 head of livestock, 6 million hides and skins, and nearly 6,000 tons of meat by rail to European Russia.[61] This trade almost certainly played a role in shifting the composition of Kazakhs' herds, as cattle were more sought after by Russian consumers.[62]

St. Petersburg had promoted the steppe trade as early as the reign of Catherine the Great. It was profitable—Russian officials obtained livestock cheaply and they offloaded grain that otherwise would have been difficult and expensive to transport back to Russia—and it was believed that the growth of this grain trade would help "civilize" Kazakhs, making them into loyal subjects of the empire.[63] During the period of intense peasant settlement (1896–1916), officials continued to link the consumption of grain with the adoption of sedentary life. In 1907, as part of a survey of Kazakh land use practices in Akmolinsk oblast, the statistician V. K. Kuznetsov proclaimed: "The mass of the population has gradually given up a purely nomadic way of life, relying on new means of existence and at the same time adopting the habits of a more cultured way of life, particularly the consumption of grain."[64] The "ancient Kazakh diet [*starinnaia Kirgizskaia pishcha*]" of meat and fermented mare's milk (*qïmïz*), he predicted, would soon give way to a diet based primarily upon grain and cow's milk.[65]

Previously, Kazakhs' diet had been based upon meat and milk products from sheep, horses, and to a lesser degree, camels. Meat was the primary focus, and the consumption of meat depended upon one's social position, with better-off nomads consuming more meat. During the late fall, Kazakhs would slaughter a portion of their herds. This meat would then be preserved for use during the winter, a process known as *soghïm*. During the warmer months (from April to October), when their animals produced fresh milk, Kazakhs shifted their diet and relied to a greater extent on milk products. They consumed soured milks, such as *qïmïz*, *shŭbat* (fermented camel's milk), and *ayran* (fermented sheep's milk), as well as a variety of cheese and butter products.

Kazakh nomads supplemented this meat- and milk-based diet in various ways. Some hunted or fished, while others consumed limited amounts of grain.[66]

There is evidence to suggest that Kazakhs' diet began to change as economic practices in the steppe shifted in the late nineteenth and early twentieth centuries. After a series of food crises in the 1830s and then again after the disastrous 1891 famine, the Russian state had become increasingly interested in managing the food supply in the Russian countryside. Officials began to discuss the relationship between food and health, in an effort to halt further outbreaks of famine. By the 1880s local government (zemstvo) officials began to collect data on peasants' food consumption in an effort to understand exactly how much food peasants needed to survive. Armed with this data, these officials sought to develop a consumption model that would be applicable to all of Russia and indicate what types of foods needed to be produced and how much of this food could be sold.[67]

By contrast, nomadic nutrition was far less well understood. The Kazakh Steppe, like many other non-Russian parts of the empire, did not have zemstva. However, the various statistical expeditions tasked with understanding how much land nomads needed to survive also collected limited data on nomads' consumption habits, which they sought to compare to those of "standard"—that is, settled—households. In 1907 Kuznetsov found that Kazakhs in Kokshetau *volost* (canton) consumed less food than they had in 1896. In addition, Kazakhs consumed a substantial amount of grain, but their consumption of meat had declined dramatically (see Table 3.1).[68]

Although Kuznetsov tried to explain these declines by pointing to various inaccuracies in the data, he concluded: "the nourishment of nearly half of the Kirgiz [Kazakh] population does not reach an average level."[69] Though Kuznetsov's data is taken from an area of intense peasant settlement, Akmolinsk oblast, during a year when parts of the Russian empire suffered from famine, it hints at the various ways that Kazakhs' diet had begun to change. Overall, Kazakhs consumed less food, making them more vulnerable to famine. Grain had come to constitute an important part of their diet. Some Kazakhs grew this grain themselves, but most acquired it through trade with settled populations.

During the last decades of the Russian Empire, the Kazakh Steppe's human and animal population grew dramatically. In the course of twenty years (1896–1916), the number of people living in the steppe increased by over two million, and three-quarters of these new arrivals were Slavic settlers.[70] In one decade (1906–1916) the number of livestock in the northern provinces of the Kazakh Steppe grew by nearly three million, an increase of 79 percent. In Akmolinsk oblast, the situation was particularly pronounced. Livestock numbers increased 135 percent during that same ten-year period.[71] Several factors help explain this tremendous growth in human and animal population, includ-

TABLE 3.1. The Consumption of Food Products in Kazakh Household Budgets

NAME OF THE GOOD	THE CONSUMPTION OF FOOD PRODUCTS IN HOUSEHOLD BUDGETS	
	1907	**1896**
Grain	1	119.6
Meat	55.3 kg	144.1 kg
Butter and cheese	12.7 kg	16.2 kg
Fermented mare's milk	17.4 buckets	19.7 buckets
Milk (cow's milk, sheep's milk, goat's milk)	19.5 buckets	30.2 buckets

ing the influx of new peoples, Kazakhs' growing involvement in the livestock trade, and the relative absence of warfare on the steppe in comparison with the eighteenth century.[72]

This rapid human and animal population growth placed greater stress on the steppe's environment. With more livestock, pastures were being used far more frequently, and many settlers did not practice crop rotation. Once the fertility of a piece of soil was depleted, they moved on to plow other lands, and soil exhaustion became a problem in areas of heavy peasant settlement.[73] Kuznetsov argued that the steppe's population increase had worsened the overall conditions of the region's water supply: "Everywhere the woods are thinning out and with each year they are depleted more and more, threatening the deforestation of whole large areas and increasing the natural desiccation of bodies of water."[74] As the steppe's animal population grew and these herds came into close proximity to one another, epizootics (large-scale epidemic outbreaks of disease among animal populations) became more likely. In 1890 officials in Akmolinsk oblast reported that thousands of livestock had died from Siberian anthrax (*Sibirskaia iazva*) and the plague (*chuma*).[75] In 1911 a large epizootic led to the death of over a thousand head of cattle in Ust' Kamenogorsk, part of Semipalatinsk oblast.[76]

By the early twentieth century, the most striking phenomenon was a trend toward the settlement of the Kazakhs.[77] Russian imperial officials noted the appearance of *dzhataki*, poor Kazakhs without livestock. These Kazakhs could be found near areas of Russian settlement, where they worked as hired laborers.[78] In other cases, nomads began to maintain just two types of pastures, summer and winter, instead of a pasture for each season. With this decrease in mobility, pastoralists began to rely less on their yurts during the colder months. They constructed semi-permanent adobe dwellings (*zimovki*), as well as shelters for their animal herds on their winter pastures.[79] The crisis of 1914–1924, which included the effects of the First World War, famine, and civil war further accelerated this process of nomadic settlement. But by the 1920s nomads who migrated year-round could only be found in areas that had few settlers, such as Transcaspia, populated by Little Horde Kazakhs. A minority of Kazakhs, perhaps less than 10 percent, had settled. The vast majority of Kazakhs continued to rely on the practice of pastoral nomadism. However, they had begun to shorten their migrations and, to a greater degree than before, supplement their practice of pastoralism with other activities, such as trade and agriculture.[80]

CONCLUSIONS

Although many Russian imperial observers had predicted that Kazakhs' pastoral nomadic way of life would disappear as a result of peasant colonization, the vast majority of Kazakhs still practiced pastoral nomadism on the eve of Soviet rule. Historically, pastoral nomads in Central Eurasia had relied on various strategies to adapt to changing political or environmental conditions, and Kazakhs responded to the challenge of Slavic settlement by limiting their mobility, increasing the level of their trade with Russia, and renting out some of their pasturelands (among other shifts in practice). Although this period marked a trend toward nomadic settlement, what Russian imperial officials referred to as "semi-nomadism," the disappearance of nomadic life under Soviet rule was by no means preordained or inevitable. As the history of Central Eurasia has shown, forms of "semi-nomadism" could be long-lasting and stable. Although some ecological, social, and political factors might cause some groups to settle, others might cause other groups to nomadize, with "semi-nomads" or settled groups taking up nomadic life.[81]

In a crucial development, however, Kazakhs began to consume more grain, and they became increasingly tied to the networks of Russian traders that supplied this grain. They also began to consume less food overall, making them more vulnerable to hunger. Many of these changes were magnified by the destruction of the period 1914–1924, which saw Russia's entrance into the First World War, the fall of the Russian Empire, and the onset of civil war. When the boundaries of what would become known as Soviet Kazakhstan were set in 1924, the region's economy was in a state of crisis, with agricultural yields and livestock numbers well below their pre–First World War levels.[82] In the winter of 1929–1930, Moscow imposed heavy grain procurements on the republic as part of the first Five-Year Plan, severing the grain trade networks that had become so important to Kazakh nomads' existence under Russian imperial rule. In the summer of 1930 famine began.

In its broad outlines the Kazakh case shares much in common with other histories of the ecological encounter between settlers and mobile peoples in the drylands, particularly in the Great Plains of the United States and in the Argentine Pampas.[83] In each case the establishment of European (or Euro-american) dominance led to the loss of traditional lands and water rights for Indigenous peoples. Also, ecological instability, or the tendency of all of these regions to suffer periodic bouts of devastating drought or severe cold, played an important role in shaping the interaction of settlers and mobile peoples with their environment. In the Kazakh case, however, peasant settlement was soon followed by regime change, the fall of the Russian Empire, and the establishment of the new Soviet state, offering environmental historians the chance to compare and contrast the effects of two different economic systems, one capitalist, the other socialist, on the drylands.

4 "The Scourge of Stock Raising"

Zhŭt, Limiting Environments, and the Economic Transformation of the Kazakh Steppe

Ian W. Campbell

In 1894 the editors of the official bilingual newspaper *Dala ualaiatïnïng gazetï* (Newspaper of the steppe region), published in Russian as *Kirgizskaia stepnaia gazeta* (*KSG*; Kazakh Steppe newspaper), provided a classic account of a recurring and frightful phenomenon on the grasslands of Central Asia: "Around Akmolinsk district, with the exception of southern cantons, on the night of November 10–11, rain fell, which then turned into snow, which then on November 13, turned into a blizzard, continuing for days. Then, as a consequence of the thaw on the 14th and 15th, all this snow melted, and on the night of the 16th froze, covering the land with a quite thick and strong crust of ice, which prevents the livestock's access to pasturage. This event will affect the nomadic population's welfare unfavorably; because of the dry summer, they prepared few hay reserves."[1]

Events like this were associated with the local or general failure of the fodder-grass harvest, known in the Kazakh language as *zhŭt*. Recurring with particular severity and scope every ten to twelve years, and more frequently on a small scale, zhŭt was a disaster for the pastoralists of the Kazakh steppe.[2] If human efforts to help animals obtain fodder beneath deep snow or thick ice failed, the Kazakhs who depended on the slowly starving livestock for food and clothing were forced either to take flight or to slaughter their animals so as to avert starvation. It was not always possible to avoid this fate. Outside observers reported that animals perished, at times, in unbelievable numbers. One author reported that hundreds of thousands of sheep perished during a bad zhŭt of 1855–1856, while another estimated that, depending on local conditions, pasturage failure could carry off anywhere from 10 to 70 percent of Kazakhs' livestock.[3] Despite its regular occurrence, its scope remained potentially devastating, and the task before a tsarist government that sought to remediate it was formidable.

Only a few decades earlier an official organ of the Russian Empire might not have found this event particularly newsworthy. Political and cultural factors influence the scope and scale of the impact of natural disasters on human and nonhuman populations alike.[4] The same factors shape the responses to disaster that state and nonstate actors can conceive of.[5] Even seeing a natural event as a disaster takes intellectual labor; perceiving a disaster as such depends on a historically contingent confluence of cultural beliefs and political goals. Just as humanitarian journalists in early Victorian Britain discovered chronic hunger among the urban poor, so too did tsarist officials gradually come to see zhŭt as a matter that required serious, transformative intervention in the lives of the largely nomadic and pastoralist Kazakh population.[6] This was part of a broader reconfiguration of the relationship between the tsarist state and its subject population, as well as between the imperial center and a remote borderland. By the middle decades of the nineteenth century, the tsarist state had developed many of the ambitions (although they were not always fully realized) that scholars associate with Foucauldian governmentality, attempting to minimize risk and increase the welfare of its population.[7] At the same time, the conquest of Turkestan transformed the Kazakh steppe into an internal province of the Russian Empire rather than one of the frontier, implying to many observers that more direct and integrative methods of rule were required there.[8]

The Russian Empire's *mission civilisatrice* on the Kazakh steppe (such as it was) was fundamentally a matter of economic transformation.[9] Zhŭt stood for all the problems that seemed to exist in pastoral nomadic lifeways. In this way it was akin to the rhetorical sticks with which rural people have always been beaten in modernizing polities, focusing on the peasant's ignorance and helplessness before acts of God. In particular, disasters and risk mitigation have worked, in a variety of settings, in the service of transformative, modernizing agendas, a phenomenon that historian Mark Carey has described as "disaster economics."[10] Although zhŭt fitted this broad typology, it was also a problem held to be inherent in nomadic lifeways. It is this linkage, as well as the particularities of the arid grasslands the Kazakhs inhabited, that lends this story its local specificity. Once tsarist administrators and Russian-educated Kazakhs agreed, for their own reasons, that the pastoral nomadic economy needed to become more secure and stable, their assessments of the steppe environment and the lifeways it was capable of supporting were vital to their assessment of the necessary change. This was particularly so from the 1890s on, when the resettlement of peasant agriculturalists to the steppe became fundamental to tsarist visions of the region's future. This promised, so it seemed, to both maximize the use of the Russian Empire's productive forces and to ease the steppe's transition to more stable and secure economic forms. Such an approach depended on the assumption that the steppe provinces were indeed unambiguously promising for agriculture. On the other hand, if the region's water, soil, or climate comprised a limiting environment for agriculture, this transformation was doomed to fail, and other approaches were needed. Kazakh intellectuals and tsarist administrators found a fleeting middle ground when they agreed that zhŭt represented an existential threat to Kazakh pastoralists and that, therefore, transforming the steppe's economy would benefit both local inhabitants and the tsarist state.[11]

This middle ground, though, was also suffused with the rhetoric of pastoralist primitivism. This dominant lens contributed to what ultimately resulted: a program of mass peasant resettlement that left the Kazakhs worse off, in many respects, than they had been in its absence.[12] A civilizing imperative was among the arguments mobilized in support of resettlement. Intervening on the steppe to prevent the disaster of zhŭt, in the end, set the stage for more thoroughgoing disasters in the future.[13]

ZHŬT AS NONEVENT

The pastoral nomadism that Kazakhs had practiced for centuries was closely connected to the unproductive soil and extreme continental climate that characterized the steppe. In this environment, seasonal or constant mobility with animals was the most logical means of extracting usable calories from the soil. The consequences of zhŭt were closely connected with such mobility; winter pastures were part of Kazakhs' annual circuit, and if they failed, ready alternative sources of fodder were not available, since nomadism militated against harvesting and storing sufficient fodder for an extended emergency.

Yet because zhŭt was a regular occurrence and part of the nomadic lifestyle, Kazakh pastoralists conceived of it as an unpleasant experience rather than a disaster. None of the diverse forms of nomadism practiced on the Kazakh steppe were so restrictive as to prevent Kazakhs' employing that most fundamental nomadic tool in times of need, flight.[14] Looking back, the Kazakh Musa Chormanov argued in the official press of Semipalatinsk oblast (province) during the early 1870s that zhŭt had never been a serious problem prior to the Russian conquest of the steppe because Kazakhs had been able to migrate more than a thousand kilometers in search of food, at need.[15] Without the restrictions created by borders and Russian settlements, Chormanov claimed, Kazakhs had been able to move to warm dry areas that, because they received little precipitation, were both useless under ordinary circumstances and vital to survival in a cold wet winter.[16] Absent the ability to flee or if the zhŭt in question was of a smaller scale, in case of misfortune Kazakhs could depend on a well-developed and, some claimed, persistent tradition of mutual aid within the clan (*rod*).[17] In a worst case scenario, entire families might undertake the backbreaking labor of manually shattering the icy crust over fodder grasses, accomplishing with hammers what their animals' hooves could not.[18] Although some of these sources are tinged with nostalgia, their overall thrust remains clear: zhŭt could be a miserable experience, but it was a component part of pastoral nomadism as Kazakhs practiced it and could be solved within that system.

Into the 1860s tsarist administrators approached zhŭt reactively rather than proactively, and on a relatively small scale. Two serious zhŭts in the Orenburg steppe, in 1850–1851 and 1855–1856 (the latter carrying off nearly a million animals), provoked no response more serious than small short-term loans of money (less than one hundred silver rubles per needy village, in 1851) or grain.[19] This might have been expected under the "frontier" system of administration that characterized the Orenburg steppe before 1868, but the Siberian steppe was historically no different. The Semipalatinsk-based scholar and political activist N. Ia. Konshin noted that, in a zhŭt of 1840–1841, local administrators provided only small amounts of grain to their needy Kazakh subjects.[20] Two decades later, the leadership of the Governor-Generalship of Western Siberia only found it possible to release loans totaling three hundred rubles in each district under its authority.[21]

To see such relief measures as part of a creeping, inexorable drive toward sedentarization would be to miss a critical shift in the way tsarist officials thought about pastoral nomads as subjects. The historian David Bello has argued that, for the comparable case of pasturage failure and relief on the Mongolian steppes of the Qing Empire, providing relief in cash and grain rather than livestock essentially "reliev[ed] Mongols of their pastoral identity in order to save it" long before Han settlement on the steppe exerted such pressure directly.[22] Yet for centuries the Kazakhs had access to grain as a trade good and occasionally as a supplemental crop Tsarist officials were, at most, providing an additional source of grain, not destabilizing pastoral identity.[23] For the first century or more of tsarist rule on the steppe, it was just the opposite. Military governors—most notably, V. A. Perovskii of Orenburg—had no greater ambition than to maintain the Kazakhs as nomads and to take the place of Central Asian merchants in supplying them with grain.[24]

When tsarist administrators on the steppe began to feel uneasy about having pastoralists as subjects (as most had, by the 1850s), they did so not for any humanitarian or economic reasons but for the same reasons their Ottoman counterparts did, and at roughly the same time: they perceived nomads as threats to the proper functioning of the state.[25] Pastoral mobility presented a host of problems for an undermanned and cash-strapped state, which zhŭt then exacerbated. Mass die-offs of livestock meant significant shortfalls (*nedochety*) in tax collection and made it forbiddingly difficult to assess Kazakhs with any sort of tax based on their actual wealth in livestock, which seemed neither stable nor predictable.[26] Moreover, the general staff officer A. K. Geins lamented that, as a result of a zhŭt of 1862, "hungry Kazakhs roamed about the steppe, carrying out plunder and robbery."[27] Nomads did not make good neighbors, and nomads in times of crisis still less so. A. I. Maksheev, professor of statistics at the General Staff Academy, in an 1856 work eloquently summarized the reasoning behind tsarist officials' growing support for sedentarization: "Agriculturalists, by the very nature of their occupations, are more peace-loving than nomads. Tied to a certain area, they do not eagerly cast it off and worry only about defending themselves and their fields, as the natural sources of their existence; and for attacks they have neither time nor means. Nomadic people are another matter."[28] Concern about pastoral mobility and concern about zhŭt still remained separate in the minds of tsarist officials. They had not yet developed a belief that the systemic vulnerabilities of the nomadic lifestyle required systematic solutions, nor had they developed more robust aspirations for the transformation of the steppe.

TOWARD INTERVENTION

As the Kazakh steppe became an internal province of the Russian Empire and attempts were made to introduce a more uniform and active type of administration to the entire region, such attitudes slowly faded from view. The problems that zhŭt presented seemed to threaten the new ambitions of the state (epitomized by the Provisional Statute of 1868) and to demand solutions both more humanitarian and more radical than anything practiced hitherto. By the late 1890s it was possible for one keen observer of Central Asian affairs, O. A. Shkapskii, to argue that zhŭt was at the very core of Kazakhs' impoverishment and that resolving the latter was, in turn, "one of the most substantial tasks of [Russia's] cultural mission in Central Asia."[29]

During the 1860s some officials began to develop transformative administrative agendas directed explicitly against zhŭt. The general staff officer N. I. Krasovskii proposed a radical change in the regulation of land use: offering

Kazakhs private property in land at their winter camps. This, he argued, would permit some Kazakhs to remain in one place year-round, which opportunity they could use to cut, prepare, and store up hay for the winter. Kazakhs could then draw upon these stores when animals returned to camp for winter in the event of pasturage failure.[30] State officials, unsurprisingly, rejected such recommendations; private property in land had recently been rejected as inappropriate for the peasantry and Kazakhs were hardly considered more developed.[31] The Provisional Statute of 1868 would acknowledge only a theoretical right to full individual ownership of land, rarely recognized in practice.[32]

Less politically sensitive, and thus more promising, were proposals to collect regular funds to create a permanent and growing capital to provide loans for Kazakhs in desperate situations, regularizing the earlier ad hoc system. In 1863 the governor-general of Orenburg and Samara, A. P. Bezak, asked the minister of internal affairs, P. A. Valuev, for permission to establish a micro-credit institution, a *ssudnaia kassa*, for the Kazakhs (mostly of the Small Horde) under his authority. He opened his proposal with a picture of a steppe and a population constantly beset by disaster and an administration unequipped to deal with it: "The Kazakhs of the trans-Ural region, subjected almost annually to disasters from drought, zhŭt [*gololeditsy*], locusts and so forth, constantly turn to the Provincial Administration of the Orenburg Kazakhs with requests for monetary loans, which, although they are rendered to them as possible, from free operating funds [*shtatnye summy*], but only in the most insignificant size for lack of funds, and these are far from corresponding to their needs."[33]

Officials in Akmolinsk oblast repeated this concern about the necessity of immediate financial assistance for the purchase of grain and prepared fodder in the event of a zhŭt.[34] Motivated by humanitarian concerns the regulations for the *ssudnye kassy*, once they were established, also drove needy Kazakhs inexorably closer to the reach of the tsarist state. These regulations, depending on the province, demanded that petitioners present written, stamped attestation from a local official.[35] In Akmolinsk and Semipalatinsk oblasti, this attestation was to include a statement that potential borrowers were registered in official books (*shnurovye aul'nye knigi*) as tent owners in a certain village.[36] By the 1870s, already, the rough-and-ready method of nomadic flight from zhŭt was becoming less tenable. Kazakhs who wanted real assistance could obtain it only by entering more closely into the embrace of a state that, by increasing its reach, sought to curb their mobility.

The fragmentary evidence that is available strongly suggests that Kazakhs made use of these new institutions once they were introduced. In 1881 Governor A. P. Konstantinovich of Turgai province noted that, despite severe winter conditions, the Kazakhs in one of the districts entrusted to him (Nikolaevsk) were doing fairly well, in part because of "the comparative prosperity of residents and the prosperity of the cantons of this district, where more than twenty-six thousand rubles circulate in the ssudnye kassy."[37] Later, reporting on the severe winter of 1897–1898, another Turgai governor, Ia. F. Barabash, noted that the provincial ssudnaia kassa had served the largest (eighty-five hundred rubles) of several sources at his disposal to aid needy Kazakhs.[38] It is safe to conclude that, for whatever reason, Kazakhs began to run to the tsarist state (in the form of disaster relief from the ssudnye kassy) as eagerly as it reached out to them.

The ssudnaia kassa institutionalized and regulated relief during the provisions crisis that accompanied zhŭt, providing money for the purchase of food and flour for hungry nomads and fodder for their livestock. By the 1880s it was an institution on which many Kazakhs depended in extreme situations. Still, at the same time as the ssudnye kassy began to flourish, officials at several levels of the administrative hierarchy began to think of solutions to zhŭt in more systemic and enduring terms. Krasovskii's economic transformation, focused on private property, was at odds with perceived state interests at the time he proposed it. These officials, rather, believed that ending zhŭt would be among the desirable consequences of a policy that satisfied the tsarist state's new economic and cultural aspirations: the peasant colonization of suitable areas of the steppe.

Colonization, according to the Provisional Statute of 1868, was to take place on a free and natural, rather than an organized basis, and could be agricultural or commercial (*promyshlennaia*). The proposal from Western Siberia that served as the basis for further discussion made only vague references to "great benefits both with respect to civilization of the Kazakhs and the development among them of agriculture and other trades particular to sedentary folk."[39] Later proposals originating from the same region made clear the sort of benefits that Kazakhs might expect from a program of organized, if limited, Russian colonization. Already in 1876 the governor-general of Western Siberia, N. G. Kaznakov, repeated his predecessor's claim that Kazakhs learning from Russian peasants how to grow grain, rendering them no longer dependent on grain imports during zhŭts, was one of the many benefits to be anticipated from "the cautious settlement [*vodvorenie*] within the Kazakh steppes of a sedentary Russian population, without constraint of the nomads."[40] Although the modest results of early experiments in agricultural colonization led a later governor-general, G. A. Kolpakovskii, to suspend it, the halt was only temporary, the civilizing logic inexorable.[41] Soon after Kolpakovskii's halt, in 1888, the majority opinion of a commission he formed to assess the prospects of peasant colonization in Akmolinsk and Semipalatinsk provinces made the connection between zhŭt prevention and resettlement explicit:

> The majority of the commission . . . acknowledged colonization of the Kazakh steppe by settlers from the inner provinces of Russia as desirable not only in view of the political and economic considerations expressed by the former administration of Western Siberia but also taking into consideration the latest information about the salutary influence, which the settlement of Russian villages on the steppe could render in support of the nomads during zhŭts, frequently recurring on the steppe, when stores of fodder in villages, obtained by the Kazakhs, could prevent the deaths of hundreds of thousands of head of Kazakh livestock.[42]

Thus, well before the peak of the resettlement era, when peasant resettlement was only allotted a "subsidiary role" in the matter of economically Russianizing the steppe, assumptions about the ability of peasant colonizers to change Kazakh pastoral nomadism sufficiently so as to render it less disaster prone were well, if not universally, established.[43] For those who viewed Slavic peasants as effective civilizers, these good neighbors would serve Kazakhs not only as a material resource but also as teachers of the techniques and habits that would prevent them and their stock from starving in the future.[44] For their own good, Kazakhs would have to surrender at least some land to civilizers from outside the steppe; colonization, sedentarization, and security would go hand in hand.

MITIGATING THE THREAT

A complex of negative ideas associated with pastoralism gave sedentarization a place at the heart of solutions both to the administrative concerns described above and the closely connected question of Kazakhs' impoverishment. Sedentary economies, however, may take multiple forms, and the transition to sedentarization may be accomplished by a range of stimuli. The range of positive valences associated with Russian settlers on the steppe illustrates this point nicely; mowing and storing hay was more compatible with limited pastoral mobility than a large-scale shift to agriculture. Such variation meant that the specific nature of the economic transformation that zhŭt was understood to necessitate was contested within a matrix of political priorities and knowledges of the steppe environment.

Agriculture and Adaptation

That agriculture was the preeminent and preferable means of struggling against zhŭt (and securing Kazakhs' existence, more broadly) was the editorial line of the official *KSG*, published from 1888 to 1902. The author of an unsigned editorial remarked: "One must rejoice at the development of agriculture among the Kazakhs first, because grain cultivation is in many cases a more reliable and lucrative occupation than stock raising; second, because with the development of agriculture for feeding a family much less land is needed than in the case when means for life are gotten exclusively with the help of stock raising."[45] A year later, in a review of a collection of articles about Siberia, the anonymous author approvingly noted: "Sooner or later the Ķazakhs will have to introduce agriculture to their economy, as a better means of securing life."[46] If zhŭt sometimes only loomed sotto voce in this rhetoric of reliability and security, at other times it became quite an explicit factor motivating the transition to agriculture. Nowhere was this connection clearer than in the words of an obscure Russian commentator, M. Imshenetskii, in 1891: "To have herds is a good thing but not durable [*prochnoe*], and if someone has herds of various stock, and then also starts to do agriculture, then that person's herd will also be better protected, for all fodder will be much, whole mountains of wheat straw will pile up, and with this fodder there is nothing to fear from zhŭt; during any misfortune, Mother Earth will rescue him from any need."[47] An enviable picture, no doubt, particularly in combination with the prosperous peasant of Imshenetskii's depiction and in light of the fabulous harvests he claimed it was possible to reap from steppe land.

Risk mitigation and "civilization" through sedentary agriculture thus became a matter of possible agreement between aggressive advocates of resettlement and some Kazakh intellectuals. For some, this was simply a matter of sowing where the land was good. The publishers of *KSG* gleefully reprinted letters from Kazakh correspondents who had sown grain successfully, propagandizing their success to the recalcitrant.[48] A faction of the Kazakh intelligentsia continued to propagate this line of thinking well into the twentieth century. This group, centered around the Troitsk journal *Ai qap* (1911–1915), favored agriculture as a means of increasing ordinary Kazakhs' civilization level and prosperity. Thus, like the official newspaper before them, they publicized cases of successful transitions to agriculture, such as among Kazakhs living on the Baraba steppe of western Siberia. Having taken good land, these Kazakhs were able to market an agricultural surplus and sell copious amounts of hay during the zhŭt years that threatened their nomadic neighbors. Investing their profits in cultural institutions and thus raising their own cultural level, these Kazakhs' activities were a "very pleasing thing to hear [*qŭlaq süisĭnerlĭk* ĭs]."[49]

Others believed that agricultural land on the steppe was of uneven quality, and so they had to propose adaptive measures to support grain cultivation in dry areas with thin soil.[50] Taking such measures held out the promise that Kazakhs could indeed change their occupation to one less vulnerable to natural caprice, rather than being condemned to poverty and vulnerability. A Kazakh writing to *KSG* invested much hope in a varietal of spring wheat called *chul-bidai* (desert wheat), reputed to give a good harvest even during drought years.[51] The interest in afforestation to ameliorate a continental climate that David Moon has noted for the Black Sea steppe found a place among Slavic commentators.[52] Still others saw similar promise in irrigation.[53] These adaptive measures rested more on imperial institutions and personnel (seed banks, agricultural schools, engineers, and agronomists) than the mass movement of peasant *Kulturträger* to the steppe. But the modernizing, transformative assumptions that buttressed all of them left the steppe vulnerable to a plan of peasant colonization based on similar rationales.

The institutions created to manage peasant resettlement to the steppe justified past and future peasant colonization in exactly such terms. These organizations—the Siberian Railroad Committee, in 1892, succeeded by the Resettlement Administration in 1896—justified the earlier arrival of irregular migrants (*samovol'tsy*) on the grounds that they introduced "higher" forms of economic practice where they settled.[54] After the turn of the century, when organized resettlement became a crucial part of the plans for economic modernization that developed under S. Iu. Witte and P. A. Stolypin, officials and statisticians in the employ of the Resettlement Administration repeated and extended these arguments, presenting Slavic peasants as "a kind of cultural or agricultural instructor to native easterners, particularly nomads."[55] Slavic peasants, they claimed, were necessary agents in the agricultural transformation of the steppe, without which Kazakhs' lives would remain precarious. Even local officials who wavered on the desirability of settlement were forced to admit that limited agricultural colonization was the cheapest, most efficient way of securing a steady supply of grain for Kazakhs.[56] Resettlement plans, mitigation of the risks associated with zhŭt, and the promotion of agriculture were thus closely intertwined.

Yet others argued that there were limits to adaptation; they were deeply anxious about the extent to which the steppe environment could ever support agriculture. Rather, a move to intensified stock raising—preparing and storing hay, stabling, and using shorter migrations—seemed to them to offer Kazakhs equal security from zhŭt while continuing to serve state interests on the steppe. An unlikely alliance of local scholars and Russian-trained Kazakh intellectuals made exactly this argument—though here, too, a positive role for settlers could not always be written out of the story.

Perfecting Pastoralism?

Following a severe zhŭt in 1892 the military governor of Akmolinsk oblast, N. I. Sannikov, came to extremely negative conclusions about the state of nomadic stock raising there. In a circular to district chiefs (*uezdnye nachal'niki*) and veterinary doctors, he argued that the root cause of Kazakhs' poverty and vulnerability consisted "in the very irrational management of the steppe economy, in the ignorance of the Kazakh nomads themselves, and their inability to adapt to their changing life conditions."[57] Sannikov considered this situation unacceptable, in turn, because it jeopardized the steppe region's position in the increasingly integrated imperial economy, which included supplying industrial workers with meat and cavalry units with mounts.[58] He noted that the pastoralist status quo had a deleterious influence on nomads' own well-being. In this view, improving the welfare of individual pastoralists and incorporating them into the movement of goods and people around the empire went hand in hand. What is absent from Sannikov's circular, however, is any mention of peasant colonizers as effecting this transformation. Rather, he advocated a broad-based educational program in which district chiefs and veterinarians were asked to explain to Kazakhs the deficiencies of stock raising as currently done; model Kazakh pastoralists would be presented to the steppe governor-general for awards; and *KSG* would also play an important role.[59]

In accordance with Sannikov's instructions, the following year, *KSG* began devoting additional column inches to problems of animal husbandry. Most prominent in this effort was a series of columns by V. Mikhailov, head of the veterinary personnel of Akmolinsk oblast, entitled, in the newspaper's usual didactic mode, "Conversations on Stock Raising [*Besedy o skotovodstve*]."[60] Having recently noted in the journal of the local branch of the Russian Geographical Society that Kazakh stock raising was "done in the most primitive way and wholly left to the conditions of nature," such that "all forms of adversity, such as storms and zhŭt . . . [were] evident in the well-being of the Kazakhs," Mikhailov kept to a similar line or argument in his column.[61] The flaws he identified among local pastoralists were numerous and included a relative lack of orientation toward the demands of the consumer markets of European Russia and Siberia; vulnerability to epizootics and wolves; and, above all, zhŭt, held responsible for starvation, premature births, and the deaths of newborn animals.[62] The solutions he advocated were of a piece with what Sannikov had indicated the previous year and also with what was being advocated by other observers either dubious about agriculture's potential on the steppe or keen to retain the advantages that a developed animal husbandry sector could provide, or both. Chief among these solutions was obligatory communal storage of hay for winter, stocked with hay that Kazakhs would gather themselves in nearby marshlands, to keep young and weak livestock alive.[63] *KSG* featured correspondents who were, like Mikhailov, advocating fodder storage and offering its readers detailed advice about how best to do so.[64] Similarly, in neighboring Turgai oblast, the veterinarian A. I. Dobrosmyslov saw little hope for agriculture in the soil of the province, which was lacking in humus-rich *chernozem* and quickly growing exhausted.[65] He praised the efforts of the military governor, Ia. F. Barabash, to introduce obligatory hay storage after the zhŭt of

1891–1892.[66] This too would involve at least partial sedentarization, returning to fixed points to store hay and sheltering animals in at least semi-permanent dwellings. By the 1890s, as provincial and district borders were drawn on the steppe and sedentary farmers claimed land permanently, the days of migrating from the feather-grass pastures of the northern steppe to the Syr-Darya River in search of open pasture were over.

As in the case of agriculture, many Kazakhs argued along similar lines. The ethnographer and pedagogue Ibrai Altynsarin, in direct response to the catastrophic zhŭt of 1879–1880, took to the pages of *Orenburgskii listok* (Orenburg leaflet) with a unique program for dealing with the crisis. Altynsarin rejected the connection between pastoralism and vulnerability to acts of God, noting that the district city of Turgai "bears adversity just the same as the Kazakhs do."[67] At the same time, he expressed a desire not to "condemn [*obrech'*] the Kazakhs to eternal nomadic life."[68] The way forward, in his mind, was a form of intensified and market-oriented stock raising, which would later be complemented by a system of vocational schools focused on processing locally available animal products that Altynsarin himself played a role in introducing.[69] Such a system, in Altynsarin's mind, leveraged both the natural conditions of the steppe and knowledge of stock raising among the Kazakh population. Within the second generation of the Kazakh intelligentsia, a group centered around the Orenburg newspaper *Qazaq* (1913–1918) took up a similar position: advocating intensified stock breeding and the expansion of a network of vocational schools, which, they believed, would give impoverished Kazakhs the skills they needed to recover financially from a zhŭt.[70] In advocating agriculture and modified pastoralism alike as responses to zhŭt, Kazakhs and tsarist bureaucrats could—and did—find common ground.

Nor were peasant colonists necessarily to be the agents with whose help Kazakhs modified their stock raising. Sannikov made no mention of colonization in his circular, and Altynsarin argued against it. After the turn of the century, a regional scholar in Semipalatinsk oblast, B. Benkevich, argued that Kazakhs were already perfectly conscious of the benefits of cutting hay, quite able to store it on their own, and that the best way to further improve their stock raising was "significant restriction of colonization, so as to preserve pastures for animal husbandry."[71] For enthusiastic advocates of resettlement, however, peasant colonists were precisely the cultural force the Kazakhs needed to improve themselves. For some observers they were sources of improved technology to cut hay; others saw in them a model of constancy and work ethic for Kazakhs to emulate.[72] In both respects, peasants represented a conduit for the habits that would help Kazakhs store up the fodder grass required for their livestock to survive a zhŭt. Thus members of the tsarist Resettlement Administration noted, with evident pleasure, that the proximity of settlers doubled

Kazakhs' hay mowing and consumption in one specific case, and that, in general, after extreme winter weather Kazakhs sought the "protection of Russian settlements," settling near them, mowing hay, and building more permanent dwellings.[73]

This humanitarian justification for colonization left open the possibility that colonization—and the modernization of the late tsarist economy that was to come with it—would someday become a matter of greater state interest than protecting Kazakhs and their livestock. This was even more the case because only a few isolated local voices expressed the idea that Kazakhs' stock raising could remain unchanged. In the era of peak resettlement after 1906, when there were particularly few restrictions on settlers' movement, colonization won out.[74] This was particularly the case in the cradle of resettlement on the steppe, Akmolinsk oblast. There, the 1907 statistical expedition of V. K. Kuznetsov gave mathematically precise evidence that the future of the region had to be agricultural and chiefly Slavic. Kuznetsov's count, dismissed by the statistician and autonomist politician Älikhan Bökeikhanov as a "lie [ötĭrĭk]," sharply reduced the minimum amount of land not subject to seizure from Kazakh pastoralists for the sake of peasant settlers.[75] His aggressive calculations left a bare minimum of room for the intensified stock raising and adaptation to zhŭt that Kazakhs and local agronomists alike had advocated: "There remains therefore only one possibility [for continued support of stock raising]—to strengthen hay mowing on the steppe. But this cannot be done quickly, and with the long duration of the process the quantity of steppes in the Kazakhs' use, as a consequence of land seizures, will decrease. Together with this will also decrease the possibility of expanding hay mowing on the steppe. All this indicates that the further development of Kazakh stock raising must meet with an obstacle in the form of deficiency of hay mowing lands for winter fodder stores."[76]

Stock raising, in this view, could never be intensified enough to accommodate the needs of land-hungry settlers, now a greater priority than any benefit pastoralists could provide the empire in meat, skin, and mounts. Kazakhs could neither remain as they were nor count on hanging on to enough land to support as many livestock through difficult times as they previously had. The only remaining option was to farm on whatever areas were left after settlers were fully allotted. What this meant in light of other indications about the difficulty of growing grain on the steppe was a question Kuznetsov never answered. Within three decades, as political circumstances evolved, humanitarian intervention in Kazakhs' lives had become extensive and disastrous expropriation.

CONCLUSIONS

Despite the multiple meanings that might have been attributed to it, zhŭt, the failure of the fodder grass harvest, and the starvation of livestock that ensued from it came to have just one meaning in the late Russian Empire. Under imperial eyes, it was above all a clear marker of the inherent instability of pastoral nomadic lifeways. Although this strongly implied that zhŭt also signified the need to improve the economy of steppe Kazakhs, there was less certainty about how this was to be accomplished. This impulse toward economic improvement, moreover, can be historicized, dating to the late 1860s. It was the product of a cognitive shift away from favoring nonintervention in the borderlands toward programs intended to improve the health and welfare of the nomadic population. In a region that did not even boast canton-level imperial schools until the 1880s, this was the main thrust of the *mission civilisatrice*.

The various programs mooted to prevent the worst consequences of zhŭt represented a series of attempts to force evolutionist ideas about pastoral life and unease about the developmental constraints represented by the steppe environment into a coherent plan of action. The political climate of the Russian Empire from the 1880s onward meant it was always likely that this solution would center on peasant colonization of one sort or another, but alternative views were possible and articulated in multiple spheres. Such alternatives are a useful reminder of the contingency and indeterminacy of Russian imperialism on the Kazakh steppe. All such alternatives, however, involved a transformation—the idea that Kazakh practices as they were had to be changed for the better. As an adjunct to such arguments, zhŭt contributed to changes in the steppe, not in short bursts every ten to twelve years but constantly, gradually, and inexorably. The many solutions to Kazakh vulnerability to zhŭt pointed to many visions for the future of the steppe, grounded in diverse views of its environmental potential; once tsarist officials determined that agricultural colonization was the best way forward, the rhetoric of humanitarian intervention remained, but these other ways of knowing the steppe faded from view.

Public and administrative discourses, and practices aligned with them, are only the beginning of the story, however. Cathy A. Frierson, in a study of the meanings and practices associated with fire in rural parts of European Russia, emphasizes that fire, beyond its symbolic meaning, was "a quantifiable factor in imperial Russia's and the Russian peasantry's economic condition."[77] If zhŭt's place in imperial discourse about the Kazakh steppe was in many ways analogous to the place of fire in descriptions of rural Russia, so too did its material effects have far-reaching consequences. Both fire and zhŭt were forces of nature, and in the eyes of modernizing tsarist officials vulnerability to these forces was perceived as backwardness, and the successful struggle against them promised a more prosperous and secure existence. Moreover, ice and snow on the steppe, like fire, transcended the discursive in their significance; grass froze, animals died, people moved around. In the case of zhŭt, though, the meaning that bureaucratic actors made of the natural world and the immediate consequences of the phenomenon—including poverty, temporary sedentarization, and reduced land use—formed a particularly inauspicious conjunction for Kazakh pastoralists. As a rising tide of peasant migrants (many fleeing the Russian famine of 1891–1892) flooded the steppe in the 1890s, many Kazakhs were vulnerable economically and were forced to compete for land with the very people who, in some understandings, were supposed to "civilize" them.[78] Peasant and Kazakh migrations at this time were thus entangled and mutually constitutive.[79] Climatic happenstance cast people onto the steppe whom tsarist officials were prepared to understand as civilizers, whose influence could protect needy Kazakhs from zhŭt's baleful consequences. At the same time, their growing numbers and settlement on the land may have ensured that Kazakh pastoralists could not recover from zhŭt in the ways to which they had been accustomed.

5

Desiccated Steppes

DROUGHTS AND CLIMATE CHANGE IN THE USSR, 1960s–1980s

Marc Elie

After settlers had plowed up vast areas of land in western Siberia, the Altai and Ural regions, and northern Kazakhstan in the course of the Virgin Land Campaign (1954–1960), the weather soon appeared to disrupt the efforts of these "New-Landers [*tselinniki*]," as they were known. The first half of the 1960s saw devastating droughts, dry winds (*sukhovei*), and destructive sand storms, similar to the North American Dust Bowl of the 1930s.[1] From the end of the 1960s through the 1980s, droughts plagued the steppes of Ukraine, Russia, and northern Kazakhstan, regions that were the main grain suppliers of the Soviet Union.[2] One visible consequence of periodic crop failures was that the Soviet government was forced to import grain beginning in 1963, using gold reserves, and later, petrodollars.[3] Agricultural production stagnated in the 1970s and 1980s, notwithstanding significant investment in chemicals, mechanization, and subventions.

In the 1970s scientists and politicians sensed that something was going wrong in the steppes. For many Soviet observers, the weather had turned mad, the natural conditions for farming were degrading, and steppe agriculture seemed deadlocked in a struggle against a changing environment. Research on the steppe environment then focused on droughts and correlated phenomena because they were the most blatant symptoms of the perceived changes. Temperatures were rising, soils were getting drier, water levels were declining in rivers, and dry winds were blowing away the soils. At the same time, many saw crop failures during droughts as the result of improper farming practices, leading to soil impoverishment and erosion. Russian and Soviet earth scientists considered both climatic change and agricultural policies as key to understanding the trouble with dry farming on the steppes.

In this chapter I focus on how conceptions of drought evolved among Soviet scientists from the eve of the Virgin Lands Campaign (VLC) at the end of Joseph Stalin's reign to the final years of the Soviet empire in the 1980s. I do not provide a complete review of the approaches taken by different scientific disciplines on drought. Instead, I concentrate on three disputed issues: the respective weight of human and natural factors, knowledge about drought before and during the VLC, and explanations for the drought crisis of the 1970s through 1980s. I handle the first issue not only as a discussion of whether Soviet meteorologists envisaged drought as a natural or an anthropogenic phenomenon but also as a general discussion of how approaching Soviet agriculture with the tools of environmental history can enhance our understanding of intertwined natural and social processes. Then I explain how Soviet scientists made major achievements in drought conceptualization and assessment in the first half of the twentieth century but were nonetheless helpless when the post-Stalin leadership launched the VLC because the climate of the regions where it played out had been understudied. Finally, I explore how political leaders, agronomists, and climatologists accounted for the perceived "aridization" of the steppes in the 1970s and 1980s.

TOWARD AN ENVIRONMENTAL HISTORY OF SOVIET AGRICULTURE

The impact of the weather on Russian and Soviet crop yields has been a subject of considerable study both in the USSR and abroad. Meteorologists, agronomists, geographers, economists, and historians have tried to assess the conditions that produced a long list of crop failures, crippling Soviet farms and bringing deadly famines (1921–1922, 1932–1933, and 1946–1947) and countless food crises. How responsible is moisture deficiency for Soviet agriculture's haphazard and overall poor performance, in contrast to governmental policy and technological advancement in a command economy? Collectivization hindered the performance of Soviet farms by removing all autonomy in farming decisions from farm managers and peasants, setting artificially low state procurement prices, implementing misguided melioration campaigns, and many other socioeconomic factors.[4] By contrast, in their 2005 book, Niko-

Map 5.1. The USSR, c. 1950.

lai Dronin and Edward G. Bellinger argue that weather conditions were the strongest factor influencing yields in the Soviet period. Irregular precipitation accounts for most of their considerable yearly oscillations. "In most cases, statistical data on harvests in Russia are in compliance with weather characteristics," they observe. Bad harvests occur when the year is dry, and the harvest is good in moist years.[5] In their view, the famines of 1921–1922, 1931–1933, and 1946–1947 were caused by violent procurement campaigns and so were exceptional, pure products of Bolshevism and Stalin's dictatorship.[6] But apart from these events, Dronin and Bellinger tend to emphasize climatic factors and downplay the impact of politics and economics. They hold the political leadership responsible for food crises when it set the wrong incentives via the price policy. But they ignore the question of how political decisions relative to the farm economy prepared farms for the drought or, on the contrary, weakened them; worsened drought manifestations or, on the contrary, alleviated them. They treat droughts as purely physical meteorological events, not as products of nature-society interactions.

I plead for another understanding of drought in the Soviet Union. In place of trying to parse human-made and climatic factors to explain diminishing yields, I propose to analyze these factors as interacting together, feeding one another, in one drought process. This approach leans on several observations. First, scholars have already debunked the preconception that catastrophes are natural phenomena that strike society from the outside. They criticize the expression "natural disasters," because it implies that disasters are physical events, insisting they are instead socially and historically constructed.[7] They refuse to link "nature" and "society" with a one-way relationship wherein physical events strike social processes; they show that they are deeply intertwined. Furthermore, neither nature nor society is a closed entity. "Vulnerability" is a key concept to approach how different groups within a society—or countries within an international system—are more or less exposed to natural hazards and risks, with the poor and the weak usually more subject to disasters than the wealthy and secure.[8] Analyzing disasters requires examining the fabric of society to comprehend how its members are situated on the sociogeographical scale and how marginalization increases exposure to hazards.[9] Remarkably, the Sahel drought of the 1960s–1970s provided the first case whereby scholars redefined disaster studies, shifting the focus from technical expertise to social problems and provoking a "vulnerability turn" in the discipline.[10]

In a famous book on the Sahel famine, the atmospheric physicist Rolando García deplores the "tendency to consider both climate itself and the climatic fluctuations as a given." To him, "climatic phenomena are only meaningful with reference to a certain society." He proposes therefore to distinguish between "water deficiency" and "drought," the former being the result of a hydrological and meteorological assessment, whereas the latter is "defined with reference to a certain productive system" whose expectations of the weather (expectations upon which crops, chemicals, machinery, and timing are chosen) are deceived by a stronger-than-normal climatic variability. A lack of water is not sufficient to explain a drought disaster. The vulnerability of an agricultural system to climatic variability is the result of economic and political processes. Plants need more or less water depending on the chosen species and variety, different soils have different water capacities and drought resilience, water may be or not be used for irrigation, and so on. Depending on its configuration, a farm is more or less vulnerable to drought. García proposes therefore a definition of drought as "the social perception of a water deficiency with reference to a normal condition socially defined."[11]

This provocative formulation of drought as perception should serve as a reminder that drought is never only a product of atmospheric physics and hydrology. García's definition reminds us that disastrous droughts are not sufficiently explained by a lack of water. This is why conventional definitions of drought usually refer to agricultural requirements. Consider the definition proposed by the US National Drought Mitigation Center at the University of Nebraska–Lincoln: "Drought is a protracted period of deficient precipitation resulting in extensive damage to *crops*, resulting in loss of *yield*" (my emphasis).[12] Climatologists themselves do not nowadays frame drought as a physical event only; they recognize that social vulnerability plays a key role.[13]

A second consideration for understanding drought as a process within the agricultural system comes from environmental studies. In agriculture, as in any other productive activity, natural processes are shaped by human intervention and harnessed to societal needs. Agriculture is the effort to bend biological, pedological, meteorological, and hydrological forces to the needs of human consumption. Natural processes in soils, the atmosphere, and waters cannot be considered external to production; they are integrated with human labor in one system.[14] Industrial farming in the nineteenth and twentieth centuries changed the traditional balance of natural and anthropic processes through a considerable increase in mechanical, chemical, financial, and scientific inputs such as selected seeds, heavy machinery, and plant-protecting and soil-enriching chemicals.[15] Furthermore, states have greatly expanded cultivated areas to the detriment of forests and grasslands, triggering far-reaching changes in ecosystems and in the biosphere. The human impact on them has considerably increased since the beginning of the Anthropocene.[16] The Soviet Union showed extreme voluntarism and transformism, aiming to control the immensity and adversity of the Eurasian territory and to overcome the limitations and plagues of traditional Russian agriculture.[17] A striking case of artificialization of natural elements is the modification of watercourses for irrigation in the North Caucasus and Central Asia, as Christian Teichmann and Julia Obertreis explore (this volume).[18]

But is not the weather different from other natural forces and entities like soils and waters? Farmers can't bend the weather to their needs as they do soils, fauna, flora, and water. Likewise, they can't mismanage and devastate it. However, with a growing understanding of how human activities affected the climate in the nineteenth and twentieth centuries, it is no longer considered independent of human agency. Climate has ceased to be considered the ultimate exogenous factor for societal change.[19] As a whole, agriculture is a major contributor to climate change via deforestation, cultivation, and emission of greenhouse gases. But it still holds that climate is not directly manipulable by farmers (as animals, plants, soils and water are). Therefore, there is no direct link between the production of food in farms and droughts hitting rural economies, contrary to the situations when maladapted agricultural practices, overuse of chemical fertilizer, and reduced biodiversity can backfire on farmers in the form of erosion, pollution, and diminishing yields. This said, and remembering García's warning, drought is not a natural fact imposed upon the productive system. Donald Worster has shown how quickly changing economic patterns produced the North American Dust Bowl in the 1930s: capitalistic farming replaced pasture lands with mechanized wheat monoculture on the Great Plains in the 1910s and 1920s, exposing farmers both to the negative impact of a decade of low rainfall and to the Great Depression of the 1930s.[20] Moreover, farmers must gamble on the stability of the weather. They were lulled into believing that the good wet years they had known until 1930 were a stable feature of the Great Plains climate, not a temporary opportunity that would sooner or later give way to series of dry years. The low rainfall of the 1930s defied the weather expectations on which they had built their farm economy. The same was true of New Landers on the Eurasian steppes: a wet year allowed for a bumper harvest in 1956, when the VLC had reached cruising speed, making many—foremost Nikita Khrushchev—believe that the rapacious plow-up had been the right thing to do.[21]

We see the entanglement of natural and social processes at work in agriculture generally and drought particularly when farmers augment their control of natural processes and spaces. Roughly speaking, drought is the downside of the agricultural extension of Russia and the Soviet Union into the dry steppe in the nineteenth and twentieth centuries. But we shouldn't overemphasize control. Natural processes have their own force and follow their own dynamics, which have not been entirely controlled and predictable, notwithstanding the claim to do so in the age of "high modernism."[22] For this reason, more often than not, well-intentioned agro-technical decisions brought adverse, unanticipated side effects to Soviet farms: after years of lobbying the government and only after Khrushchev's fall did Kazakhstani agronomists succeed in imposing a fallow-wheat crop rotation on the Virgin Lands, to leave soils time to recover and accumulate moisture and nutrients. Although created to conserve the soils (it was named "soil-preserving"), this system worsened erosion in the long run, because unprotected fields are more prone to deflation than cropped fields.[23]

A third motive for considering drought as a social phenomenon stems from the work of Soviet natural scientists. They hinted at the entanglement of natural and societal forces in drought formation when they discussed whether a purely physical definition of drought was possible and desirable. Meteorologists Vladimir Loginov and Anatolii Uteshev called for a universal definition of drought based on the discovery of meteorological laws. They criticized scholars who inferred meteorological conditions from yield data. Especially during earlier periods without instrumental measurements of weather (but for the Soviet period, too), many meteorologists used yield data as a proxy for weather situations. Roughly speaking, a bad harvest meant that the year was dominated by drought (as seen above, yields and weather show a strong correlation). Or they would compare yield curves (a proxy for biological production) to meteorological ones and use yields to confirm or contradict the characterization of a given year as drought-stricken and assess its severity. For Loginov, such an "agronomic" definition of drought "can't be objective" because yields do not depend exclusively on weather.[24] Uteshev added that working with a purely meteorological definition of drought would help understand its effect on other plants and animals, whereas the use of harvest data, almost exclusively for wheat, pinned the analysis to one crop.[25]

Against Loginov and Uteshev, Feofan Davitaia and Oleg Drozdov criticized the quest for a purely physical definition of drought as naïve and fruitless. For agrometeorologist Davitaia, droughts are "agro-hydro-meteorological phenomena" that cannot be studied "outside of the object which they impact and without accounting for the agricultural techniques used" on the farms.[26] The classical definition of drought given by the agronomist Vladimir Rotmistrov in 1913 stresses the link between weather, water, soil, and plant in drought: "Drought is a rainless period of a length such as to exhaust moisture accessible to plants in the root layer of the soil."[27] Drought is not the mere lack of rainfall, but a deficiency of moisture. To this we should add: not for all plants, but for crops; not all year long, but during the vegetative period; and so on. An objective meteorological definition of drought is possible, argues Drozdov, but climate data says nothing until the anomaly series are compared with that of plant development and other processes.[28] The interesting question is: What weather conditions are detrimental to water availability, to soil moisture, and to plant and animal development in a given region's particular agricultural system? Or, to reverse the question: What selection of crops and soils makes a farm vulnerable to which departures of the weather from the long-term norm?

Most Soviet meteorologists agreed that there can be no universal characterization of drought. For Davitaia, "criteria for drought . . . cannot be universal. What is drought for a type of plant raised using some agricultural practices can be optimal conditions for another type with more advanced agricultural practices."[29] Drozdov adds that a given drought level in the Moscow region can represent perfectly fit moisture conditions in central Kazakhstan. The same is true of soil drought: soils are differently affected by a lack of moisture. In Karelia in the north of Russia, pedological drought occurs even after a short interruption in spring rainfall because soils are shallow and rest on a rocky foundation. The same is true of peat bogs that have bad capillary properties, so that plants suffer from water deficit with even a small negative variation from the norm. Therefore, Drozdov observes, drought disappears as a clear-cut natural phenomenon.[30] Drought characterization is more deeply dependent on human agricultural parameters than are lightning, avalanches, or forest fires. Floods are other such phenomena that are heavily dependent, in their definition and their development, on societal factors and parameters.[31]

The regional character of drought definitions is reflected in the difficulty of transferring the operational definition of drought (often crystallized in indices) from the region for which they have been created to other regions. The famous Palmer Drought Severity Index works best for arid and semi-arid regions in the United States. As in many other arid regions of the world, it was not adopted in Russia.[32]

Practicing an environmental history of agriculture does not mean introducing "nature" into the narrative. Economic historians do not ignore nature in their studies of Soviet agriculture. They handle physical phenomena such as weather and climate, more rarely waters and soils. But, adopting the methodologies of agrometeorologists, they isolate and ponder the weather, the economy, and politics as factors to explain the outcomes of Soviet agriculture. One of their key imperatives is to find methods to distinguish natural from other factors in harvest data. In their view these natural factors obtrude on agriculture.[33] In contrast, environmental historians do not assign weight to each factor, for instance by insisting that natural factors are more determinant than social or economic ones. They analyze these factors as entangled in feedback processes and refrain from speaking of an exogenous, direct, and one-way impact of physical processes on society, because nature in agriculture is mediated by the production process.[34]

CLIMATE KNOWLEDGE AND THE VIRGIN LAND CAMPAIGN

After having discussed how environmental history can enhance our understanding of drought and agriculture, let's move to the body of knowledge on steppe climate accumulated by Soviet scientists on the eve of the VLC. The motivation for this review comes from the bewilderment of historians when they observe that the New Landers had no usable knowledge of droughts when they rolled into the eastern steppes, although Soviet science had made major achievements in climatology, agrometeorology, and drought study in the first half of the twentieth century. To account for this paradox is the goal of this section.

When Stalin's heirs launched the VLC in 1954, politicians and scientists had long recognized that drought was the major calamity of Russian and Soviet agriculture. The famine of 1891 had triggered state reaction on an unprecedented scale in Russia and had led many scientists to look for solutions: the father of modern pedology, Vasilii Dokuchaev (1846–1903), went on an expedition to the drought region. His widely read book *Our Steppes: Past and Present* (1892) analyzed the causes of the crop failure and set up a plan to render Russian agriculture resilient to drought.[35]

In the first decades of the twentieth century, Russian climatologists and agronomists made several breakthroughs in measuring drought intensity and explaining their causes. Aleksandr Voeikov (1842–1915) and Petr Brounov (1853–1927) founded agrometeorology, a discipline crossing plant physiology, pedology, and weather data with harvest statistics. Drought theory and prevention is a major field of study for agrometeorologists.[36] On the eve of the First World War, Brounov, chief of the Meteorological Bureau of the Ministry of Agriculture, designed the first drought probability maps. In 1915 Boris Mul'tanovskii (1876–1938) of the Main Physical Observatory in Petrograd gave a comprehensive explanation of atmospheric drought formation over European Russia and Ukraine. Linking drought to a peculiar type of atmospheric circulation, he showed that air descending from the North Pole southward along the Volga meridian dries out along the way and forms anticyclones over eastern Europe and southern Russia. Later meteorologists corroborated and refined his theory, showing that severe droughts occur when highs remain stationary for several weeks, "blocking" moist westerlies.[37]

During the civil war, in April 1921, the young Bolshevik state created an agrometeorological service for the whole Russian Republic, the first of its kind in the world. In early Soviet times, drought research progressed unabated, with the terrible 1921 famine triggering more meteorological investigations. The network of agrometeorological stations, already large before the war, increased tenfold in the 1920s.[38] To measure drought magnitude, Georgii Selianinov (1887–1966) proposed a simple index in 1928, the hydrothermal coefficient (HTC; *gidrotermicheskii koeffitsient*), which is still widely used today to map drought in former Soviet countries.[39] The HTC represents precipitation deficit as the relationship of precipitation to evaporation. It is defined as the quotient of precipitation in millimeters (multiplied by a coefficient of 10) to the sum of temperatures above 10°C for some period of time, usually the three main growth months in the steppe region (June, July, and August).[40] The weaknesses of this index are that it does not account for soil moisture prior to the beginning of the vegetation and that temperature is a very rough approximation of evaporation.

The industrializing Stalinist state lent support to meteorologists in the form of an expanding network of meteorological stations covering all time zone and geographical regions. Motivation for this investment was agronomic, but weather forecasting is essential also for war strategy and planning and for flying airplanes and guiding missiles, key tasks during the Second World War and the Cold War. As a result, drought specialists benefited from an influx of data on rainfall, temperature, and soils from the 1930s to the 1950s.

However, the commitment of the Soviet state under Stalin to grow a large station network did not extend to seriously financing meteorological and climatological research. After the Great Patriotic War, the interpretation of drought data lagged badly behind their production and collection. Meteorologists lacked the technical means to build upon this data. In education, meteorological knowledge and climatology did not enter the agronomic curriculum.[41] Worse still, from 1941 until at least 1952, Soviet higher education had no program for agrometeorology, which created a severe shortage of qualified personnel.[42]

After the drought of 1946 Soviet meteorologists saw in Stalin's "Plan for the Transformation of Nature" an opportunity to deepen work on droughts in the southeastern part of European Russia.[43] The idea behind the Plan and other "great construction projects of communism"—first and foremost the massive hydroelectric and canal schemes on the Volga, Don, and Dnepr Rivers—was to transform the nature (climate, water, plants, and soils) of the drier regions of southern parts of European Russia, northern Caucasus, southern Ukraine, and the Crimea to make it wetter and more amenable to agriculture.[44] The Soviet Hydrometeorological Service was called in for expertise and investigation on the effect of forest shelterbelts on climate, on the impact of dams on air humidity, and other practical issues.

This positive effect on research notwithstanding, the Plan favored the diffusion of erroneous conceptions of drought. Two ideas were especially detrimental to understanding the nature of drought and imagining effective ways to mitigate it: first, the preconception that droughts occur on the Volga and in

Ukraine when dry winds (*sukhovei*) bring warm air from Central Asia. In 1959 Davitaia published a paper in an agronomic review in which he debunked the idea that drought was imported from the deserts, an idea that he claimed was still profoundly entrenched in agronomic circles.[45] Mariia Zavarina had shown that sukhovei were driven by the drought-producing anticyclone itself and were not coming from outside the drought region. But because the air masses move clockwise in an anticyclone, the sukhovei blow from the east and southeast in the Lower Volga when a drought-bringing high nests above Stalingrad, giving the impression that they come from Kazakhstan.[46]

Second, the theory of the "breath of the desert"—which was fueled by a vague, crooked parallel to nomadic incursions into historic Muscovy—led to the central phony idea in the Plan: that to prevent the desiccating easterlies from Central Asia and Kazakhstan from entering the steppes of European Russia and Ukraine, the state had to build "ramparts [*zaslony*]": forest shelterbelts, up to one hundred meters deep and nine hundred kilometers long, extending on both shores of the Ural, Volga, and Don Rivers to stop the winds.[47] The Plan and other "great construction sites of communism" focused on immense infrastructures (shelterbelts, networks of canals, cascades of hydroelectric dams) built using Gulag prisoners and extracting corvée labor from famished peasants, diverting financing and attention from long-known effective measures to mitigate drought at the level of the field and the farm. The Plan misrepresented and misused the recommendation to plant trees to protect fields, which Dokuchaev and Vladimir Sukachev had theorized.[48]

When the post-Stalin leadership scrapped the plan to transform nature and launched the VLC, scientific understanding of drought in the eastern part of the Soviet Union and meteorological training among agronomists were insufficient. Although equipped with a large observation network and state-of-the-art theorization of atmospheric circulation and global heat and moisture balance, Soviet meteorologists and climatologists had no explanation for and no systematic data on droughts in the eastern part of the USSR (east of the Ural mountains), that is, in the regions where the VLC was to unfold. For their part, agronomists were mostly ignorant of contemporary meteorological research.

The campaign started without serious preparatory climatological studies about the regions to be put under the plow. Geographical study of the lands began with plowing, too late to influence the campaign. The main institutions for climatic research sent out expeditions of postgraduate students and seasoned scientists to accompany and advise settlers: the Main Geophysical Laboratory, the State Hydrological Institute, the Central Forecasting Institute, and Academy of Sciences' Geography Institute were present from the spring of 1954. Their first major scientific report came out at the end of 1955 when the campaign had already plowed up thirty million hectares. It was a general description of northern Kazakhstan's and western Siberia's physical geography without practical information about drought origin, occurrence, and patterns in these regions. A short chapter on drought written by M. S. Kulik followed a chapter on frosts—a real, although far lesser problem in this region—and gave little data or advice. The book did not acknowledge the centrality of drought in the Virgin Lands (VL), comprising dry and hot winds, dust storms, and periods of deficit rainfall over several seasons or years.[49] The knowledge about drought accumulated by local farming communities already living in the VL before 1954 was ignored by both the New Landers and the scientists.[50]

I have shown elsewhere that knowledge about specific natural conditions and the risks they posed to wheat monoculture in the VLC regions developed only in the 1960s, after the settlers had plowed forty million hectares in the steppes and after the first serious setbacks and harvest failures at the turn of the 1960s began endangering Khrushchev's rule. Although wind erosion (deflation) and especially its extreme manifestation, dust storms, had long blighted steppe agriculture, research began only when Moscow perceived significant political ramifications.[51] The same is true of drought research. Four years after the start of the VLC, leading Soviet meteorologists published state-of-the-art studies on droughts in an important edited volume. None of the papers handled the eastern USSR and Central Asia where the campaign was at full throttle. In his paper, the leading drought researcher Selianinov explained that he had no data for droughts east of the Urals Mountains.[52]

The situation improved only at the turn of the 1960s, when Kazakhstani climatologists Anatolii Uteshev and M. Kh. Baidal tackled statistical and physical studies of drought and their work became known at the Soviet level. Uteshev published the first systematic characterization of Kazakhstan's climate in 1959. One chapter analyses three meteorologically typical situations during which drought occurs in Kazakhstan. One drought condition strikes central Kazakhstan when a latitudinal type of atmospheric circulation dominates in Russia and Kazakhstan: highs drift slowly from west to east over central Kazakhstan, squeezed between lows to the north (northern Urals and the Kara Sea) and to the south (over Lake Balkash and the Tian Shan range). Second, like the Lower Volga, western Kazakhstan is subject to drought during the East European type of circulation, if an anticyclone from the White Sea descends toward the Caspian Sea, as Mul'tanovskii had shown. Last, droughts occur in eastern Kazakhstan when the western Siberian type of circulation brings cold dry air masses from the northeast to the Altai Mountains (the rest of Kazakhstan then is under strong cyclones).[53] Uteshev published the first fundamental monograph on droughts in Kazakhstan in 1972, two years before the VLC's twentieth anniversary, after more than a decade of devastating dust storms and several crop failures.[54]

The first half of the 1960s had been marked by a major erosion crisis in the VLC, which contributed to Khrushchev's ouster in October 1964. The steppes of northern Kazakhstan and western Siberia had turned into a "Soviet Dust Bowl," contributing to food shortages across the country.[55] The party first secretary, who had built his legitimacy on finding a definitive solution the bread issue, had to import wheat from Canada in 1963 and announce it publicly, a major humiliation for the Soviet leadership which probably saved the population from a serious food crisis.[56] In Kazakhstan, yields per hectare diminished by 90 kilograms (–14.5 percent) in the first half of the 1960s compared to the second half of the 1950s.[57]

The insufficient understanding of the steppe climate and the ignorance of the knowledge that was already available facilitated the party's insistence on raising production every year regardless of the economic situation of the farms, the state of the soils, or the availability of seeds, labor, machinery, and chemicals. The will to quickly produce extraordinary results—to make the Virgin Lands the new breadbasket of the country, to solve the problem of meat production, and to show to the world that Soviet agriculture was the most productive and effective—led to relentless expansion of the sown surface. Wind erosion and dust storms damaged millions of hectares because marginal lands, easily salinized soils, and light, pulverizable soils had been plowed under political pressure. Until at least 1963, the policy imposed by the party hierarchy on the farms was to acquire more lands to compensate for diminishing yields per hectare because of erosion and drought, leading to still more erosion and vulnerability to low rainfall. The "Soviet Dust Bowl" was clearly human-made. Water deficiency in the first half of the 1960s could not account for the crop failures and resulting food crisis. Erosion provoked by soil mismanagement explains the bad harvests and Khrushchev's unfortunate decisions explain the food crisis.[58]

SCIENTIFIC EXPLANATIONS FOR THE DROUGHT CRISIS OF THE 1970s

In the 1970s, many Soviet scholars were convinced that the climate was becoming more arid. A series of severe droughts challenged climatologists, pedologists, and agronomists for explanations. The political leadership under Leonid Brezhnev needed to explain to the population why food provisioning was faltering. Leading agronomists, climatologists, and soil scientists came up with theories to interpret the succession of droughts.

After Brezhnev replaced Khrushchev, a series of years with favorable weather, high investments, and better central management of agriculture allowed the Virgin Lands to recover from the crisis. However, the second decade of Brezhnev's rule was marked by an unusual series of dry years and a protracted food crisis. Drought hit central Russia in 1972 and provoked well-remembered peat and forest fires. The Virgin Lands were hit by drought two years in a row, in 1974 and 1975. Three years in a row—1979 to 1981—were disastrous across the country, and 1984 was very dry.[59] The droughts of 1975 and 1981 rank among the most severe of the century on Soviet territory.[60] The severe droughts of 1963, 1975, and 1984 encompassed all main grain regions.[61]

Khrushchev and Brezhnev hid behind "natural disasters" to draw attention away from the consequences of their agricultural policy. Khrushchev said at a Central Committee plenum in 1953 that first there was a "severe winter, and then a fierce drought has caused damage to the most important agricultural regions of the country."[62] Again in 1963, to justify importing Canadian wheat, Khrushchev blamed the dry weather.[63] Speaking with Kazakh comrades in Alma-Ata, Brezhnev exaggerated boldly: "Last year [1972] due to the drought that encompassed three-quarters of our country, we had to overcome many difficulties. According to scientists' data there wasn't such a drought in the last hundred years."[64] In fact, Kazakhstan and the eastern USSR were spared by the 1972 drought, which concentrated on the European part of the USSR.

Scapegoating low-rank party functionaries was another technique used by party leaders to account for crop failures. After a decade of droughts, Brezhnev in 1982 accused farm managers and party functionaries of not sticking to the right agricultural practices. The Politburo threatened them with retaliation if they would not respect the prescribed timing for sowing.[65]

Many agronomists did not buy into the "natural disaster" rhetoric. For the academic Alexander Baraev from Vsesoiuznaia akademiia sel'skokhoziaistvennykh nauk imeni V. I. Lenina (the All-Union Academy of Agricultural Sciences of the Soviet Union; VASKhNIL), who led a major research institute on grain farming in northern Kazakhstan, disastrous crop failures were by no means inevitable. He wrote a lengthy memo to the government in 1976 promoting the soil-protecting farming technologies his institute had tested for steppe agriculture. He refused to define drought as a disaster: "Often one tries to account for the low crop capacity in grain and the strong reduction in gross grain yields in certain years by ascribing them to a natural disaster [*stikhiinoe bedstvie*]: drought and sukhovei. However, such explanation is not objective. In fact extremely dry years in the steppe regions of our country are an inevitable phenomenon, independent from the will of man and, we could say, normal."[66]

Remarkably, Baraev refuted the climatic argument just one year after a drought had devastated both the Asian and European parts of the country. Dry years in the Kazakh steppe were not droughts, he wrote, but a regular feature of the semi-arid and sub-humid climates. Distinguishing between drought and aridity, the former was a human-made problem.

In his memo, Baraev defended the soil-conserving farming system he had devised in the first half of the 1960s under the influence of Canadian agronomists and farmers. The basis of the system was to leave fallow one-fourth of the tilth each year. To Khrushchev, removing such a significant amount of land from sowing (and so from harvest) was unacceptable. After two clashes in Kazakhstan over the fallow issue, he almost removed Baraev from his post as director of an agronomic institute near Tselinograd (now Astana). After Khrushchev's removal Baraev's farming system received Moscow's blessing: he was named an academician of VASKhNIL and received two Orders of Lenin; his institute was promoted to be the central scientific hub for dry farming in the steppes east of the Urals.

Despite Moscow's support, however, Baraev's set of rules to spare soils under wheat monoculture was not unanimously enforced even in northern Kazakhstan. The fallow requirement posed a threat to farm managers and regional leaders laboring to meet the production plan. During the drought years of the first half of the 1970s, some Kazakh grain regions decided to cultivate all fallow lands in order to maintain production. Kustanai oblast forbade bare fallows in 1974–1976, and Severo-Kazakhstanskaia oblast did the same in 1975. Unfortunately, cultivating fallows proved disastrous: yields fell in these two regions, despite their having the richest soils in Kazakhstan. Recovery would take years, Baraev warned.[67] He tried to convince the political leadership that rotation with 25 percent of land laying fallow would yield more grain than land cultivated without fallow, given the yearly oscillation of production and the quantity of seeds.[68] For Baraev the arid conditions of the steppes challenged agronomists and peasants to look for coping strategies, especially agro-technical systems; blaming the weather for the dry years was infantile and irresponsible.

In contrast to Baraev many climatologists had begun to reject the view that climate was a stable and unchanging feature of the environment. For Baraev, steppe climate was stable in the sense that it was best understood as a succession of dry and wet years, even if science could not yet explain or predict such oscillations. He theorized that two of every five years were highly dry in Kazakhstan. The occurrence of two droughts within five years (1972 and 1975) in the steppes west of the Urals was for him proof of the regularity of steppe climate.[69] But Soviet climatologists, in tune with their colleagues around the world, worried that the climate itself was becoming more arid. Contrary to Baraev, the sequence of droughts in 1972 and 1975 shocked Drozdov as abnormal. He classified them as the worst of the century and calculated the probability of two such severe droughts in a half decade was one in thirty-eight thousand years. For him climate warming could not sufficiently explain the growing occurrence of droughts and the aridization of steppe climate.[70] He found that farming had modified the climate of the steppe. In comparison to the original grass steppe, evaporation is reduced in the spring on fallow land and on sown surfaces until sprouting. Consequently, precipitation is reduced on millions of hectares of recently cultivated lands, increasing the probability of droughts.[71]

For meteorologist Mikhail Budyko, the industrial and agricultural activities of humankind brought changes in the global climate.[72] He forecast a global increase of temperatures in the 1970s because of human activity.[73] Since the middle of the twentieth century, economic activities had had the same quantitative impact on temperature change as its natural oscillation, he wrote. The warming process, boosted by energy consumption and the greenhouse effect, materialized in rising surface temperature at the end of the 1960s. Two severe droughts had struck the USSR in the first half of the 1970s alone, confirming the prediction of decreasing humidity and increasing temperatures, Budyko claimed in 1977. Temperatures would further rise, and rapidly, he warned, bringing about economic disruption.[74] Within just a few decades the earth would return to the carbon dioxide concentration it last reached in the Tertiary Period, millions of years before. The temperature increase, already destructive in the 1970s, had the potential to bring about a "global ecological crisis" like the one announced in the famous report to the Club of Rome, *Limits to Growth*, published five years earlier.[75] For the dry steppes in particular, climate change promised duress: one could expect more regular and marked anomalies in temperature and precipitation. The effect of the reduction of precipitation on agricultural zones with irregular humidity like the steppes—and on bodies of water—increasingly worried Budyko.[76]

Some meteorologists pointed out one probable positive consequence of global warming for Soviet agriculture: with the increased concentration of carbon dioxide, the physiological capacities of plants would grow, and rising temperatures would extend and warm the vegetative period, leading to increased yields. However, these meteorologists warned that the anticipated diminution of precipitation would probably bring about disturbances in the distribution of yields in time and an increase in extreme dry events.[77] The main forecast for the steppes and forest steppes remained pessimistic among meteorologists in the 1980s. The prognostication made by Budyko at the beginning of the 1970s seemed to have become reality: according to his aridity index, the conditions of humidification in the main crop-growing regions deteriorated markedly in 1975–1985 in comparison to 1964–1974.[78] From the second half of the 1960s, wheat yields stagnated at 1,400–1,500 kilograms per hectare, but with considerable anomalies and amplitudes.[79]

Pedologists shared the climatologists' concern about the drought wave of the 1970s. They argued that the growing number of droughts and crop failures and the destruction of vegetation and soils were linked. Among them, Viktor Kovda, renewing a century-old debate among Russian geographers, elaborated on the idea that the steppes of the Soviet Union were desiccating. Kovda had played a key role in the international desertification debate triggered by the 1972 Sahel drought. At the turn of the 1980s, Kovda was unsure whether the high climatic variability and repeated droughts he witnessed in the Northern Hemisphere meant that the climate was warming or cooling. But he was certain that a general arid trend animated the climate, which in his understanding could be linked to either a colder or a warmer climate.[80]

In a monograph published in 1977 and translated into English three years later, Kovda built on George Kukla, Hubert Lamb, and a 1975 report of the US Academy of Sciences on climate change, all warning that a new global cold era was on its way.[81] He shared Kukla's and Lamb's views that the Northern Hemisphere was probably entering a new Little Ice Age, but he rejected the latter's sun spot theory.[82] Instead, Kovda used geological cycles that had been uncovered by A. Shnitikov in 1957: ice, peat, climate, and ocean data revealed moisture cycles of approximately eighteen hundred years for the Northern Hemisphere. The last of these cycles began in the thirteenth century and had entered its cooling and dessicating phase in the nineteenth century. He insisted that in the Holocene, periods of warming were coupled with greater precipitation; conversely, cool periods and aridization—or xerotization—due to decreased precipitation were linked. To him, the diminishing global air surface temperatures of the 1960s and 1970s signified that the multi-secular cooling and aridizing trend had resumed after a short interval in the 1930s–1950s. He diagnosed a slow tectonic rise of water-accumulative plains around the world with severe consequences for soils, waters, and vegetation: subsoil waters became less accessible to the plants, the hydrological network cut deeper, and the snow limit receded.[83] Although the main force behind aridization was geological, Kovda stressed how human activity reinforced the trend toward a more arid climate: heightened groundwater usage, the acceleration of runoff, deforestation, and the destruction of grass vegetation for agriculture all "contributed to the lowering of the water table."[84]

The droughts of the 1970s and the bad results of Soviet agriculture prompted earth scientists to find explanations. Those reviewed here represent some of the most important among them at the time. Their differences testify to the diversity of scientific approaches to drought and climate and to the lack of a convincing explanatory scheme for the perceived growing occurrence of heat waves. Underpinning their theories were both a century-old thread of thought about climate change among Russian thinkers and the participation of Soviet scientists in international debates about global warming, global cooling, and desertification.[85] Remarkably, all of them made room both for natural and human factors, letting them act at different chronological and geographical scales—from microclimate and pedological change via farming to global climate change via greenhouse effect.

The alarm rung by Soviet earth scientists in the 1970s and 1980s might have been exaggerated. Thirty years later climatologists find that farming conditions actually improved in the last quarter of the twentieth century in the grain-growing regions of Russia and Kazakhstan. Oleg Sirotenko and Vera Pavlova find that the climate became less arid in the Russian steppes after 1975.[86] Like Budyko and Drozdov, they see the temperatures growing, but this growth went hand in hand with more rainfall—a correlation reminiscent of Kovda's view on aridization. All in all, Russian climate has become less continental, with the vegetative season gaining two weeks in the second half of the twentieth century.[87] In a reversal of what their predecessors and colleagues concluded, they find that everywhere except in central Russia a better climate has contributed to rising yields.[88] Retrospectively, it appeared that, when comparing 1966–1995 to 1936–1965, climate change had actually led to a warmer vegetative period and to increased precipitation in the main grain-growing regions of the former USSR (Kazakhstan, Russia, and Ukraine). Thus, the climatic conditions for growing crops improved markedly, contrary to the diagnoses of the 1970s and 1980s.[89]

CONCLUSIONS

Russian and Soviet earth scientists have demonstrated a strong interest in drought and climate change since the nineteenth century. They played an important role in describing global atmospheric circulation and theorizing the climate system.[90] In this article I have shown that their theories of climate change were linked to their observation of droughts in the main grain baskets of Russia, Ukraine, and Kazakhstan. Crop failures triggered research on droughts. Studies of atmospheric droughts nurtured scholarly reflection on changes to the steppe climate. In the 1970s repetitive droughts resurrected a century-old supposition that the steppe climate was becoming more arid. Scientists devised several theories to account for the desiccating change in the climate of the steppes. They diagnosed the temperature increase as a product of global warming, as Budyko demonstrated. But trends in precipitation are more difficult to distinguish and assess. It appears now that levels of precipitation have increased in most steppe regions. Thus, Russian meteorologists nowadays estimate that—notwithstanding the droughts—the climatic conditions of Soviet agriculture have improved since the 1960s, not worsened.

These meteorological findings tend to indicate that the stagnation of agriculture and bad harvest results from the second half of the 1970s and in the 1980s cannot be adequately understood from climatic data. Instead, the causes should be sought in the agronomic and economic functioning of dry farming in the steppe *sovkhozy*. Drought and crop failures are the result of complex interactions between human and natural entities and processes. The soil—heavily and widely altered by wheat monoculture—is an especially important component of this system. It is time for an "ecologically informed history" accounting for soils, climate, plants, technologies, and humans to reassess the constraints and results of Soviet dry farming in the steppes.[91] Soviet agriculture remains understudied. We lack convincing explanations of its challenges and outcomes, which the study of climate alone does not provide. A localized study of how *kolkhozy and sovkhozy* survived—or not—the natural, economic, and political constraints could well advance our knowledge in this field.

PART II

Water Engineering

6 Leviathan on the Oxus

WATER AND SOVIET POWER ON THE LOWER AMU DARYA, 1920s–1940s

CHRISTIAN TEICHMANN

COMRADE LENIN understood little about irrigation. Yet he knew that it "is more necessary than anything else and changes the country more than anything else; it leads to its rebirth, buries the past and strengthens the transition to socialism." Joseph Stalin's protégé, Lazar Kaganovich, called irrigation a "key factor" necessary to "fully reconstruct the agriculture of our country" and hailed its "revolutionary importance for the foreign Orient, where water is still the source of tribal feuds and hatred." Uzbek party leader Usman Iusupov glorified the construction of irrigation canals, which in his view served as "an indicator for the imperturbable movement of the Uzbek people toward communism."[1] According to Soviet leaders and communist believers, environmental change and social change were mutually reenforcing. With the revival of ancient canal networks and the construction of new dams, they aspired to undo tsarist Russia's history of colonial suppression in its Central Asian peripheries and strove to begin building a socialist utopia. State intervention in the natural environment, they rightly expected, was to radically change the human condition.

The Soviet pattern of thought was neither original nor unique to modernizing political regimes. As the environmental historian Joachim Radkau has pointed out, from antiquity onward irrigation was a "cornerstone of state formation" and "water issues gave rise to regulatory forms that went beyond the house economy," because "ecological necessities often go hand in hand with opportunities for intervention from the top."[2] Although the establishment and pursuit of state power in the Soviet Union has long intrigued historians, environmental and climatic factors have been surprisingly absent from wider debates about the nature of the Soviet state.[3] On the contrary, many historians have tended to view the construction of the Soviet political space as an ideology-driven venture constrained by factors such as feeble infrastructures, bureaucratic sluggishness, popular resilience, and mass terror. While drawbacks, failures, and violence are frequently portrayed as anomalies and aberrations, Soviet state building is depicted as a continuous and ultimately successful progression toward the normality of a modern bureaucratic dictatorship. If not for ideologically driven state intervention, how was it possible, one could ask, to pursue those grand construction projects of the Stalin era that so tremendously changed the face of the Soviet natural environment?

Still, the modernist view of the Soviet state has some shortcomings when seen through the lens of environmental history. In fact, Soviet ideological discourses on capturing traditional peasant societies and controlling natural spaces were deeply divorced from both the realities of local administrative practice and the inability of the regime to control the powers of nature.[4] Stalinist construction projects were carried out not in the manner of modern grand planning schemes but by makeshift improvisation. Modern technologies of engineering and administration were, moreover, not socially embedded by the Soviet state. While successes in construction were achieved mainly through forced mass mobilization and manual labor, Soviet officials created administrative utopias in their reports and planning documents that hardly allowed their superiors (or their historians) a meaningful sense of the realities on the ground. Only cases of gross mismanagement, planning failures, and natural disasters drew the attention of Soviet higher-ups (and their historians) to the fundamental workings of the Soviet state.[5]

An environmental approach allows us to shed new light on the nature of the Soviet state. Karl August Wittfogel's study *Oriental Despotism* famously linked the construction of monumental irrigation works to the rise of powerful state bureaucracies. In Wittfogel's "hydraulic societies," the "total power" of the state is based on the mastery of complex administrative tasks on one hand and control over land property and natural resources on the other. According to Wittfogel, the power accumulation of the state is enforced by ambitious construction projects to build imposing canals, palaces, and temples. While Wittfogel's hypothesis was much criticized during his time, it comes as a surprise that James Scott's well-received study *Seeing like a State* strikes a very similar tone. Here, not ill-willed totalitarian despots but powerful authoritarian rulers plan and implement grand development schemes that are built at tremendous social and environmental cost. In the age of "high modernist ideology," Scott claims, states transform into opaque administrative machineries that "simplify" territories and chart populations on "state maps of legibility." Similar to Wittfogel's view that a strong state is able to suppress weak civil society, Scott finds a "prostrate civil society that lacks the capacity to resist the state's transformative plans."[6]

Although Wittfogel succeeded in establishing tentative connections between environmental conditions and the rise of despotic forms of governance, Scott aptly spelled out the risky implications of modern administrative techniques and the environmental impact of new technologies. Yet, both authors generalize about "the modern state as a homogenous and all-seeing entity" and overestimate the state's actual capabilities.[7] Modern states are much less modernist and much less homogeneous than Scott and others propose. As Frederick Cooper pointed out, "the supposedly modern apparatus of rule was itself laced through with particularistic mechanisms" such as patron-client networks and elements of personal rule, the authority of local power brokers, and the establishment of second economies based on state funds.[8] Since states are not monolithic and rarely develop into unified bureaucratic structures, as Max Weber suggested they did, discussions about state building in the Soviet Union should engage with administrative practice rather than subsuming historical inquiry to meta-narratives such as bureaucratization and modernization. In regions such as Central Asia, where irrigation is the basis of agriculture, water distribution and canal maintenance are key political issues that drive the interaction between state actors, local environments, and rural populations.[9] Since water allocation and river regulation determine the survival of oasis communities, intervention into water management is particularly complicated and fraught for state administrators who arrive as outsiders to a colonized or occupied region. One Russian engineer captured the tensions and struggles arising over water distribution during the 1920s when he reported that "the water administration faces the overwhelming task of how to stop the theft of water, which here has the value of gold and is the sweat and blood of the working population."[10] Soviet European officials in Central Asia faced this and other dilemmas. They not only had the duty to create viable administrative structures but were obliged to do so in the face of an unforgiving natural environment. Additionally, they were regularly confronted with hostile local populations who disregarded their claims to cultural superiority and political legitimacy. For officials and administrators, therefore, building the Soviet state in Central Asia proved hard and dangerous work.[11] To observe the peculiar patterns of power that formed within the Soviet state in a natural and social setting that was challenging, changing, and unsettling, the region of the lower reaches of the Amu Darya River provides a good test case.

Amu Darya, Capricious River

When trying to understand the obstacles to Soviet state building in the 1920s and 1930s, a closer look at the ecology of the lower reaches of the Amu Darya is indispensable. The lower stream of the Amu Darya and the delta region at the Aral Sea can be described as a natural unity because of their connected landscape and geology. Still, the region consisted of diverse ecological spheres where intensive irrigation agriculture, nomad cattle herding, and fishing went hand in hand. One can distinguish the lower stream with its canal oases from the delta region, where the river divided into numerous arms and streams that constantly changed paths and beds. Descriptions of the region from the first half of the twentieth century mark the beginning of the lower section of the Amu Darya River at the point where the river enters the oasis of Khorezm at Pitniak, situated 380 kilometers south of the former Aral Sea (see map 6.1).

Whereas in its middle reaches the Amu Darya streamed evenly and formed terraces into its fixed riverbed that enabled the population to cultivate a small strip of land along the shores, in the river section from Pitniak to the Aral Sea the Amu Darya started to deposit great amounts of sand and silt that it had collected on its way downstream. These fertile alluvial deposits formed the agricultural basis of the oases of the lower Amu Darya. Fed by the river waters, the canals of the Khorezm oasis were up to 160 kilometers long (the Shavat canal) and had internal canal networks that were as long as 900 kilometers (the Palvan-ata canal). The canal oases were not small strips of fields but large and densely populated agricultural zones covering up to 500 square kilometers. Still, the alluvial deposits (amounting to 1–1.5 kilograms per cubic meter) could quickly lead to the congestion of canal heads and irrigation systems. Peasants were therefore required to clean the canal heads and the canal annually so as to alleviate siltation. In some places the alluvial deposits formed steep banks at the river and canal shores that made the use of waterwheels mandatory.[12]

The Amu Darya had other ecological peculiarities. Fed by its tributaries, the Panj and the Vakhsh Rivers, with melting waters of the Pamir and Tien Shan glaciers, the water levels of the Amu Darya varied seasonally. In the lower region of the Amu Darya, water levels were low from November to March. When the first water arrived in spring, it was eagerly awaited by the population. The sluices at the canal heads were opened, and the crops—which included the local pulse variety *dzhugara*, wheat, alfalfa, cotton, rice, grapes, and honey melons—were watered several times between April and July. In the middle of the agricultural season, the water levels started to rise further. Now canals and riverbanks had to be protected from floodwaters and erosion. The floodwaters

Map 6.1. Soviet Central Asia, Showing the Decrease in Size of the Aral Sea.

arrived at very high speeds and washed away everything in their path. The water could not be used for irrigation purposes and the canals had to be kept closed. Furthermore, dams were built to protect the land and the towns from flooding. The high water levels and scouring (*deigish*) continued into November and sometimes December. Moreover, during the cold months of the year, the river was covered with drifting ice (*shuga*). Given these conditions, it is unsurprising that Soviet sources described the work of irrigation as "a bitter struggle with the caprices of the Amu Darya."[13]

Ecology, irrigation, and communal life were intertwined, and irrigation structured the social order of the populations living along the canals. Despite its ecological fluidity and the technical challenges it posed, the social organization within the irrigated oases were characterized by remarkable stability in the late tsarist and early Soviet periods. The ways in which canals and irrigation systems were constructed contributed to this fact. Main irrigation canals took their beginning at the riverbanks where the inflow of water was regulated at the canal head with moveable wooden barrages and dikes. Many canals had several heads in order to forestall the risk of water shortages during the irrigation season. The sluices at the canal heads had to be renewed regularly. Water was distributed from rivers and the main canals into smaller canals (*jab*) and eventually arrived in small and deep ditches from where it could flow or be manually lifted onto the fields. A local canal overseer (*mirab*) planned and organized canal maintenance works and arranged water allocations to water users. Often single clan groups or extended families settled along one side canal and ran its maintenance.[14]

In the Khorezm oasis, throughout the 1920s, the basic unit of administration was not the single rural settlement but the canal community, which unified all settlements that lined one canal (*jab*) or along several parallel canals. The head of the canal community (*aksakal*) controlled the fulfillment of seasonal work obligations (*kazu*). Maintenance of irrigation systems was organized according to field size and the amount of water needed by a family, group, or canal community. But social divisions among canal communities loomed large. The closer the fields were located to riverbanks or canals, the more reliable was the water supply even in dry years. Popular wisdom had it that it was "better to have a field at the head of the canal than a brother who is a canal overseer" because water supply at the end of canals was irregular and could quickly lead to soil salinization and crop failure.

The fabric of social networks within canal communities could hardly be penetrated by outsiders. In 1925 the oasis of Khorezm proved a difficult field of action not only because of the "remoteness of the location, rebel activity, and the absence of housing that would be suitable in a certain way." Russian irrigation engineers—sent to the region ostensibly to improve and modernize water systems and their administration—lived isolated from the local communities, were detested both by the rural population and the indigenous administration, and had to cope with nonacceptance and a lack of authority. Conflicts were frequent. One official wrote in 1926 that "technicians either leave their work after a few months or they take to drink and end their lives tragically. In Kunia Urgench district, for example, the first assigned engineer shot himself, the second poisoned himself, and the third asked to be removed after he was severely beaten up in the local police station and fled to Tashauz."[15]

The river ecology created further problems for Soviet water management efforts. In the delta region, the Amu Darya constantly changed its course and shifted riverbeds, river channels appeared and disappeared, and sandbanks built up at the canal heads. Although the settled populations lived clustered along canals, these could disappear and reappear very quickly. The ever-evolving water and sand systems led to ever-changing settlement patterns. Soviet administrators complained that "as a result of these circumstances, when statistical inquiries are conducted every three to five years, in places that had been cultivated lands formerly you can see only water, whereas in places that before had been river arms and lakes agricultural settlements have emerged."[16]

The state's administrative network was, however, not very dense. In 1925 the *volost*, or county, of Khanki in the Khorezm oasis—fifty kilometers from the town of Khiva and composed of twenty-five canal communities with a population of approximately twenty-one thousand people—was administered by a single indigenous official. The duties of this district chief included the organization of canal cleaning, tax collection, and agricultural campaigns as well as the gathering of information for land statistics. Additionally, he mediated conflicts that arose over water distribution, property divisions, payment of credits, and marriage problems. The power of the district chief was restricted, though, and relied on the support and goodwill of the heads of the canal communities and their support networks: "When an aksakal does not want to give information about the real situation of his village, the district chief cannot obtain this information: the peasants are used to the fact that the district chief comes to their village and leaves again while the aksakal stays behind. For them, it does not make sense to go against the aksakal."[17]

THE CREATION OF A PERIPHERY

State-building Soviet administrators in the lower Amu Darya region grappled with both environmental conditions and human nature. They saw the irrigation system as a key site of their struggle, their "battle mission."[18] Consequently, during the latter half of the 1920s, state interference in everyday irrigation management became more and more common. Soviet officials endeavored to establish the presence of the state not only in the towns and administrative centers but also along the canals and within local communities. But contrary to official proclamations, activist state policies in the irrigation sector were accompanied by a parallel process of cultural marginalization and systematic economic neglect of the region.

The political and social upheavals that accompanied the transition from the old to the new regime in Central Asia presented the Bolsheviks with a range of immediate state-building tasks. By the mid-nineteenth century, the chancellery of the Khan of Khiva annually mobilized the population for seasonal work obligations and was able to collect taxes that were based either on land ownership (*salguta*) or the size of herds (*zekat*). The khan's rule was stable, especially in the oases, and even extended to some of the nomadic peoples of his realm.[19] After Khiva was captured by the tsar's army in 1873, the territory of the khanate was divided along the Amu Darya. Still, the Khanate of Khiva, even if reduced in size, was left intact under the auspices of a Russian protectorate.[20] During the late 1870s and 1880s tsarist officials organized several expeditions to explore possible sites for the construction of a canal connecting the Amu Darya to the Caspian Sea. But on the Amu Darya, plans remained plans and dreams remained dreams.[21]

Under the old regime Russian military and state officials lacked both the interest and the resources to interfere in the local irrigation system. When Soviet power was formally established in Khiva in 1920, Soviet rule over the region remained conflict-ridden. The main task of the Bolsheviks and their sympathizers was to fend off rebel attacks and negotiate truces with local strongmen and community leaders. Violent conflicts between the "Uzbek" settled population and "Turkmen" nomads were rampant throughout the 1920s.[22] Moreover, all state institutions had disappeared during the period of war, famine, and massive outmigration between 1915 and 1925.

Bolsheviks emissaries saw it as their first obligation to ameliorate the conflicts among the local populations. They came to the conclusion that "nationality" was at the root of the violent struggles between settled populations and nomads over water rights and land titles. During national delimitation (the creation of the Central Asian Soviet republics Kazakhstan, Uzbekistan, and Turkmenistan in 1924–1925), the region of the lower Amu Darya was divided into three administrative units. The northern part of the region was incorporated into Kazakhstan (as Karakalpakstan oblast, or province), the western part into Turkmenistan (as the Tashauz oblast) and the central and eastern parts into Uzbekistan (as the Khorezm oblast).

The division did not lead to appeasement but, rather, created new conflicts over economic assets and natural resources, particularly water.[23] In the oasis of Khorezm, national delimitation had created "an extraordinarily inconvenient situation," one engineer reported in 1926 when he described the situation in Tashauz. The district was irrigated by nine major canals, but only three had their canal heads on Turkmen territory, which were not as important as the six other major canals that had their heads on Uzbek territory. Turkmen territory covered only the ends of these six canals where water supplies were irregular and depended on the goodwill of the Uzbek side. Additionally, the three Turkmen canal heads were disputed, because the Karakalpak government was quick to claim that these canals were located on Karakalpak territory. The Turkmen-Karakalpak conflict could not be resolved because no reliable maps were available and a new land survey had not yet been conducted.[24]

The national delimitation of 1924–1925 set in motion a process of marginalization and peripheralization of the lower Amu Darya region. With the creation of the new borders, water conflicts were not resolved but became institutionalized. Moreover, all new administrative centers of the new Central Asian republics (Kyzylorda, Ashkhabat, and Samarkand) were far from the Amu Darya region. In the case of the Karakalpakstan, the closest administrative center was the then-Kazak capital of Kyzylorda to the north. This was a distance of eighteen days of travel until a road was built in 1927, making it a two-day drive. The situation was similar for Tashauz in Turkmenistan and Novyi Urgench, the administrative center of Uzbekistan's Khorezm oblast. Here, officials could either arrive by airplane or overcome the distance by railroad and boat. When Izaak Zelenskii, the chairman of the Central Asia Party Bureau, visited Tashauz in 1925, he made his trip through the province in an open horse carriage that had been left behind by the Khan of Khiva.[25] The administrative distances added to the isolation of the region. Moreover, throughout the 1920s and 1930s administrative borders frequently changed all over the Soviet Union. Districts were created and abolished, internal borders were changed, and even the status of big administrative territories did not remain untouched. An extreme case was the Karakalpakstan. The territory was part of Kazakhstan between 1925 and 1932, then transferred to the Russian Federation (RSFSR), and finally, in 1936, incorporated into Uzbekistan.

The peripheralization of the Amu Darya region was accomplished not only through inconvenient and fluid administrative divisions but also by a process of cultural marginalization. Again irrigation played a key role. Since the revolution and civil war, views held by Russian engineers and administrators about the irrigation system had radically changed for the worse. Tsarist administrators believed the irrigation systems of Central Asia were a treasure of ancient culture and civilization that should be revived and restored. They viewed the nomadic lifestyle negatively because of the "low reproduction rate and low culture of the population" as well as the "instability of the economy that depends almost entirely on the well-being of the herds." However, they hailed the irrigated agriculture of the settled peasants because of the "significant reproduction rate of the population, its comparatively high cultural level and the stability of its economic situation, the abundance and variety of agricultural and gardening products as well as the intensive system of the agricultural economy."[26] This rosy view of irrigation disappeared in the course of the Bolshevik conquest of Central Asia and was substituted by an overtly negative assessment: "The experience of the local population with irrigation is significant but the system works uneconomically; there is no intensive economy but a colossal waste of water and physical labor."[27] Pessimistic views of the irrigation system were the new "common sense" among the engineers running the irrigation system during the 1920s and 1930s.[28]

For Soviet officials and engineers, irrigation became synonymous with the "backwardness" and "darkness" of the local populations of the Amu Darya region. They judged the amount of physical labor needed to run the irrigation system, the irregularity of the irrigation canals, ineffective flood protection, and constant water shortages to be clear signs of the "chaos ruling the irrigation systems." The way in which Khivan peasants worked in their fields underlined the "extreme primitiveness" of the local agriculture, "based exclusively on elementary hand labor and colossal amounts of physical labor." War and anarchy in Khiva had led to both massive salinization in many parts of the oasis and damagingly high groundwater levels. There was a shortage of draft cattle and, as a consequence, of fertilizers. Private land strips were tiny and fields were often located at great distance from homesteads.

Similar to the technical flaws of local irrigation, the populations of the oases came to be viewed as inactive and unknowable: "Because they lack sufficient technical means and knowledge, they somehow got used to the regular flooding and made peace with a situation when a district, town or village is stuck in floodwaters for one and a half months every year." Given the habitual passivity and fatalism attributed to the local people, it did not come as a surprise that one could "get the impression that the population is slowly but truly dying out from malaria and other swamp diseases."[29]

Soviet disregard of local technical knowledge and the population's apparent cultural backwardness were two key elements in the marginalization and peripheralization of the region. The rhetoric of cultural superiority allowed Soviet administrators to take action and to cope with frequent drawbacks. However, the same rhetoric also undermined the building of trust with the local population. Russian engineers who worked in local water administrations along the Amu Darya knew this dilemma very well. Local power networks often involving indigenous party secretaries, police, and government officials remained stronger than the small groups of European officials and specialists who had recently arrived in the region. Throughout the 1920s reports indicated that the "broad masses of the rural population" ignored the demands of the water administration, stole geodetic signs (because they were made from wood, a good burning material), and did not participate in mandatory canal cleaning works. Moreover, farmers boycotted the water distribution plans of the water administration, operated canals as they pleased, and built new canals without approval.[30]

Russian engineers saw it as their duty to fight against tradition and overcome local arrangements in the irrigation sector. The test case was the fulfillment of seasonal work obligations, the annual working obligations to clean and maintain canals: "The psychology of the peasants is based on the conviction that they can work today as they did for hundreds of years only because they are not paid for their labor. But one has to fight against this mental dullness (*kosnost'*) and engineers should not passively watch how the work is proceeding and stand idly by the construction sites."[31] In 1926 the amount of mandatory work obligations was increased by 100 percent. In this year, peasants in the Chimbai oasis of Karakalpakstan had labored 1.0 million days to fulfill their obligations; they had moved 3.9 million cubic meters of earth—an average of thirty-five working days per water user—but the water administration was far from happy with the results. Because the Amu Darya shifted its riverbed westward and away from the canal heads, the sanding up of canal heads in the district continued and caused massive water shortages in 1927. Engineers realized that the population was as unsatisfied with the massive but unsuccessful effort as they were. The way in which engineers managed the canal construction "made a bad impression on the population and will have serious consequences for future mandatory construction works; it leads to a total loss of confidence with the work of the water administration because the population has no comprehension whatsoever about the real causes of the current situation."[32]

As a consequence engineers demanded that the Karakalpak government purchase construction machinery and issue a decree easing the work obligation, but money was in short supply and machines were not available to make canal construction a success. Instead, political campaigns and brute force were used to enhance peasant participation in construction work. Starting in the fall of 1928, under the slogan of "Sovietization of water and labor obligations," mandatory works were organized according to the class principle (*samooblozhenie*). This huge change from past labor-water practice meant that labor obligations were not assigned according to land size but, rather, according to ascribed class status. Households of "rich people, religious authorities and traders" had to send more workers to the construction sites and relieve the poorer families. Reports confirmed the apparent success of these measures: the local population enthusiastically helped to "unmask" the privileged "because it sees that the state is coming to their help in the question of how more workers can be mobilized and, thus, how the time line of the construction works can be shortened." The campaign was accompanied by appropriate measures to ensure that it "would not be conducted in the framework of formality": show trials, fines, and police force provided for "best results." Despite snow, rain, and strong winds, the population's participation in construction had supposedly doubled or tripled in comparison to previous years.[33]

RURAL DESPOTS

Overcoming what Soviet officials and engineers saw as the challenges of backwardness and tradition proved as difficult as handling the unforgiving nature of the Amu Darya. Facing a problem typical for state officials in many European colonies, Soviet administrators had three options to deal with the situation: first, to rely on arbitrary violence to secure their rule; second, to use intermediaries and local power brokers; third, to build up a functioning bureaucratic apparatus.[34] During the 1920s all of these strategies were tried interchangeably. Armed resistance was suppressed by the military and secret police, local intermediaries were tolerated because reliable administrative personnel remained scarce, and formal government institutions were established in the major towns. Yet the failings of the Soviet state were visible to all: rebel attacks continued; local power brokers remained unreliable; and Soviet officials created fictitious administrative worlds that had few connections to the realities outside the papers, reports, and discussions of government offices and party meeting rooms. Although the central state administrations gained comparatively more leverage and resources, they were unable to take root locally. Instead, the successes of state building relied on traveling emissaries, control commissions, plenipotentiaries, and special brigades who were sent out to obtain information, check local conditions, and exercise intermittent control.

The collectivization of agriculture partially changed this situation. At the onset of the campaign, it was Stalin himself who called for the establishment of the Soviet state in the villages when he reminded party members that "we have all the power we want but we lack the ability to exercise our power" and that "we are not a party of agitators but the party who is in power and who determines the laws of the country. If we cannot enforce our laws the peasant will ignore them."[35] In the case of the lower Amu Darya region, collectivization met with strong resistance. When the first campaign was started in early 1930, famine raged in the Khorezm oasis. Because the campaign came at a time when canal cleaning works had to be conducted, many of the 240 people who were sent out to organize collective farms (*kolkhozy*) restricted themselves to forcing peasants to work on the canals while they randomly confiscated property from private homes. At same time, the majority of plenipotentiaries had "no idea how to organize the everyday functioning of a kolkhoz" because no one had told them. Moreover, the first wave of collectivization in spring 1930 proved that local party organizations were unable to write the appropriate resolutions and that its leadership needed to be "strongly thrashed so that all districts feel the necessity of a true Bolshevik reorganization of their work."[36]

In 1930 and 1931 violent methods were used to force peasants to work under the conditions of collectivized agriculture. Water distribution was an officially sanctioned tool to pressure peasants to join the kolkhoz.[37] In 1932 and early 1933 purges of the kolkhoz membership and mass deportations of peasants served to increase the pressure. In the 639 kolkhozy of Karakalpakstan, 1,246 members were excluded as "alien and dangerous elements." Amid hunger, violence, and uncertainty the number of kolkhoz households decreased from 33,948 to 31,061 in the same two years. Workers were in short supply on the cotton fields, and sluggishness and absenteeism posed major problems. In 1934 only 50 percent of the households of Karakalpakstan were kolkhoz members. The peasants did not understand how a kolkhoz was supposed to function, and kolkhoz activity was mostly restricted to irrigation and cotton cultivation. But Soviet planners did not agree on cotton plans beforehand, leaving these proposals administrative works of fiction. Work obligations were not assigned, and there was no registration of working days. In the assessment of the Karakalpak party leadership, the local kolkhozy were "artificial economic units that were built up as a result of administrative pressure and for this reason do not have the strength and necessary preconditions for a normal economic development."[38]

The kolkhoz system was economically unviable, but its internal structure and organizational hierarchy significantly changed the balance of local power within the village communities. Kolkhoz chairmen from the local population could institute themselves as an intermediary administrative layer between the local governments and the rural population. Remote from the centers and with virtually unchecked power, they could establish personal fiefdoms where their word counted as law. When government commissions inquired about local conditions and regularly uncovered the misdeeds of kolkhoz chairmen, the higher-ups regularly chastised the behavior of their village representatives: "They kept old traditions in the kolkhozy, especially the feudal relics that are a dark heritage in our region from the times of the Khan. They cultivated administrative measures, organized mass beatings of peasants to force them to take part in seasonal work obligations, they cultivated bribery and saved the custom of *mul'deki*, forcing kolkhoz members and individual peasants to perform personal labor obligations for them by working on their private fields, they lived as parasites on the account of the kolkhoz members by forcing them to slaughter muttons for them, and they maltreated the best kolkhoz people."[39]

Administrative despotism by kolkhoz chairmen and other rural notables was the rule, not the exception, on the Amu Darya during the 1930s and 1940s.[40] Since administrative despotism appears in a wide range of different colonial and postcolonial settings, not just Soviet, and has a unique power logic, it serves to appreciate its implications. According to anthropologist Gerd Spittler, administrative despotism is a way to exercise power that combines arbitrariness and physical violence.[41] Arbitrariness implies that each and every administrative decision is taken individually. Whether it concerns the amount of cotton or the sum of taxes to be paid by a peasant household, state officials collect resources and money on the basis of case-by-case decisions. Hence, their actions are unpredictable for individual peasants and village communities. Because peasants are likely trying to evade payments and the delivery of produce to the state, the threat of violence has to be used by state officials to enforce their arbitrary decisions. As a result, peasants and state officials lack trust. On the contrary, the deficit of information (on the side of state officials) and the lack of security (on the side of the peasantry) prevent each side from building trust relations that would enable them to engage in meaningful/sustainable economic activity and to plan for the future.

During campaigns to collect taxes and revenues, state officials have to constantly uphold the threat of violence and use violence to enforce their arbitrary decisions. Power exercise by a despotic state administration is, therefore, closely tied to the physical presence of officials in the villages. This is a noteworthy restriction on administrative despotism: while it absorbs a great deal of personnel and demands constant physical effort, it cannot forestall strong tendencies toward decentralization and autonomous action by local officials. Reluctance and passivity toward orders and instructions of the state bureaucracy are, thus, common features of the rural administration. Since this control is usually lacking, local officials can act as they please. When control from above is actually superimposed on them, only failures and shortfalls become visible for higher-ups. While local officials impose their despotic rule, they themselves are threatened by sanctions and punishment from above. Administrative despotism generates not only a frail (but violent) local administration but also a frail (but violent) central state: "In despotism the area in which authority is actually exercised is extremely limited. Arbitrary rule involves a great deal of work and is therefore inevitably restricted."[42]

STATES OF MOBILIZATION

On the Amu Darya, administrative despotism had severe implications for the day-to-day management of the irrigation system. Water distribution, maintenance works, and mass mobilization for canal construction posed overwhelming challenges for administrators and engineers during the 1930s and 1940s. Instead of obeying orders, peasants remained passive as long as decisions were not imposed by force. But the use of violent methods was necessarily restricted. Consequently, locals neglected canal networks that went to state fields, irrigated these lands carelessly, and cultivated cotton crops sluggishly. While water distribution plans were officially fulfilled, kolkhoz peasants used abandoned canal systems and nighttime hours and arranged small-scale accidents in order to water their private land plots.[43] Despite continuous efforts, state officials were unable to gain control of the irrigation system. Moreover, the threat of violence and the use of force hampered larger construction efforts. Because of passivity, distrust, and fear, peasants were unmotivated to participate in the large-scale hydraulic construction to which Soviet leaders aspired.[44]

Starting in 1932 flooding and the erosion of the riverbanks became an ever more serious problem in Karakalpakstan as the Amu Darya changed its banks in its lower reaches. Every year during November and December, the fast floodwaters and ice drift washed away parts of the riverbanks and came closer to the villages and towns on the river's right-hand banks. During July and August, the summer floods continued the destructive work with water speeds of up to four meters per second. In some places, the river washed away as much as fifty meters of agricultural land per day. Flooding became so devastating that the capital of the Karakalpakstan had to be moved from Turtkul to Nukus in 1936.[45] In late 1934 serious works were started to protect the riverbanks by improving the existing dams and by covering the dams with concrete elements. But even when protection works were taken up, building materials (concrete elements and wood) were always short in supply, preventing engineers from finishing their work. As a result, the local peasants were spontaneously mobilized for several days to collect bushes and other wooden plants that could be used to protect the dams. The efforts made to protect the riverbanks and the agricultural land "had the character of improvised and experimental measures."[46]

After the flooding destruction had worsened over the years, the Moscow government ordered building materials sent to Turtkul in February 1937.[47] However, because Moscow's decision to grant subsidies was taken only in mid-February, by the time local authorities could start construction work, it was already too late to have any lasting effect. There was only time for minor improvements to the existing fortifications before the beginning of the new flood season.[48]

Moreover, construction relied far more on improvisation than on planning. Engineers and managers at building sites asked Moscow only for formal approval of technical projects. When construction works were actually implemented most of the work relied not on the approved plans but, rather, on local knowledge. Decisions were then taken on the spot. Because of the belated arrival of construction supplies, traditional methods and building materials were used.[49] Hastiness and improvisation negatively affected both the provision of building materials and the recruitment of an appropriate workforce; transport was never available in time, and there was never enough food for the workers. Construction machinery was barely used at the building sites. If and when machinery arrived, it was always late, and usually sufficient fuel supplies were unavailable. The construction management frequently ran out of money because the central agencies did not remit it in time.[50]

Mobilization of the workforce, however, was also a major problem on the construction sites. In late 1937, even after several calls on the local government to implement the working schedule in the Khorezm district, only 1,020 instead of 2,660 peasants as planned showed up for canal construction. They had to be guarded by 255 policemen.[51] In early December 1937, only 37 percent of the official working plan had been accomplished and the district government had to be asked again "to forge the participation of the workforce in order to speed up the canal cleaning work." When mid-December temperatures hit minus 15 degrees Celsius and the surface of the Amu Darya River was covered with drifting ice, only 25 percent of the workers worked according to the plan.

Because the icy waters threatened to destroy dikes and cause flooding, local authorities were yet again asked to "take decisive measures and fulfill the workforce quotas."[52] The mobilization of workers for canal construction posed a problem for Soviet administrators because their everyday exercise of power relied on administrative despotism. Canal works were only added on top of other work obligations the peasants had to fulfill in the kolkhozy. Consequently, peasants were forced to work by police and were fined and punished if they chose to walk away or ignore the working plans. District party bosses, kolkhoz chairmen, and local police singlehandedly made arbitrary decisions about who had to participate in the demanding canal works and who could stay behind.

Faced with the power and unpredictability of the river's floods, combined with the Soviet command economy and administrative framework in the region, officials based their approach to canal construction in the 1930s and 1940s on arbitrary mobilization by violent means. Construction continued to rely on physical labor, since machinery was only available in exceptional cases. Few technical innovations were introduced. Irrigation construction showed, on one hand, that "the state" had gained more room to maneuver and, on the other hand, that the space of its authority remained limited and restricted. The ever-changing "capricious" Amu Darya River reminded authorities of the limits of their power and the daily difficulties they had to face.

Arbitrariness and violence determined the relations between the agents of the state and the populations. The consequence for the everyday working irrigation system was passivity on the side of the population and, as a result, the lack of technical innovation and improvements that would have met with the acceptance of the population. The passivity of the population challenged Soviet rule as much as did the complex ecology of the Amu Darya River. The "capriciousness" of the river could not be ignored or overcome. Despite the introduction of economic planning and administrative technologies, canal construction and flood protection works were conducted as if the irrigation system was in a permanent state of emergency.

7 Soviet Irrigation Policies under Fire

Ecological Critique in Central Asia, 1970s–1991

Julia Obertreis

As a region on the periphery of the Russian Empire and, later, the Soviet Union (USSR), Central Asia has long been seen as a mere colony or as developing republics. These perspectives, held by the imperial center and scholars alike, have perhaps degraded the region to an object of necessary central interference. Yet the fact remains that major sectors of the economy and of social and cultural life were historically shaped by the state's modernization efforts, especially during the Soviet period. Policies governing irrigation and cotton growing had particularly grave consequences for the environment. The siltation of the Aral Sea as a consequence of the expansion of irrigation for cotton cultivation is a well-known phenomenon but only the tip of the iceberg.[1]

While geographers, ethnographers, and other scholars working on post-Soviet Central Asia have long dealt with environmental aspects, especially concerning water management, historians have only more recently begun to include environmental perspectives. Existing surveys of Central Asia's modern history have largely concentrated on cultural, political, and religious aspects.[2] At the same time, environmental perspectives on Soviet history have focused on Russia and Siberia and particularly Moscow.[3] The further development of environmental perspectives on Central Asia's history should be embedded in general political history, including modernization, Sovietization and nation building, thereby combining subjects as yet treated separately.

The time period covered here is the 1970s and 1980s. In environmental history, a turning point around 1970 has been identified as a breakthrough of a modern environmental activism and consciousness distinct from earlier forms of nature protectionism. John McNeill articulates two phases of modern environmentalism, the first one beginning in the mid-1960s and continuing into the late 1970s. This is when, in Western industrialized states, environmental movements and green parties emerged and governments began to establish environmental agencies. The second phase began around 1980. The "poorer countries" began to establish their own environmental agencies and grassroots movements, like the famous tree-planting initiative in Kenya, which could influence national politics.[4] In Germany's literature on environmental history, the early 1970s are described as an "ecological turn."[5] Joachim Radkau characterizes the "Age of Ecology" as being marked by two turning points around 1970 and 1990.[6]

Communist-governed states do not figure prominently in international environmental history of this period because of language barriers and the early stage of this research. McNeill does not attempt to characterize socialist environmental politics and environmentalism under the conditions of communist regimes, although he stresses the environmental consequences of the "grand social and ideological systems that people construct for themselves."[7] Radkau examines the USSR and the German Democratic Republic (GDR) in some detail, especially regarding the 1980s, but he does not lay out generalizable features of environmentalism in socialist-governed countries. We are just at the beginning of finding out more about environmental activities under socialism, although seminal works laid the groundwork some time ago. Naturally, because of its geographical and political size, the Soviet Union should be regarded foremost. It was not only a socialist state but also a modern socialist empire with a mission to civilize and modernize its peripheries.

In this context Central Asia is especially interesting as the only undisputed example of a colonial possession of the Russian Empire and because of the grave environmental problems that have shaped its Soviet and post-Soviet histories. The reasons for these problems have to be detected in the context of a Soviet technical modernity agenda that legitimized Soviet rule and was realized by technical and planning experts as well as state functionaries. Their spheres of activity and worldviews were removed from the cultural and religious questions that have been highlighted in existing Western historiography. The severity of environmental problems was a test case for Soviet modernity in Central Asia, provoking heated debates, gloomy statements, and faction building among intellectuals and experts. In the realm of irrigation and modernization of agriculture, the Soviet regime sought legitimation in its southeastern peripheries. Thus, fast-growing ecological, sanitary, and social problems in the countryside became a parting of the ways for those who challenged established modernization patterns and those who clung to them.

This contribution shows that a specific form of environmentalism emerged in allegedly backward Central Asia during the 1970s and 1980s, which included outspoken criticism of modernization policies and their representatives. By elucidating this example, it is the general aim here, as depicted in Nicholas Breyfogle's introduction to this volume, to overcome the "ecocide" thesis that stems from a specific time and political context and has proved unsatisfactory.[8] The "ecocide" narrative has one-sidedly highlighted the environmental problems of the Soviet Union and has depicted the mid-1980s as a sudden outburst of environmentalism. Although this narrative and this chronology have to be corrected, environmental problems and their consequences should not be played down.

In examining the emergence of ecological criticism and environment-related activities during the 1970s in Soviet Central Asia, in this chapter I find that 1975 was a turning point. I examine different actors and forums with a focus on press discourses and scholarly journals. During the 1980s criticism expressed since the mid-1970s grew more sharp and expansive, especially in the era of Perestroika and Glasnost during the second half of the decade. What consequences, if any, did this "ecological critique" have? The experts responsible for the planning and realization of Soviet modernization policies in irrigation and agriculture were a multiethnic group. As a result of Moscow's "affirmative action" policies beginning in the 1950s, more and more Central Asians joined the ranks of the cadres in almost all professional and political-economic spheres.[9] Did they take a special position in the debates examined? Did they promote "national" points of view?

SOVIET IRRIGATION POLICIES

The expansion of irrigated areas, the construction of canals, dams, and reservoirs, and the opening up of new lands were cornerstones of Soviet success in Central Asia. The main aim of all these efforts was to expand cotton cultivation, which was mostly dependent on irrigation. Cotton was the most important agricultural product delivered by Central Asian republics to the western parts of the empire. Uzbekistan was frequently referred to as the "cotton supplier [*khlopkovaia baza*]" of the USSR and maintained the far largest share of the country's cotton production.[10]

Since the end of the 1930s and with the introduction of so-called people's construction (*narodnoe stroitel'stvo*), large-scale projects in irrigation were realized, including the Great Ferghana Canal in Uzbekistan in 1939 and the Qara-qum Canal in Turkmenistan, construction of which began in 1954. The Qara-qum Canal derived huge amounts of water from the Amu Darya River, one of the two main water depositories in the region and one of two feeding the Aral Sea. Both canals were celebrated in Soviet mass media and academic writings as victories of the Soviet (Uzbek, Turkmen) people over nature. Besides these mega projects, many other canals were built, especially beginning in the 1950s. Accordingly, the usage of irrigation water from the main rivers grew ever more intense and the rivers' horizons declined.[11]

Besides the construction of irrigation systems, the opening up of agricultural lands was also a prominent goal of Soviet development policies. Khrushchev's Virgin Land Campaign (VLC) in the Kazakh Republic, in western Siberia, and in the southeast of the European part of Russia is regarded as an important element in his struggle to achieve and maintain power in Moscow's leadership circle in the mid-1950s.[12] In the Uzbek republic a major showcase project he supported was the opening up of the Hungry Steppe (see map 7.1), more than a million hectares southwest of Tashkent believed since tsarist times to contain fertile soils.[13]

In 1956 a new period of development of the Hungry Steppe in Uzbekistan began. An expanse of three hundred thousand hectares was to be irrigated, and a special trust, the Golodnostepstroi (Hungry Steppe Construction Trust), was created to realize the ambitious plans. A few years later, another major agency called Glavsredazirsovkhozstroi (Central Asian Irrigation and Sovkhoz Farm Construction Administration) was founded to supervise development programs in all of southern Central Asia.[14] In official statements, the opening up of the Hungry Steppe was depicted as an "attack on the desert [*nastuplenie na pustyniu*]." The turning of barren, "dead" land into a "blossoming oasis" was a much-used metaphor since tsarist times. In general, the idea of man conquering and defeating nature now reemerged as a characteristic both of late Stalinism and the Khrushchev period.[15]

In the late 1960s and the 1970s, Soviet irrigation and development policies as sketched above continued with new canal building and VLC projects.[16] Cadres of bureaucrats in the Ministries of Water Management in the republics and in Moscow mostly supported expanding irrigation systems and opening up new lands because these costly projects secured their high status. More and more water was derived from the major rivers for irrigation. Simultaneously, the long-term consequences of the expansion of irrigation turned into major economic and social problems. The Aral Sea was shrinking (see map 6.1) as both its feeders, the Amu Darya and Syr Darya, were heavily reduced by the redirection of water. The siltation of the Aral Sea had been anticipated in the

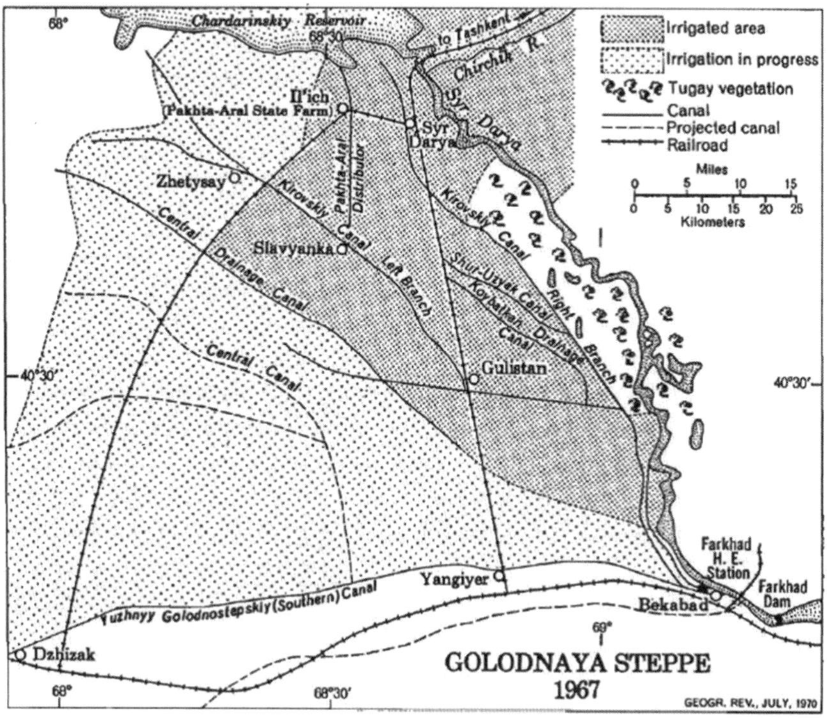

MAP 7.1. The Hungry Steppe in 1967. From Ian M. Matley, "The Golodnaya Steppe: A Russian Irrigation Venture in Central Asia," *Geographical Review* 60, no. 3 (1970): 343.

tsarist period already and was detected in the 1960s, but Soviet officials, experts, and scientists did not regard it as a problem at first. Now, its rapid shrinking evoked more concern among nature protectionists and journalists.[17] Desertification went on with rapid speed, changing the landscapes of Karakalpakstan and other areas. As the coastline retreated, cities were stripped of their primary economic basis, the fishing industry.[18] Another problem was the general scarcity of water, known as *malovod'e*. Irrigation water became a deficit in many places, which affected cotton harvests.[19] Drinking water was polluted by fertilizers and pesticides, which was a serious problem since large parts of the rural population drank unpurified water from the canals.[20] As a solution to the water scarcity, the idea of diverting Siberian rivers to Central Asia (known as *perebroska*) was very much *en vogue* in the 1970s and 1980s. Most officials and technical experts in Central Asia heavily supported this project. They set high hopes on the arrival of the "Siberian water."[21]

ECOLOGICAL CRITIQUE IN THE 1970S

Against this background, the treatment of natural resources and development policies came under fire in these decades. In the genesis of ecological thinking and criticism and in the strengthening of environmental issues, different actors and forums participated. One of these—admittedly a rather fragmented one—was constituted by academic and scholarly journals. Not accidentally, the earliest of some keen publications were to be found in the journal of the Uzbek Academy of Sciences, *Obshchestvennye nauki Uzbekistana* (Social Sciences of Uzbekistan), a scientific journal edited by the Uzbek Academy of Sciences that covered a wide range of disciplines and subjects. As was the case with the Moscow Academy of Sciences, the one in Tashkent held a prominent position as a research institute.[22] Its active engagement in nature protection issues had already become a Soviet tradition by then but it was noticed before only the central Academy of Sciences in Moscow had been.[23] As the publications analyzed show, the Uzbek Academy also gave leeway for the introduction of the ecological perspective and thus contributed to a reorientation, at least in the intellectual sphere. Another important sphere of growing environmentalism was legislation—despite the fact that it has often been dismissed in existing scholarship as having proved ineffective.[24] The Obshchestvo Okhrany Prirody (Society of Nature Protection) was another relevant player. It existed not only on the USSR level but also in the republics. It was more than a mere propaganda organ of the government and contributed to putting environmental issues on the agenda in its own way. In addition, there was an interesting case of environmental activity in terms of specific public spheres—an expedition organized in 1978 that combined Moscow and Central Asian expert public spheres and activism.

Serious criticism and ecological thinking emerged not in the context of a special organization but in scholarly articles and books. Beginning during the 1960s, when economists and engineers addressed the wastage of irrigation water on the local level, this discourse developed out of an economic criticism and became a discourse on nature protection and ecological thinking. In the 1970s this criticism intensified when economists and others introduced new environmental and ecological perspectives.

A discourse on the "(ir)rational usage of natural resources" with regard to irrigation management had been well established since at least the early 1960s.[25] It addressed the problem of water wastage in terms of economic harm. In general, waste of scarce resources (and capital) was a central issue in the Soviet economists' debates beginning in 1959.[26] In Central Asia, the ubiquitous problems of salinization of soils and rising groundwater could be addressed in this economic context. This discourse did not usually address the economic system as such and thus the foundations of socialism. Instead, state and party officials blamed the kolkhoz and sovkhoz farms for wasting water.[27]

In the early 1970s aspects of water pollution were more frequently addressed than before.[28] In this period, established topics of nature protection—such as the preservation of nature in reserves, the preservation of endangered species, and the fight against poaching—were usually related to the existing laws and were addressed as "nature protection [*okhrana prirody*]."[29] Only in one article on "bio-methods" of pest management, published in 1971, was the term "environment [*okruzhaiushchaia sreda*]" used.[30]

In 1975 several articles united these different strands and introduced ideas of ecology. Therefore, the year 1975 can be regarded as a caesura. Several articles in *Obshchestvennye nauki Uzbekistana* added new elements to the existing discourse on "rational usage of natural resources," using new expressions like "pollution of the environment [*zagriaznenie okruzhaiushchei sredy*]" and "the ecological problem."[31] One author demanded an "ecological style of thinking" and an "ecological education and ecological ethics of scientific and technical cadres."[32] The problem of the old-style mentality of scientists and technical experts working on the large infrastructural projects was thus addressed directly—an important precursor of the criticism that would intensify in the following years.

These changes in established discourse and this introduction of new ideas fulfilled several functions: the established subject of resource usage was connected with an environmental perspective that was highlighted by linking it to the "welfare of the present and succeeding generations of Soviet people."[33] Moreover, environmental problems were put in a global context. In the following years, authors commonly regarded pollution as a worldwide problem, implying that socialist and capitalist countries were experiencing the same problems. Nevertheless, the basic differences between the two blocs in treating environmental problems and the much better chances of the socialist system to overcome them were highlighted by almost all Soviet authors, probably as required lip service.[34]

The articles implied new and provocative thoughts on mankind's responsibility to future generations and on the need to harmonize technology and nature and to care about the consequences of industrial production and consumption. Where did these thoughts come from? To answer this question in detail, further research will be necessary.

It can be assumed, however, that some of these impulses came from abroad. The Soviet Union took part in a UNESCO program titled Man and the Biosphere (*Chelovek i biosfera*).[35] Through contacts at Soviet and international conferences, ideas could spread relatively quickly. There were also translations of single Western books on environmental questions, such as Barbara Ward's

Only One Earth.[36] Another possibility of the spread of contributions of Western authors in the Soviet Union was the translation of Western language articles in the proceedings of joint conferences.[37] As the example of Vladimir Vernadskii's notion of "biosphere" shows, however, the exchange between the Soviet Union and the West was by no means a one-way street.[38]

Compared to the *Obshchestvennye nauki Uzbekistana*, the journals *Narodnoe khoziaistvo Turkmenistana* (Agriculture of Turkmenistan) and *Narodnoe khoziaistvo Uzbekistana* (Agriculture of Uzbekistan) were to a lesser degree a forum for new ideas. But even here, nature protection issues gained more significance, in part because of the intensification of the state's environmental protection legislation. In the early 1970s several central decrees were issued. No USSR nature protection law existed at this time but there was legislation on soils, waters, and health care both on USSR and republic levels that touched upon environmental issues, and there was a new wave (after around 1960) of nature protection legislation in the republics.[39] Additionally, a new nature protection decree was issued by the central government in December 1978.[40] Nature protection boomed in the sphere of legislation. Certain terms such as "reproduction" of resources or "environment" were introduced to legislation. "Ecological," though, was not.

To be sure, all these laws were weakly realized and large-scale pollution continued. Priority was on the fulfillment of agricultural and industrial plans, and existing protection authorities were badly equipped and staffed.[41] But at least legislation in this sphere existed and could be cited in demands to raise environmental protection activities and effectiveness.

Nature protection was also mentioned in the new Soviet Constitution of 1977. In public discussion of the document, many Soviet citizens sent letters to state and party bodies demanding an ecological reorientation of politics.[42] Although not in very prominent places, the constitution made reference to environmental concerns.[43] The reference to the constitution, in turn, became a widespread pattern of argumentation in the following years: nature protection was mentioned as an aim of state politics in the constitution, so it was requested and desired by the regime, and it was something one could make a claim on. The effect was, it seems, comparable to that of the Helsinki declaration for the sphere of human rights.[44]

At this point, another player entered the scene. The Turkmen *Obshchestvo Okhrany Prirody* (Turkmen Society of Nature Protection) took advantage of the topicality of the subject and published several articles promoting environmental protection. As in other republics, this society was a mass organization. In 1979 it had 277,000 members (almost 10 percent of the population of the Turkmen Republic), most of whom were children and students, pioneers, and Komsomol members.[45] Membership was often a formality, and the society was mostly a transmitter for the state's environmental policies with no critical agenda.[46] Nevertheless, its cause had received a strong impulse from the constitution and from the new decree of 1978, so the society could claim, "To protect nature is a mandate [*nakaz*] of the party," as the title of one article ran. In recent years, the society had organized or attended several conferences, among them the first conference on nature protection in Turkmenistan with representatives of the other republics' nature protection societies and a conference on the protection of ecosystems of arid areas attended by American specialists.[47]

In 1978 an expedition called Zhivaia voda (Living water) visited Central Asia. It was a small group of journalists and scientists from Russia aimed at ecological and economic problems related to water usage in different parts of the Soviet Union. This time, they had chosen the Amu Darya as the focus and destination of their expedition. The expedition visited different places along the river and talked to local party and Komsomol organizations and many scientists and experts who were often complaining about the lack of water.[48] Such expeditions were a special form of non-state organization. They too linked Central Asia to Moscow by providing information on the current situation in the periphery. They also enhanced personal contacts in an ethnically mixed community of scholars and activists.

In 1980, against the background of a serious water crisis, the critical discourse on environmental issues intensified. Water was so scarce that people in Karakalpakstan and other regions used more and more groundwater extracted from illegal artesian holes.[49] Officially, cotton harvests were steadily improving; in 1980 there was a record harvest of six million tons in Uzbekistan. But in reality, yields were stagnating.[50]

Authors from different disciplines criticized not only the wastage of water or the endangering of animal species but also the one-sided orientation of agriculture on cotton growing and the pushing back of crop rotation.[51] Three agronomists linked the rising salinization of soils directly to the decline of cotton yields and thus linked ecological to economic problems.[52] The use of "bio-methods" in pest management, propagated in various articles since the early 1970s, was now demanded more energetically.[53] It was nothing unusual at this point to bring forward arguments for nature protection, the protection of the environment, or ecological correlations. Even a continuing education course on "Agrarnaia teoriia i politika KPSS [Agrarian theory and the politics of the Communist Party of the Soviet Union]" addressed environmental problems.[54]

An outstanding article was published in 1980 by a female researcher of the Institut Filosofii i Prava Akademii Nauk Uzbekskoi SSR (Institute of Philosophy and Law of the Uzbek Academy of Sciences), I. Khashimova, in *Narodnoe khoziaistvo Uzbekistana*. It was a direct criticism of the gigantomaniac development policies that had been alluded to in one of the articles published in

1975. Khasimova wrote: "One of the serious ecological problems of today is the attack on the desert. . . . In conjunction with the opening up of new lands the water deficit will rise more and more."[55]

This was the earliest article I have found by an author from Central Asia that attacked directly the opening up of virgin lands and the idea of subjugating nature that went with it. The author regarded the VLC as responsible for the water scarcity that had amounted to a serious water crisis by 1980. Khashimova's article remained an exception for a few years; at the least, it does not seem to have triggered similar statements. The amassing of criticism in special journals did not influence politics at this time. But Khashimova opened a new phase of criticism that was to burst out after 1985. Her criticism of the development of virgin lands at the expense of a regular water supply then became an often repeated topic.

To sum up, the mid-1970s can be seen as a turning point. The environmental discourse that emerged was not a broad one and was not connected to a mass movement, as in the West. But there was a noticeable change in what could be said in the rather fragmented public sphere of expert and special journals, and the discourses that developed in this forum laid the groundwork for more intense and broader criticism in the second half of the 1980s. An expedition with journalists and scientists drew public attention to Aral Sea siltation and excessive irrigation in southern Central Asia. Other driving forces were the expanding environmental legislation and the changing agenda of the USSR and Turkmen/Uzbek societies of nature protection.

THE PERESTROIKA YEARS

The political history of southern Central Asia differs from that of the Soviet Union in general. Important events and caesurae were the death in October 1983 of the Uzbek party leader Sharaf Rashidov, who had been in power since 1959, followed by the "cotton affair [*khlopkovoe delo*]," which began in 1983–1984. The cotton affair is well known as a large-scale punitive campaign directed by Moscow against corruption. During 1984 and 1985, hundreds of high officials of the Uzbek Republic were dismissed. Ten out of thirteen province secretaries lost their offices; five of them were arrested and one was sentenced to death. Half of the Uzbek *nomenklatura* was exchanged.[56] Some of the most serious charges concerned falsifications of cotton production figures. Moscow reacted to what it perceived as large-scale deceit from the Central Asian side with mistrust and Russification. This, in turn, affected ethnic relations in Central Asia and provoked Uzbek nationalism and chauvinism.[57] The underlying reasons for the cotton scandal were the pressure and demands by the government to deliver ever more cotton to the center. The record six-million-ton harvest of 1980 was an unrealistic goal to achieve and was partly made up.[58] There were serious environmental and ecological reasons for lower cotton harvests, among them the salinization and exhaustion of the soils and the lack of irrigation water, but these problems were not addressed by representatives of the central government and institutions in the course of the affair. The cotton scandal overshadowed the beginning of Perestroika in Central Asia. Perestroika and Glasnost were short-lived in Uzbekistan and Turkmenistan and were limited in their effects because the new power holders repressed non-state groups by limiting their access to the media and their ability to register as organizations or parties.[59] Nevertheless, new groups emerged, most prominently the group Birlik in Uzbekistan, which expressed cultural and national but also ecological demands.[60]

Building on the ecological critique of the 1970s, ecological and environmental criticism gained momentum after 1985 and the fronts became more clear-cut. The broadening of leeway for public discourse in the press intensified debate on many crucial issues. Despite the strained relations between Central Asia and Moscow, experts and journalists continued to cooperate in drawing public attention to intensifying water problems. Another large expedition called "Aral-88," headed by Grigorii Ivanovich Reznichenko, visited the Aral Sea basin in the fall of 1988.[61] This expedition was organized by two journals, *Novyi mir* (New world) and *Pamir*. Since 1986, *Novyi mir* had been edited by Sergei Zalygin, an influential opponent of the diversion project.[62] Among the group arriving in Central Asia from Moscow were seventeen journalists, writers, and scholars from different disciplines, one representative of the prosecution and one of the USSR Ministry of Water Management. Upon arrival in their first destination, Dushanbe, eleven others from Central Asia joined in.[63] Aral-88 received some media coverage, organized several press conferences, and helped direct unionwide attention to the problems of the Aral Sea basin. Like the expedition, the journal *Zvezda Vostoka* (Star of the East) represented a platform of exchange of opinions and findings of research between Moscow and Tashkent. It was edited in Tashkent and contained many articles by authors from Central Asia but also from Russia and translations from prose by foreign authors. It had readers both in Russia and in Central Asia and linked the discourse of the two regions.

In these expert scholarly and publicist debates, no segregation on ethnic lines can be detected. The intellectuals taking part in them were builders of Soviet modernity and were raised in the spirit of the "friendship of the peoples [*druzhba narodov*]." Many of them had for decades supported the basic politics of the Soviet regime. But they could also become critics of long-standing Soviet policies including water management and excessive cotton growing. An example is the agronomist Mirza-Ali Mukhamedzhanov who had supported Soviet agricultural politics since the 1930s and later occupied high positions in ministries and in the Academy of Sciences. He also represented the republic abroad as an agricultural expert.[64] By the 1980s, he had become a critic of the Soviet agricultural politics in Central Asia. In 1985 he published a book entitled *Berech' zemliu, umnozhat' ee plodorodie* (Protect the soils, multiply their fertility), in which he heavily criticized "cotton monoculture," as a factor in the exhaustion of the soils and environmental pollution.[65]

The most prominent debates in the 1980s, however, did not center on cotton growing but, rather, on water management, the diversion of Siberian rivers, and the fate of the Aral Sea. Debates were influenced by new information delivered by mass media. Revelations on the state of health and sanitary conditions had a shock effect. Before, health statistics had been subject to massive falsification. Rates of infectious diseases proved to be at unimagined levels, especially in Karakalpakstan and Turkmenistan, and were partly caused by the usage of polluted water.[66] The connection between water pollution, water scarcity, and the Aral Sea siltation on the one hand and rising illness and mortality rates on the other were obvious. This spurred the debates on water management.

A decree by the central government dated August 14, 1986, stopped the river diversion project.[67] This was a shock and a big disappointment for its advocates, especially in Central Asia. The decision evoked anger among the ranks of officials and technical experts who saw it as the result of a press campaign directed against them by Sergei Zalygin and other writers. Indeed, as Douglas Weiner has shown, criticism of the diversion projects also came from scientists in a broad range of institutions. Among the Village Prose writers, the debate had nationalist overtones.[68]

Moscow's ban did not end the debate over the diversion projects. On the contrary, scholars and journalists from Central Asia continued their heated argument. The debates and published opinions during Perestroika should be seen as extending the critical discourses that existed before. What had begun in Uzbekistan in the 1970s with ecological critiques by economists and others now evolved into a broader debate involving journalists, writers, and scholars.

The front lines were clearly established. On one side, there were the *vodniki* (roughly translated as water construction experts), officials, and engineers responsible for large-scale irrigation construction. The vodniki's adversaries were journalists, scholars, and writers. Neither side pulled any punches. The central and republican Ministerstvo melioratsii i vodnogo khozaiastva (Ministry of Reclamation and Water Management; known as Minvodkhoz) was tackled both from the public and from within the state leadership.[69] The achievements concerning the opening up of the Hungry Steppe did not have the magic of the 1960s anymore. Instead, it seemed now that the above-mentioned organization, Glavsredazirsovkhozstroi, had run down the areas developed since then. Sovkhoz farms in the development areas were operating at a loss financially.[70]

The organization had neglected to provide sufficient infrastructure other than irrigation systems.[71] The opening up of new lands was viewed by critics as an enterprise that had used up huge resources and had led to little positive results. The Minvodkhoz on both the USSR and republican levels was held responsible for the gigantic development projects and the excessive expansion of irrigation systems and all their negative consequences. It was also attacked as an authority with extreme power and giant financial means.

In the rows of the adversaries of the vodniki, Irina Aliab'eva took a prominent position. She was one of the editors of *Zvezda Vostoka* and was the only woman on the expedition Aral-88 when it arrived in Dushanbe.[72] In an article published in 1989, she expressed a radical, relentless criticism. She concentrated on the reclamation projects in the Hungry Steppe and elsewhere that had contributed to the catastrophic situation in the region adjacent to the Aral Sea. Aliab'eva castigated the dolce vita of the Soviet nomenclature and juxtaposed it to the poverty of children in Karakalpakstan who had to drink water from the irrigation canals "intoxicated with pesticides and infected with agents of typhoid fever." She criticized the convergence of interests of the party and the Minvodkhoz. The "marathon opening up of virgin lands" in the Hungry Steppe and other steppe regions, "if it has added to anything, then to losses for the already poor village, and it added to the number of medals, awards, and other signs of fame for those people reclaiming the land and for Rashidov himself."[73]

Central Asian writers, too, engaged in criticism of the regime's policies. They tried to act as spokesmen for the rural population suffering the most from water scarcity, desertification, and a polluted environment. Some writers were engaged in the newly emerging Komitet po spaseniiu Arala (Committee for the Salvation of the Aral Sea).[74] Writers' publications, speeches, and nongovernmental activism were not directed against Russians or Slavs. Rather, the writers, at least the ones who were well-known representatives of their republics, appealed to their Soviet colleagues for support. For example, the well-known writer Tulepbergen Kaipbergenov from Karakalpakstan used his speech at the congress of the Sojuz Pisatelei (the Soviet writers' union) in May 1987 to draw attention to the "ecological catastrophe" of the Aral Sea. He appealed to the established role of Soviet writers as defenders of ecological issues—in the case of Lake Baikal, for example.[75] As persons and institutions responsible for the siltation of the Aral, he named scientists, economic managers (*khoziaistvenniki*), and the Minvodkhoz. He also referred to the "attack" on the desert and steppe lands, the opening up of ever new lands for irrigation.[76]

The debates also contained a controversy about human attitudes toward nature that had, of course, begun much earlier. Criticism of the idea of subjugating nature that had dominated politics and infrastructural projects in the past decades was often repeated. The publicist Aliab'eva, for example, wrote about an "unrestrained passion for the subjugation of nature" that had led to the well-known negative consequences.[77] She juxtaposed it to her own viewings and ascribed to nature both reason and agency: "I believe in Nature's reason [*Ia veriu v razum Prirody*]" and "The rise of dangerous diseases, high mortality rates and the mass degradation of lands—isn't this a riot of nature against its unbridled subjugation?"[78] The technical expert I. Chervanev put it more mildly and concluded: "nature is more organized than we have thought."[79]

Among the defenders of the diversion project, two authors—B. Atchabarov and T. Sharmanov (professional positions unknown)—polemicized in 1990 against Zalygin and the project's critics. They ended their contribution with an interesting statement on humans' relations to nature in which they repeated the long-standing Soviet paradigm of nature's submissive role: "Nature is maybe severe, but this is to make man courageous. It is closed [*skrytna*] to make him not lose its curiosity. It is unbounded to not let him be limited. It is in a way imperfect to not let the genius of creation and transformation [go] extinct in him. Nature has even given man power over it to not let him be anybody's slave."[80] Such ideas of the complete domination and regulation of nature by humans lived on, up to the end of the Soviet period.

Despite broad public criticism, the Uzbek government officials did little to advance new technologies or enforce a reduction of water usage. In a letter to two central government bodies that had severely criticized Uzbek politics in this sphere a year before, they defended the Uzbek government in 1988. Allegedly it had done a lot to encourage the use of "intensive technology" in irrigation agriculture. Moreover, a government commission for the Aral Sea problem had been founded recently. At the same time, government plans were still aiming to cultivate new areas.[81] All measures taken by the USSR and republic governments came too late for any genuine reversal of Soviet policies. Several decrees were taken by the central government to improve the situation in water management and to counteract the Aral Sea crisis. Usually, the decrees were not put into effect.[82]

There were administrative and institutional changes that took place, however.[83] Glavsredazirsovkhozstroi was abolished in 1988.[84] Its sovkhoz farms were integrated into the newly founded Gosagroprom structures. In 1990 the USSR Minvodkhoz was abolished, partly because of the ongoing protest against its power position.[85] In Central Asia, though, the Minvodkhozes on the republics' level continued to exist and maintained their strong position even after the breakup of the Soviet Union.[86] Regardless of the abolition of the central Minvodkhoz, the fronts between its adversaries and defenders remained stable.

As these insights into the debate on the river diversion and the Aral Sea siltation have shown, the old establishment of the engineers and officials involved in large-scale reclamation and irrigation projects braced themselves against hostilities from journalists and scientists. The vodniki and the Uzbek government tried to maintain the flow of investment in this sector. Although the cadres in the Uzbek republic had been replaced as a consequence of the cotton scandal, the government did not dare to break the mold. There was a strong power of persistence in the government but also in the agencies responsible for planning. Most importantly, the state planning authority Gosplan proved a reactionary force, simply because it continued to set ambitious, but finally unrealistic, plan targets. For the vodniki, the government, and Gosplan, a reduction of irrigation norms and cotton-growing areas was unthinkable.[87]

CONCLUSIONS

The analysis of the debates and fights over water management in Uzbekistan and Turkmenistan in the 1970s and 1980s allows for several conclusions. First, it shows how environmental history can challenge an established chronology based on a conventional political history. The ecological turn, if we can speak of one, took place since the mid-1970s and thus undermines a chronology oriented on the periods in power of first secretaries. It can help us change our idea of the 1970s and early 1980s: they were not just a time of "stagnation" and crisis but also a time of different forerunners for the reforms of the Perestroika period.[88] The ecological critique as presented here also undermines the usual chronological approach of environmental history that saw the beginning of the Perestroika years in the mid-1980s as a caesura.[89] Changes began about a decade earlier, and there is a strong continuity between the 1970s and the 1980s.

To be sure, this environmentalism differed greatly from that in the West. There was no broad social movement as in some Western states where non-state environmental groups sprang up like mushrooms beginning around 1970. Environmental activism in the case of Soviet Central Asia—and, it seems, in the Soviet Union in general—was more elitist and more formal. Major forums were academic and specialized journals in which representatives of different disciplines introduced new ideas such as an ecological thinking or an understanding of environmental problems as global problems. The Academy of Sciences, both central and local, provided an academic location for (critical) environmentalism. This is a role that it seems to have played in other socialist societies in Europe as well.[90]

An important element of the ecological criticism that emerged was a challenge to the established politics of water management with the latter's compulsive expansion of irrigation areas and its neglect for environmental aspects. The "attack" on the desert as a rhetorical figure and as an idea was a marker of this politics and was openly criticized from 1980 onward. An alternative view of nature emerged that became more clear-cut in the debates of the Perestroika period: nature not as an enemy fought by humans or subjugated by humans but as a complex system or something quasi-sacred. The intellectual, scholarly criticism implied ideas of nature protection and modern environmentalism. What is more, it was directed against a certain lobby, the lobby of the vodniki, which was—rightly—held responsible for the severe ecological and economic problems that plagued southern Central Asia from the 1970s. In the Perestroika period this criticism grew stronger and played out more strongly, although conditions for Glasnost in Uzbekistan and Turkmenistan were less favorable than in other parts of the Soviet Union.

These observations help to describe a new type of environmental development in the "era of ecology." Hitherto, focus has been on the West with its early, broad social movements and on "Third World" countries with their grassroots movements such as tree-planting initiatives. The socialist-governed countries represent a distinct type and have to be included in this spectrum.

Scholars have debated the contribution of environmental activism and protests to the transformation of political culture in the Soviet Union and, ultimately, to the downfall of the communist regime. This is an important question for global environmental history because it relates to all other communist regimes in Europe and elsewhere as well. Today, it is largely common sense that environmental protests were important not so much as an immediate cause of the system's collapse but as a transitional phenomenon.[91] In the words of Joachim Radkau, it was "a catalyst for democratic change and the creation of new nation-states."[92] In the Soviet Union, environmental activism was at first not regarded as a political threat by the regime and thus could unfold relatively freely.[93] Environmental activism pushed the development of the nongovernmental sector.

Ecological, cultural, and national demands of newly emerging groups were interrelated, and the case of Uzbekistan follows this general pattern. At the beginning of the Perestroika years, when no or only a few non-state organizations existed, environmental groups served as melting pots for different kinds of political activism.[94] This is true for Central Asia to a lesser extent, however, because of the limited leeway for Glasnost. Although criticism of existing water management politics and the reckless treatment of nature and natural resources became sharper and could reach out to a broader public, no sweeping changes in politics could be achieved. The old elites remained in power and the much-chided ministries of water management continued to exist.

Central Asia is, at the same time, a very good example for another important observation: ecological problems as such, whether made an issue in public debate or not, were clearly one of the factors that led to the downfall of the Soviet and, more general, of the communist state system. In Central Asia the downfall was, above all, because of the water crisis and the decline of cotton yields as severe social and economic problems with environmental roots. Ecological problems were intimately intertwined with economic problems, and these led to a sharp decline in living standards, especially in the countryside. This, in turn, resulted in an intensification of ethnic, regional, and interrepublic conflicts that advanced the breakup of the USSR.[95]

At the same time, Soviet Central Asia is a good example for a (post)colonial context in which representatives of indigenous peoples supported the advance of a modernity that did not treat natural resources with care. Central Asian agronomists and engineers as well as functionaries built their careers on the expansion of cotton-sown areas and the irrigation of ever more hectares.[96] Some of them, however, like Mukhamedzhanov, turned into critics of this old-fashioned, reckless kind of modernity. The eco-critics of the 1970s and 1980s were a multiethnic and diverse group. Rather than allowing for national tensions between Slavs and Central Asians—which could have easily gone along with making the center responsible for the misguided modernization politics—the environment-related discourses reflected the multiethnic composition and rather peaceful cohabitation of ethnic groups in Central Asia's elites. The children of Soviet modernity engaged in a serious questioning of the dominant politics. No nationalistic views were expressed by them. Ironically, the only national overtones related to the source of environmental problems: cotton appeared as a symbol of "national pride" in the eyes of some Central Asian scholars.[97] All of this shows how strongly Soviet politics and discourses had shaped the professional elites. They were too Soviet to engage in "eco-nationalism."[98]

PART III

Land, Rocks, Soil

8 Models of Soil and Society

The Legacy of Justus Liebig in Russia and the Soviet Union

Mieka Erley

In this article I examine the cultural reception of the ideas of the German chemist Justus Liebig (1803–1873) in nineteenth-century Russia and consider the far-reaching effects of his model of soil chemistry on Russian and Soviet attitudes to agriculture and soil management as social questions. Founder of one of the most important chemical research centers in Europe, his chemical institute at the University of Giessen, Liebig would become a pioneer of modern organic chemistry and one of the most significant promoters of science in the public interest in nineteenth-century Europe. Liebig revolutionized agriculture with his assertion that plant life relies on minerals for nutrition, which laid the foundation for the development of artificial fertilizers and the later Green Revolution. Liebig's 1873 obituary in the English journal *The Chemical News* summed up the chemist's renown among the general public: "The application of chemistry to agriculture, and to many of the wants of daily life, received so powerful an impulse from Liebig, that the popular mind has taken him for the representative of the science in its application to practical purposes."[1]

Liebig's ideas on soil fertility were at least as powerful an influence on the Russian "popular mind" as on the English or European mind, and arguably more so, in view of the pervasiveness of the "soil question" in nineteenth-century Russian political and intellectual history. By the mid-nineteenth century Russia possessed its own distinctive discourse of soil, and when Liebig's ideas began to filter into Russia, they met with a charged political and cultural environment. Beginning in the 1840s and intensifying in the 1860s, Liebig's work was widely discussed in Russia's press and literature, among gentlemen farmers and amateur scientists, novelists such as Ivan Turgenev and Fedor Dostoevskii, and Marxist and other social revolutionaries. Liebig's ideas became a site of contest between the Romantic anti-reductionist trend in science and the radical positivist-materialist trend that would follow, and between the social and political postures associated with these two tendencies. Liebig's name became a rallying cry for groups with different political agendas and a symbolic object of contention between intellectual generations. Above all, Russian discussions of Liebig and his materialist soil chemistry spiraled around questions concerning Russian society and how Russia's distinctive national spirit arose from its "native soil." In addition to tracing Liebig's direct impact on Russian discourses of agriculture, history, and society, I will consider how creative reworkings of Liebig's ideas by such figures as Karl Marx, Friedrich Engels, Nikolai Chernyshevskii, and Aleksandr Engel'gardt had a far-reaching impact on Russian and, later, Soviet attitudes to soil and society. The transfer of Liebig's ideas across epistemic domains—from soil science to social science—had a profound impact on discourses of soil, social formations, and agricultural policy for a century to come. This case study of the reception of Liebig in Russia establishes a genealogy of Soviet understandings of the relationship between soil and society; in particular, the Soviet political concept of the *smychka*—the unification of the city and the countryside—is interpreted as a political ecology with roots in Liebigian soil economy. I also examine how cognitive models and metaphors travel across the scientific, social, and cultural domains, and how Liebigian agricultural chemistry became a symbol and cypher for materialist philosophy and radical politics in an era of political repression.

Before I discuss the Russian cultural reception of Liebig's ideas, it is necessary to provide some background on Liebig and his ideas. Liebig's renown, like that of many great popularizers of science, was far greater in his own time than it is today. At his peak of productivity from the 1840s to the early 1870s, his name was constantly before the European public in the ephemera of the age: an extensive body of journalistic writing including his popular *Letters on Modern Agriculture*, public and professional polemics around his work, product testimonials, and his own commercial ventures (of varying success) involving everything from chemical fertilizer to meat extract. Liebig's books on plant and animal chemistry, moreover, were popular among lay audiences and translated into a number of European languages by his numerous students. Even if Liebig's works have been called too technical for the lay reader (Karl Marx complained of having to "wade" through them), it did not prevent a broad public from attempting them.[2] In a widely circulated anecdote, it was said that the passport official examining Liebig's documents on his arrival in London in 1842 shook his hand and affably chatted with him about his first major work translated into English, *Organic Chemistry in Its Application to Agriculture and Physiology*.[3] In addition to his direct outreach to the public, Liebig was enormously important in building the institutions that supported the development of chemistry as the editor of a major journal of the day, through his professional relationships with chemists across Europe, and in his long teaching career, which produced several generations of students who would go on to form the core of Europe's great chemists and industrialists.[4]

Liebig's mineral theory of plant nutrition supplanted the prevailing "humus theory" of his German predecessor A. D. Thaer, which asserted that humus, the "living" component of soil, nourished plants, while minerals, the "non-living" component of soil, were not necessary to plant nutrition.[5] Liebig argued that, in fact, all life has a chemical basis and that "nonliving" substances such as minerals interpenetrate soil, plant, animal, and human organisms. Liebig wrote that minerals from the soil "assume the form of corn, flesh, and bones; they pass into the bodies of men, and again assume the same form which they originally possessed."[6] As Liebig notes, this mineral economy depends on the recycling of minerals back into the soil as fertilizer, a process that, he argued, was not only technologically but also economically and socially mediated. Liebig's original German term for this process of exchange across the mineral, vegetable, and animal kingdoms was *Stoffwechsel* (metabolism), translated into Russian as *obmen veshchestv*, literally "the exchange of substances."[7] Liebig narrativizes the movement of minerals and imagines the exchanges between plant, animal, and human life through the medium of soil as chemical equations that must be balanced.

Liebig was a transitional figure between German romantic *Naturphilosophie*, exemplified by the works of Goethe and Friedrich Schelling, and the new philosophical materialism. After attending lectures by Schelling in his student days, Liebig wrote that "Schelling possessed no thorough knowledge in the province of the natural sciences, and the dressing up of natural phenomena with analogies and in images, which was called exposition, did not suit me."[8] In his later works, Liebig would call this Romantic tendency in science "a dead tree, which bore the finest leaves and most beautiful flowers, but no fruit. With an infinite sagacity, only pictures were created."[9] Despite his rejection of Naturphilosophie and his endorsement of rigorous experimentation, reliance on empirical data, and the practical application of scientific knowledge, Liebig himself would come under fire for antireductionism and excessive "literariness." Liebig's mentor and erstwhile friend, the chemist Jöns Jacob Berzelius, ironically made a similar insinuation about the literary nature of Liebig's own work, writing of the latter's *Agricultural Chemistry*: "This kind of facile physiological chemistry is created at the writing table."[10] A younger generation of positivist materialists in Europe took up Berzelius's critique of Liebig, but in Russia the chemist was venerated in materialist thought and discourse for several decades to come.

LIEBIG IN RUSSIAN INTELLECTUAL HISTORY

Liebig's Giessen Chemical Institute has been called a "chemist breeder," and indeed a significant number of Russia's chemists and scientists in the mid-nineteenth century were mentored by Liebig or one of his disciples.[11] By the 1860s Liebig's Russian students had returned home to establish themselves as major figures in their fields and to propagate Liebig's legacy and materialist vision of the world in both the scientific and cultural domains. This cohort formed the first generation of Russian chemists: Nikolai Nikolaevich Zinin (first chairman and cofounder of Russkoe khimicheskoe obshchestvo, or the Russian Chemical Society), Nikolai Nikolaevich Sokolov (who published the first Russian journal of chemistry, *Khimicheskii zhurnal*, from 1859–1860 together with the agronomist and journalist Aleksandr Nikolaevich Engel'gardt [see below]), Aleksei Ivanovich Khodnev, Nikolai Erastovich Liaskovskii, Fedor Fedorovich Beilstein, and Aleksandr Aleksandrovich Fadeev, among others. Liebig's famed student Aleksandr Abramovich Voskresenskii, called the "grandfather" of Russian chemistry, trained the succeeding generation of Russian scientists—among them Dmitrii Ivanovich Mendeleev, the physical chemist Nikolai Nikolaevich Beketov, and Nikolai Aleksandrovich Menshutkin, the discoverer of the Menshutkin reaction.[12] Beketov and Mendeleev, in turn, mentored the founder of modern soil science and soil classification, Vasilii Dokuchaev, who was, nonetheless, critical of Liebig's limited chemical vision of soil.[13] Aside from transmission of influence through these direct lineages, Liebig shaped Russian chemistry on an institutional level, as the first Russian chemistry labs were modeled on the Giessen research center.[14]

Liebig's students of chemistry also popularized his ideas by translating his works into Russian. The earliest translations, mostly excerpted in journals, were published in the 1840s and 1850s, but it was not until the 1860s that Liebig's works exploded in Russia, a trend coinciding with the professionalization of his first generation of Russian students and the coming of age of a generation of radicals known as the "men of the sixties."[15] Between 1860 and 1863, at least nine full Russian translations of Liebig's books were published; Liebig was such a sensation that Aleksandr Engel'gardt remarked on the phenomenon in an editorial in *Sankt Peterburgskie vedomosti* (Saint Petersburg news) in

1863.[16] The first full Russian translation of Liebig's *Agricultural Chemistry* was published in 1862 by Liebig's student Pavel Antonovich Il'enkov (1821–1877), who would become professor of chemical engineering at St. Petersburg University.[17] *Brockhaus-Efron*, the most authoritative Russian encyclopedia of the prerevolutionary period, hailed Liebig's work a "blessing to mankind" because it promised to increase crop yields and end hunger.[18] As Liebig's works became widely available in Russian, and with the memory of the 1833–1834 famine fresh in their minds, many Russians had high hopes that Liebig's ideas could provide a solution to the multifaceted "soil question" in Russia.

The reach of Liebig's ideas in Russian society extended, then, through scientific institutions, print publications, professional relationships, and finally, social-intellectual circles (*kruzhki*). Kruzhki had become an important institution of intellectual life during the repressive reign of Nicholas I, and they served as a forum for the intermixing of politics, ideology, and science. Not only in organized kruzhki but also in informal circles of intelligentsia sociability, Liebig's disciples brought their scientific interests, education, and expertise to the troubling questions of Russian society.[19] There were many close social and family ties between scientists and artists during the mid-to-late century. For instance, the chemist Nikolai Beketov and his brother Andrei Nikolaevich Beketov, a celebrated botanist who wrote extensively on soil conditions, were close friends of Fedor Dostoevskii, who criticized the abuse of Liebig's name as an empty cultural signifier.[20] These social networks conditioned the movement and contestation of metaphors and images across cultural and scientific domains.

LIEBIG AS A RUSSIAN CULTURAL ICON

The "soil question" in Russia was of concern not only to scientists and agronomists but to the entire Russian intelligentsia of the mid-to-late nineteenth century. Soil was an endless topic of discussion, as both a material and a symbolic object. In Imperial Russia's agrarian society even the most disinterested landowner was aware that soil was the foundation of his estate; at the national level soil was the source of national wealth and the object of contention in an age of political and land reform, imperial expansion, and borderland colonization. It was also a symbolic source of national identity in the tradition of Romantic nationalism and a nostalgic proxy for the Russian people among Slavophiles, who contended that Russian soil and the character that arose from it were distinct from those of western Europe.[21] Most importantly, soil as material became conflated with soil as symbol in theories of geographical determinism and primordial nationalism. *Pochvennichestvo*, the "native soil movement," fetishized soil as a source of national identity. One of its chief proponents, Dostoevskii, wrote in his announcement for the journal *Vremia* (Time): "We have finally seen that we, too, are a distinct nationality, highly original, and that our task is to create a new form, our own native form, drawing on our own soil, drawing on our national spirit and national source."[22] A fellow *pochvennik*, Apollon Grigor'ev, further articulated the link between *pochva* (soil) and identity: "The fact of it is that the soil here is completely virgin, untouched, it has not been planted with anything that would interfere with the natural growth of this organic product."[23]

Commenting on the excessive rhetorical and symbolic inflation of "soil" in the discourse of Dostoevskii and the pochvenniki, Maksim Antonovich, a journalist and amateur geologist, wrote: "'Soil' [*pochva*] was the philosopher's stone for us, a journalistic elixir, an inexhaustible goldmine, a cash cow, in a word, everything."[24] Dostoevskii and the pochvenniki participated in a much broader discourse of soil that had been developing in Russia over the preceding twenty years as the word *pochva* had become highly politicized and symbolic, its usage more distinctly marked than the ubiquitous and polysemous *zemlia* (earth, soil, land). *Pochva* evolved from *podoshva*, meaning "the sole of the shoe," which acquired the more figurative meaning of foundation, ground, or the top layer of the Earth.[25] *Pochva* had also entered scientific usage by the eighteenth century in discussions of geognosy and the attempt to systematize the study of earth. But the most transformative semantic shift in *pochva* occurred in the 1830s–1840s, under the influence of the French *terrain*. A number of new phraseologies from French entered Russian, considerably expanding the figurative uses and associations of *pochva*. Linguist V. V. Vinogradov notes several of these, including *naiti podkhodiashchuiu pochvu* (to find common ground) and *na pochve chego-nibud'* (on the grounds of something).[26] These new phraseologies further shifted *pochva* into a predominantly symbolic or figurative register, placing the material within the semiotic, symbolic, and social orders.

When the journalist Antonovich critiqued the symbolic use of soil, it was part of the codified discursive performance of a materialist and social radical—a "man of the sixties." Materialism as a philosophy was nearly inseparable from radical politics in the Russian 1860s, and Liebig, as an icon of materialism in Russia, also became an icon of radical politics. In Europe Liebig's works emerged during a period of political revolution, but in Russia these works were received in an era of political repression. Michael Gordin notes that the "German states perceived the revolutions [of 1848] to be at their roots *agricultural* disturbances caused by instability in crop production" (my emphasis) and that Liebig himself believed agricultural chemistry could help stabilize the social order.[27]

While German states and scientists were promoting agricultural chemistry as a utopian cure for revolutionary unrest, Nicholas I was reacting to the events of 1848 with repressive measures, including curtailing foreign travel and study and eliminating philosophy from the university curriculum.[28] Nearly all of Liebig's most important Russian disciples had studied in Giessen before 1848, and they were a key force in importing to Russia not only Liebig's ideas but materialist philosophy in general. Victoria Frede notes that Nicholas I's censorship "delayed the arrival of the new materialism into Russia."[29] However, agricultural chemistry had become a sanctioned way of discussing new materialist ideas during the repressive 1840s and 1850s in Russia. Crucial to the strategic interests of the Russian state, Liebig's works managed to escape the censor, yet they carried potentially subversive materialist ideas and placed soil in a broader social and political economy in which serfdom was an inescapable question. It was but a short step from Liebig's mineral theory of nutrition to a materialist ontology of human life, and as Jacob Moleschott would suggest, cognition as well (exemplified by the slogan "No thought without phosphorus").[30] Thus, it was precisely because of the political repression of mid-nineteenth century Russia that Liebig's works on agriculture became so symbolically charged: they offered a means of talking about new materialist philosophies and of critiquing the Russian sociopolitical order grounded in an unbalanced economy of soil. By the 1860s, while Liebig's ideas were being critiqued as vitalist and "literary" by a younger generation of positivist materialists in western Europe, many young Russians continued to hail Liebig as an icon of scientific materialism and a broader materialist philosophy as well as radical politics.

If Liebig exposed a generational divide in Russia, this divide was exemplified by the polemic between the gentry belletrist Ivan Turgenev and the revolutionary writer and critic Nikolai Chernyshevskii. Liebig appears in the works of these authors as a symbolic object of contention between two ideological camps and intellectual generations in mid-century Russia. In 1861 Turgenev had dissolved his relationship with the journal *Sovremennik* (The Contemporary) over ideological conflict with Chernyshevskii and a younger, politically radical generation of contributors to the journal. This debate was playing out publicly just as Turgenev was finishing work on his novel *Otsy i deti* (Fathers and children [1862]), and Turgenev's portrait of the young nihilist student Evgenii Bazarov was regarded as an unflattering reflection of Chernyshevskii's generation of positivist materialists. The novel has been widely regarded as a "concrete social picture" of the generational conflict between the men of the forties—liberals like Turgenev who favored gradual reform of Russia's political institutions and a holistic vision of *anthropos*—and the succeeding generation of radicals, "men of the sixties" such as Chernyshevskii.[31]

Turgenev's *Otsy i deti* is a compendium of the intellectual debates of the age, and Liebig appears in Turgenev's story as an empty idol of the younger generation.[32] In the novel the positivist materialists Bazarov and Nikolai Petrovich excitedly discuss a number of contemporary works of science and social theory, and a dilettante "lady chemist" named Kukshina refers to Liebig to legitimate her amateur enthusiasm for chemistry. At their first meeting Bazarov and Kukshina turn their conversation from lunch to Liebig:

> [Bazarov] "A piece of meat's better than a piece of bread even from the chemical point of view."
>
> [Kukshina] "Are you are studying chemistry? That is my passion. I myself have even invented a new sort of glue."
>
> [Bazarov] "Invented glue? You?"
>
> [Kukshina] "Yes. And do you know for what purpose? To make dolls' heads so that they won't break. I'm practical, too, you see. But it's not quite ready yet. I've yet to read Liebig."[33]

Kukshina herself (whose very name suggests *kukla*, or doll) applies Liebig's chemical theories to the formulation of better glue for dolls' heads. In his savage parody of the lady chemist, Turgenev declares Russia's men and women of the 1860s to be puppets uncomprehendingly replicating Western materialist discourse. Kukshina and her dolls also constitute a parody of the mechanist and materialist view of the human body—a doll held together not by vital spirit but by chemical glue.

Chernyshevskii responded to Turgenev's provocation in his work of the following year, *Chto delat'?* (What is to be done?), a major influence on the architects of the Russian Revolution, including Vladimir Lenin who famously called it his favorite novel. In *Chto delat'?* Liebig's ideas concerning soil appear as the basis for an extended metaphor in the famous dream of the novel's heroine, Vera Pavlovna. The evening before her dream, Vera's dinner guests steer the conversation to "the current debate about the chemical basis of agriculture according to Liebig's theory, the laws concerning historical progress—an unavoidable subject of conversation in such circles at that time."[34] The association Chernyshevskii makes between theories of "agricultural chemistry" and "laws of historical progress" shows the post-Darwinian tendency to extrapolate universal models of history, including models of human historical development, from specific bodies of natural-scientific knowledge. That night Vera Pavlovna's dream logic fuses agricultural chemistry and historical development into an extended metaphor of Russian soil and revolution. In the dream the young positivist materialists Lopukhov and Aleksei Petrovich contrast plants grown in "real" soil with those grown in "putrescent" soil. "Real" soil is composed of healthy elements (*elementy*) which form "complex chemical arrangements" through energy from the sun. The main "element" of this healthy soil is human labor. Putrescent soil, on the other hand, is characterized by "stagnation."

This discussion of healthy and putrescent soil mobilizes a dialectic between *pochva* and *griaz'* (dirt). Chernyshevskii marks *griaz'* as "real" soil, a lexical choice that underscores his desire to reject the lofty and encumbered symbolism of *pochva* and restore soil to a material condition. The putrescent and "fantastic" soil of Vera Pavlovna's dream is the symbolic *pochva* of Russian discourse: an emptied sign, filled with abstract cultural symbolism that is stagnant and inert.[35] Liebig's writings on soil provide a metaphoric structure, then, for Chernyshevskii's proposal for revolutionary social transformation. Chernyshevskii suggests that minerals circulate freely in a healthy economy of soil. In this utopian vision, labor cures the defects of the "stagnant" Russian national character and improves the quality of the mythologized Russian "native soil."

MODELS OF SOIL AND SOCIAL ECONOMY

Chernyshevskii's vision of the imbrication of soil economies and social economies echoes Karl Marx's extension of Liebig's agricultural models into sociopolitical theory. While formulating *Das Kapital* (*Capital*), Karl Marx was keenly interested in new developments in agricultural chemistry, a field that was already making important social and economic changes throughout Europe. Liebig in Germany and John Bennet Lawes in Scotland were developing fertilizers that could raise crop yields significantly and mitigate the effects of the declining fertility of Europe's soil under the *Raubsystem*, or "robbery system" of agriculture.[36] In his letters to Engels, Marx wrote that "the new agricultural chemistry in Germany, especially Liebig and [Christian] Schönbein . . . are more important than all the economists put together."[37]

As Joan Martinez Alier, John Bellamy Foster, and others have shown, Marx extended Liebig's diagnosis of the source of soil depletion into a broader socioeconomic critique of the unhealthy relationship between city and country.[38] In Liebig's model of mineral exchange, the agricultural cycle depends on the return to the soil of all extracted minerals in the form of food waste, animal waste, and human waste (and even human remains). Liebig states: "In the large towns of England, the produce both of English and foreign agriculture is largely consumed; elements of soil indispensible [*sic*] to plants do not return to the fields—contrivances resulting from the manners and customs of English people . . . render it difficult, perhaps impossible, to collect the enormous quantity of the phosphates which are daily, as solid and liquid excrements, carried into the rivers."[39] Liebig considered it a major problem of agriculture and society alike that minerals were not returned to the soil of the countryside but were "wasted" in cities, poisoning water and spreading disease.[40] These were conditions of both social and soil imbalance between the country and the city. Marx explains how Liebig's agricultural metabolism is sealed into the larger interdependent processes of "social metabolism": "Large landed property reduces the agricultural population to a constantly falling minimum and confronts it with a constantly growing industrial population crowded together in large cities. It thereby creates conditions which cause an irreparable break in the coherence of social metabolism [*Stoffwechsel*], a metabolism prescribed by the natural laws of life itself. As a result, the vitality of the soil is squandered, and this prodigality is carried by commerce far beyond the borders of a particular state (Liebig)."[41] Marx assures his reader that "social metabolism" is prescribed by nature, as he transfers a "natural law" from soil science to social science. Marx interprets what Foster calls the "metabolic rift" as a broader form of economic exploitation of the country by the city.[42] Friedrich Engels, in *The Housing Question* (1872), also appeals to Liebig as an authority in his articulation of the "rift" between town and country:

> The abolition of the antithesis between town and country is no more and no less utopian than the abolition of the antithesis between capitalists and wageworkers. From day to day it is becoming more and more a practical demand of both industrial and agricultural production. No one has demanded this more energetically than Liebig in his writings on the chemistry of agriculture, in which his first demand has always been that man shall give back to the land what he takes from it, and in which he proves that only the existence of the towns, and in particular the big towns, prevents this.[43]

In his discussion of the "antithesis" between town and country, Engels frames Liebig's theory of soil metabolism in dialectical terms and suggests that Liebig's work points to an effective means of synthesizing town and country into a single economy, founded on material and mineral exchange.

The impact of Marx and Engels's interpretation of Liebig on later Soviet agricultural and social policy is unquestionable, but the concept of imbalance between city and country also evolved independently within the Russian intellectual tradition by way of Liebig's Russian interpreters. In his "Ocherki iz istorii truda [Sketches from the history of labor]" (1863), the radical critic Dmitrii Pisarev echoed Liebig's critique of robbery of the soil:

> Only in one case does human intervention weaken the productive forces of nature; this occurs when a person takes out the raw products of the earth to distant markets, and thus permanently deprives the land of its known constituents and does not return any fertilizer in exchange. Such a course of action is possible only in places where there are few people, and where consequently there is no industrial activity. If there were a lot of people, enterprise would be necessary,

> factories would grow, the raw products would be recycled [*pererabotyvat'sia*] and assimilated on the site. The remains of processed products would provide rich fertilizer, and the soil rather than being depleted, would continually become more fertile.[44]

Pisarev implicitly argues against the law of diminishing returns; in his view, the only condition under which agricultural labor reduces the fertility of soil instead of enhancing it is one characterized by unbalanced exchange between raw products from the country and processed products from the industrial center. Pisarev focuses on the need for the products of agriculture and industry to be consumed and then recycled (from waste to fertilizer) within a single location and community.

Liebig's model of the robbery of nutrients from the countryside also drew the interest of Aleksandr Nikolaevich Engel'gardt, a distinguished chemist, popular journalist, and active figure in the Russian Populist movement of the 1870s. Engel'gardt had studied with Liebig's disciple Nikolai Zinin before attaining the position of chemist and rector at the St. Petersburg Agricultural Institute.[45] Engel'gardt wrote that as a student he was "very attracted to Liebig's scientific genius" and interested in Liebig's observations about "soil exhaustion."[46] Engel'gardt himself developed phosphate fertilizers and worked together with Pavel Antonovich Il'enkov, Liebig's student and translator, to develop a chemical process to dissolve bone for fertilizer using alkalizing compounds, a method that Il'enkov discussed in detail with Liebig himself in correspondence.[47] Unsurprisingly, given the links between materialist philosophy and political radicalism, Engel'gardt was arrested during his tenure at the Agricultural Institute and went into internal exile on his estate near Smolensk in 1881, where he continued his agricultural experiments and wrote prolifically on agronomy, farm management, and soil enrichment.[48]

Beginning in 1872 Engel'gardt had published a regular column on agriculture, "Pis'ma iz derevni [Letters from the Country]" in the popular journal *Otechestvennye zapiski* (Notes of the Fatherland). In one of his letters, Engel'gardt applies Liebig's theory of the "robbery of the soil" to the description of peasant and gentry estates in his own region:

> In our parts, both the landowners and the peasants fertilize their land with manure. The need for fertilizer has entered everyone's consciousness, so that the landowner devotes all his attention to building up stores of manure. . . . But at the same time that the landowner, who is selling grain and cattle, renting out his meadows in part, and leasing land for flax and grain, depletes the soil because of the removal of soil particles [*pochvennikh chastits*] (most importantly—phosphate salts) through the grain, cattle, and hay. On the contrary, the peasant,

> by bringing in grain, straw, etc., improves and humifies his own land, bringing in soil particles from outside.[49]

Engel'gardt describes the peasants' practice of skimming hay, straw, flax, grain, and firewood from rented gentry land to enrich their own small plots. Peasants also consume the fruits of their own land, and their "excrement remains on their own plot." "Thus," he writes, "the peasant brings soil particles as hay from other places, and the soil particles remain on his allotment, increasing the amount of nutrients in his land."[50] Large gentry farms continually lose minerals through export of agricultural produce: "The situation is completely different on the farms of landed gentry. There the soil is always exhausted, and farming further exhausts the land. Under serfdom, the landed gentry produced enormous quantities of grain, which they sold from their holdings and which carried off with them masses of valuable soil particles taken out of the earth, carried them across the sea to the Germans and the English, carried them off to the cities, from which these particles flowed down into the rivers."[51] Engel'gardt's description of local Russian agricultural conditions closely models Liebig's description of the unbalanced social metabolism of soil—but with an attention to distinctly Russian social conditions. The influence of Engel'gardt's "Letters" ran deep among the Russian intelligentsia at the turn of the century. Lenin was a careful reader of Engel'gardt and devoted an entire section of *The History of Capitalism in Russia* to Engel'gardt's life and work.[52] Although Lenin rejected Engel'gardt's Populist valorization of agriculture over industry, Engel'gardt was another link between Liebig's conception of the incomplete rural-urban cycle and Lenin's ideas on the disconnection between city and country.

The unification of city and country (*smychka goroda i derevni*) became a crucial slogan of Soviet politics in the 1920s. The smychka can ultimately be traced to Liebig's system of soil metabolism, but perhaps not only by way of Marx and Engels.[53] There is a distinct genealogy of the smychka that drew on the highly developed Russian discourse surrounding Liebig and Russian soil directly, as seen in Chernyshevskii, Pisarev, Engel'gardt, and many other commentators. Scientific understandings of soil and mineral economy were transferred into sociopolitical paradigms that would later be reified in Soviet environmental and social policies. The state-directed effort to achieve the smychka resulted in policy decisions with real environmental and human impact. For example, Lenin echoed Liebig's prescription that mineral exchange between the city and the country should be balanced: in his 1919 document "O perevozke udobrenii po zheleznym dorogam [On the transportation of fertilizers by rail]" he declared that shipments of chemical fertilizers to the country should be precisely equivalent to shipments of grain to the city.[54] Where Liebig had recommended managing the exchanges between city and country as one would balance a chemical equation, Lenin brought those exacting equations into the realm of social policy.

Lenin's innovation in interpreting Liebigian social and soil metabolism, then, was the centralized control of these metabolic functions within a state apparatus, managed by the first generation of professional Russian scientists and technicians who were themselves products of the nineteenth-century tradition of cultural and scientific materialism and, in many cases, claimed direct professional descent from Liebig through a chain of apprenticeships. The smychka—grounded in a unified field of social and soil metabolism—would offer a model in the later Soviet period for the exchanges not only between city and country but also between center and national republics within the Stalinist "all-union division of labor," which coordinated the material flows of the Soviet economic regions.

The reception of Liebig's ideas in nineteenth-century Russia and their afterlife in Soviet culture offers a case study in how metaphors transfer across scientific and cultural domains. This transfer was facilitated by the first professionalized Russian chemists, agronomists, and soil scientists, who were also frequently active in political circles and intelligentsia cultural discourse. These figures popularized Liebig's scientific works, translated his ideas into idioms that were familiar to the reading public, and suggested social implications of his scientific theories.

Within this pervasive discourse of soil, many Russian commentators, from Antonovich to Engel'gardt, had already anticipated Ellen Stroud's question: "Where is the dirt?"[55] Liebig's own exposition of mineral economies and the relationship between agriculture and culture relied on structural metaphors of the sort that some considered literary, produced, as Berzelius scornfully remarked, "at the writing desk" rather than in the laboratory.[56] But knowledge of the soil was generated not only at the writing desk or in the laboratory but also in the field. This was the project of the founder of Russian soil science, Vasilii Dokuchaev, who synthesized German organic chemistry and practical Russian peasant folk knowledge. Dokuchaev argued that soil was an "independent natural-historical body" formed in specific natural environments and characterized by much greater complexity than mineral composition alone. Dokuchaev's theories of natural zones and soil formation offered a more nuanced understanding of the relations between living and nonliving matter and pointed to the field of biogeochemistry developed by his student Vladimir Vernadskii.[57]

All of the metaphors and social models adapted from Liebig's mineral economy might have remained in the realm of discourse and theory but for the October Revolution. Although the materialists may have sought to disassociate material from lofty metaphor, they were likewise inspired to extend Liebig's work into social models, as in the case of Chernyshevskii, and into "natural laws" of social and historical development, as in the case of Marx. The Soviet utopian project to remake human nature and the material world according to Marxist "laws" of historical development resulted in a reification of ideology at many levels, including agricultural policy. Policy was not always consistent with ideology, nor was it only a reification of it. As many of the articles in this volume show, policy also responded to pragmatic needs rather than ideology, and scientists and citizens adapted to natural environments. However, Soviet environmental and land policies were products of a complex intellectual history that drew on both European and Russian nineteenth-century materialist traditions. As a totalizing method, Marxist dialectical materialism was particularly susceptible to the free transfer of metaphors and models across various domains, and its materialist analysis of history and society—in part influenced by Liebig's soil chemistry—would shape the material transformation of the Eurasian environment for many years to come.

9 How a Rock Remade the Soviet North

NEPHELINE IN THE KHIBINY MOUNTAINS

ANDY BRUNO

WHAT COULD a rock have to do with it? It turns out that the right rock in the right place can have quite an influence on human history. A rock can excite and destroy, facilitate and undermine, or create value and costs. In the Khibiny Mountains in northwest Russia rocks played a significant and multifaceted role in the environmental history of industrial development and pollution. In particular, a form of inert inorganic matter aided the development of a model of industrial operations that aimed at minimizing environmental wreckage and then later precipitated considerable damage to a fragile ecosystem.

In this essay I address a fraught question of Soviet environmental history—the origin of extreme pollution—from the perspective of a rock. The mineral nepheline was a main by-product of the phosphate fertilizer industry in the Khibiny region of the Kola Peninsula (see map 11.1). Its significance varied in the region's environmental history. Nepheline first served as inspiration for quasi-conservationist schemes to eliminate industrial wastes and then became a chief source of environmental pollution in the region. By tracing this material and efforts to manage it, I show how human engagement with physical elements of the natural world affected what the Soviet system tried to and was able to accomplish. I tell the story of an Arctic mineral here, but the case for the material features of Eurasia's influencing Soviet history could also be made about the Siberian permafrost from Pey-Yi Chu's chapter (this volume) or, in an earlier period, the soils discussed by Mieka Erley (this volume). My aim is to offer a fresh account of how the Soviet Union came to despoil its environment in which nature is treated as an actor. This viewpoint augments and enhances the human-centered explanations dominating the debate.

EXPLAINING ENVIRONMENTAL DEGRADATION IN THE SOVIET UNION

What accounts for the extreme levels of environmental degradation in the Soviet Union? Answering this question was once the province of social scientists, but over the past decade and a half historians have been paying more attention to the USSR's inauspicious environmental record.[1] Two of the leading environmental historians of the Soviet Union, Douglas Weiner and Paul Josephson, have weighed in on this question. Both scholars seek to pinpoint the specific causal dynamics of the Russian case, while also acknowledging worsening environmental conditions as a global phenomenon.

Weiner argues that the practices of a "predatory tribute-taking state" that has ruled the Eurasian landmass since at least the era of Mongol rule accounts for the "treacherous patchwork quilt of poisoned lands, poisoned air and poisoned water we encounter today."[2] A succession of militaristic regimes in Russia "saw the population and the land over which they ruled as a trove of resources to be mined for the rulers' purposes."[3] This argument for deep continuity puts Ivan the Terrible's declaration of Muscovy as his patrimonial demesne, the seventeenth-century military encroachment on the steppe, Peter the Great's forest exploitation, the late imperial gentry's approach to estate management, Stalin's agricultural and industrial policies, and the appearance of state gangsterism in the post-Soviet era into an abiding pattern of shortsighted resource predation with nefarious environmental and social consequences. The pretense of "planning" in the Soviet Union, the "fig leaf" of Marxist ideology, and the heedless embrace of economic development have only been subsidiary issues in creating an "acute ecological situation" over 15 percent of the Russian Federation by the time of the Soviet collapse. The "insidiously plunderous" practices of the state ultimately bear the blame.[4]

Josephson's approach to this question of environmental deterioration in state-socialist countries centers less on an exploitative process than on a specific form of degradation. The Soviet Union shared with capitalist countries the Enlightenment impulse to tame the natural world with science in order to turn it into a commodity machine and an ideological attachment to large-scale technological systems. Its distinction came more from the predilection of authoritarian socialism to produce "industrial deserts." This tendency resulted from the technological giganticism of state socialism, a disregard for consumer preferences, the strong momentum toward new projects in the absence of public opposition or market constraints and with the imperative of full employment, and the wide-ranging use of slave labor for industrialization.[5] Not simply aggregate anthropogenic damage to ecosystems, industrial deserts are concentrated zones of environmental destruction "where intense, virtually unregulated industrial activities poisoned waterways, forests, fields, and the cities and towns in which factories are located." They are "denuded landscapes in which mountains of slag, mining waste, and heavy metals seem the only distinguishing marks" with proximate urban populations suffering from the pollution.[6] The production of these hazardous geographical spaces more than environmental damage writ large was the legacy of Soviet socialism.

Some historians have more recently reconsidered the question of environmental pollution from what might be called a revisionist angle. Stephen Brain provocatively argues that the negative assessments of environmental stewardship in the Soviet Union may be attributing too much weight to the Stalin era. His research shows that Soviet forest policy from the 1930s until the early 1950s established "levels of protection unparalleled anywhere in the world." In his assessment it was Stalin's passing that allowed for hyper-industrialism to infiltrate the forest after decades of utilitarian conservationism.[7] Labor historian Donald Filtzer's detailed discussion of river pollution in central Russian cities in the decade after World War II suggests that Stalinist environmental protection did not extend to all other realms.[8] But even in this case the impact on ecosystems, as opposed to public health, became much greater in later decades as the scale of industrial activities ratcheted up and prolonged periods of pollution took their toll. Finally, Kate Brown's scholarship on the nuclear facilities in the Urals and my own on the nickel industry in the Far North stress the importance of a deeper comparative perspective when accounting for the environmental legacy of the Soviet Union.[9]

As these reassessments make clear, environmental deterioration is a complex and heterogeneous process that often lacks unified coherence. Disastrous environmental destruction can coexist with more laudable instances of environmental performance. Sometimes efforts to solve one environmental problem create new ones, as in the case of plans to save the Aral Sea through river diversion. A comparative assessment of the United States and the Soviet Union by Natalia Mirovitskaya and Marvin Soros counters the claims that the USSR had an overall environmental legacy that was degrees of magnitude worse than the West. The authors demonstrate, for instance, that levels of air and water pollution in the late Soviet era were not obviously worse than in the United States, though the degradation of forests and farmland more closely resembled the experience of Third World countries.[10] The official Soviet statistics used in their study are subject to skepticism. However, Jonathan Oldfield's 2005 overview of the environmental situation in Russia, based largely on post-Soviet data, comes to a similar conclusion.[11]

Furthermore, efforts to "de-exceptionalize" the Soviet environmental experience (to cite Pey-Yi Chu's apt phrase) compel a greater appreciation for the influence of global processes and contexts in bringing about the conditions that earned the USSR a reputation for ecocide. First off, the twentieth-century world witnessed unprecedented levels of population, energy use, and economic growth that fundamentally transformed the relationship that humans on the aggregate had with the natural world.[12] The Soviet population, for instance, rose from just over 172 million in 1945 to more than 290 million in 1991, albeit after the major demographic shock of the Second World War.[13] Meanwhile, electricity production in the country (to take one sphere of energy use) grew from 48.6 billion kilowatt-hours in 1940 to 1,722 billion kilowatt-hours in 1989, while the economy grew by several orders of magnitude during the seven decades of Soviet rule.[14] The attendant increase in the use of natural resources helped generate significantly more pollution than Russia had ever experienced before.

Additionally, the specific context of the 1970s and 1980s should be acknowledged. It was in this period, and not the earlier Stalinist or even Khrushchev eras, when the situation with pollution in the Soviet Union became extraordinarily destructive for ecosystems on a large scale. Whereas at earlier moments territories such as the Urals, Magnitogorsk, and the Donbas resembled concentrated industrial centers in the German Ruhr and the Hanshin region of Japan in terms of air pollution, by the late Soviet period a combination of accelerated production and ineffective mitigation of emissions led to an especially dramatic proliferation of toxic landscapes.[15]

THE ROLE OF ROCKS IN HISTORY

On top of the insights offered by a comparative perspective for understanding environmental degradation in the USSR, historians would do well to pay heed to the influence of the actual environment in the generation of pollution. Nature is not just a stage for human action, a passive background that can be destroyed or protected. As numerous theorists of the intersection between the categories of the social and the natural insist, the environment can function as an actor in human stories, shaping history in unexpected and unacknowledged ways.[16] Following historian Ellen Stroud, I am convinced that showing how material nature—the "dirt" of history, in her words—affected human experience affords environmental history its strongest potential to transform scholars' thinking about the past as a whole.[17] "Nature" and "environment" are also constructed and contested concepts that can be and have been put to different political purposes. But a focus on nature's materiality in all its grit and glory best demonstrates its potency in human history.

Two perils appear immediately in any attempt to take the autonomous materiality of the natural world seriously. They relate to a long-standing concern in social theory about the balance between agency and determinism. On one side, arguments that climatic conditions, local biota, landscape features, and microbes define the history of certain peoples have been advanced and received with justified skepticism for centuries. The notion that Russia's cold, for instance, sufficiently accounts for abiding forms of authoritarian rule there can quickly be discounted by comparisons with other circumpolar countries.[18] It is important to avoid relying on such crude forms of geographical and environmental determinism in emphasizing the influence of material nature. The other danger comes from excessive anthropomorphism that attributes agency and intentionality to elements of nature. Influenced by the actor-network-theory and critical political ecology, some scholars have gone as far as suggesting that they figured out what some nonhuman organism or artifact "wanted" in a given situation.[19] Regardless of the complex philosophical and theoretical problems that emerge in assessing what is agency and where it might reside, the capacity of people to communicate their own perceived desires and intentions makes human agency accessible in a way that nature's agency, if it exists, cannot be.

At the heart of this conundrum sits the question of how to use knowledge from the physical sciences in historical narratives. A divide seems to exist between environmental historians who rely on insights from science and scholars of the production of scientific knowledge and expertise. The former often use interdisciplinary and necessarily simplified understandings of the behavior of some part of the natural world and incorporate them into historical assessments. The latter more frequently interrogate the limits of science and the borders between human cognition and the natural objects being studied.

Contrary to some, I believe that a constructive synthesis between these two approaches can be found by simply setting aside the epistemological debate about science and concentrating on examples of how elements of material nature actually acted in history. Scholars can tentatively trust scientific knowledge about the natural world and still analyze environmental transformation as an interactive process of reciprocal feedbacks between human and nonhuman actors. The key is to accept that the influence of material interactions also extends to the development of scientific ideas.[20]

A quick detour into research on the history of mosquitoes can flesh out this common ground. In dealing with the methodological problem of environmental determinism in his book on the partisan role of mosquitoes carrying yellow fever and malaria in the greater Caribbean, John McNeill states his intent "to provide a blended perspective that emphasizes the mutual and reciprocal impacts of geopolitics and ecology. . . . [A]lmost all human history is really a co-evolutionary process involving society and nature. But the degree to which this is true varies greatly from context to context."[21] Although McNeill keeps his analysis rooted in his understanding of ecology and epidemiology, it is also possible to treat moments in the production of scientific knowledge as also co-evolutionary. With different terminology this is essentially what political theorist Timothy Mitchell does in his essay "Can the Mosquito Speak?" He stresses the entanglement of the categories of nature and society through an exploration of a deadly malaria outbreak in Egypt in 1942–1944. His main point is not to assert an affirmative response to the essay's provocative title but, rather, to demonstrate how human thought and activity remain fully interwoven with the exterior world and do not function in isolation. "What is called nature or the material world moves, like the plasmodium, in and out of human forms, or occurs as arrangements, like the river Nile, that are social as well as natural, technical as well as material. The world out of which techno-politics emerged was an unresolved and prior combination of reason, force, imagination, and resources. Ideas and technology did not precede this mixture as pure forms of thought brought to bear upon the messy world of reality. They emerged from the mixture and were manufactured in the processes themselves."[22] In Mitchell's conception, parts of the nonhuman material world influence their own treatment by interacting with humans who are engaged in scientific and technological endeavors. Nature, therefore, plays a productive role in the development of modern schemes to transform it. By blending McNeill's and Mitchell's approaches, scholars can treat the dirt of history as a nondeterminative actor without ascribing agency to it.

NEPHELINE AS INSPIRATION

One way to get at the productive role of the material world in the generation of a polluted environment in the Soviet Union is to examine a local case of the complicated influence of a particular rock. The rock is a material called nepheline, which was a by-product in the mining and enrichment of apatite into phosphorous fertilizers in the Khibiny Mountains. Over time the role of nepheline in the region switched from potentially being environmentally helpful to unquestionably causing environmental damage. Nepheline inspired plans to eliminate industrial wastes but then became a pollutant despoiling air and waterways. That is—in contrast to official ideological confidence about the ability of the Soviet state to control nature—inorganic matter influenced both the specific forms of Stalinist utopianism and the undermining of this fantasy.

The Khibiny Mountains are located on the Kola Peninsula (the Murmansk region) in the northwest corner of Russia. A sparsely inhabited territory before

Soviet rule, the Kola Peninsula became one of the most built-up parts of the Arctic in the twentieth century. Industrial interest in the area first emerged in the 1920s when teams of exploration geologists discovered large veins of apatite embedded in the Khibiny Mountains. Apatite is chemically composed of calcium phosphate with an extra ion of fluorine, chlorine, or hydroxyl. As a source material for the manufacturing of phosphate fertilizers, apatite earned the moniker the "stone of fertility" in Soviet culture.[23]

During the rapid industrialization of Stalin's first Five-Year Plan (1928–1932), state economic planners decided to transform proposals for the limited extraction of apatite ore into a grandiose project to build a new socialist city north of the polar circle. The chaotic establishment of the new industrial town of Khibinogorsk (later Kirovsk) and the mining and enrichment enterprise of Apatit occurred with a heavy dose of forced labor and an array of setbacks and mishaps that plagued the Soviet Union in this period. It also featured a local scientific infrastructure that supported industrial activities. By the late 1930s the Apatit enterprise was extracting over two million tons of ore and manufacturing over one million tons of apatite concentrate a year, while the Kirovsk region possessed more than forty thousand inhabitants.[24] In the postwar decades the industrial civilization of the Khibiny Mountains renewed its rapid growth in population and production, including the development of a larger second city called Apatity.

As geologists and boosters of northern development began to promote the possibility of apatite operations in the 1920s, they simultaneously addressed the question of what to do with the massive amounts of nepheline in the area. The bulk of the Khibiny Mountains was a large igneous intrusion of this rock form, making the reserves of nepheline in the region seem "inexhaustible" to commentators.[25] This nepheline rock gave rise to many of the rare minerals of the Khibiny Mountains, including crystalline pegmatite rocks, which helped initially attract scientists to the region. Nepheline in the Khibiny occurs as nepheline seynite, an alkaline rock containing sodium, potassium, aluminum, and silicon. Unlike the deposits of apatite that existed in isolated slivers along a narrow ring that could quickly and fairly cheaply be transformed into fertilizers, nepheline did not have an industrial significance that would have warranted building a settlement from scratch. However, its potential status as a subsidiary material prompted scientists and industrial planners to contemplate its versatile uses.[26]

From the very beginning the leaders of Apatit attempted to organize industrial production so that it would be maximally efficient with raw material and would minimize wastes. Even before the basic plans for the first year had been decided and before making his first visit to the construction site, the new manager of the Apatit trust, Vasilii Kondrikov, pressed the issue of the potential applications of nepheline. At a meeting with Apatit's oversight organization in December 1929, Kondrikov diverted from the slated agenda about the possible use of prison labor in the Khibiny. Having already received coaching from the geochemist Aleksandr Fersman, Kondrikov insisted on the need to invest in scientific research on the reuse of nepheline tailings (the solid residue of processed mineral ore) during the 1929–1930 accounting year.[27] "Today the apatite problem is clear," said Kondrikov, "but nepheline is an entirely new product."[28]

Four different forms of nepheline interested Soviet scientists working in the Khibiny. The mass of urtite (nepheline seynite) outside the apatite deposits could in theory be used to make a whole array of nepheline products, but project leaders considered this economically inexpedient. The nepheline sands on the eastern shore of Lake Imandra (see map 11.1) could serve the glass, ceramic, and porcelain industries. However, the supply of these sands came to only five million tons, a relatively small total. The main forms of nepheline that attracted attention were the sections of mined apatite-nepheline ore that were not suitable for processing as phosphates and the tailings from apatite enrichment that contained 70–75 percent nepheline. Breaking down the chemical structure of Khibiny nepheline to 21–22 percent alkaline, 34 percent alumina (aluminum oxide), and 44 percent silicon dioxide, Fersman elucidated twenty-three different uses for these molecular agents. Least importantly, silicon dioxide could serve as a versatile industrial product in sludge-based cement, leather, rubber, processing wood, and roadwork. Nepheline's alkaline substance could help manufacture "glass, ceramics, enamel, paper, liquid glass, ultramarine, kaolin, and nephelinized soil." The latter entailed using nepheline as a potassium fertilizer. Most significant was the potential use of nepheline as an alumina source. This compound could produce "metallurgical aluminum, aluminum chloride, abrasives, coagulants (sulfuric acid salts of aluminum), alum, water resistance in textiles, and a mordant for textiles."[29]

The processing of nepheline into alumina for aluminum production was entirely new. Bauxite was and has remained the primary source material of the aluminum industry globally. Rich deposits of bauxite rock contain up to 40–60 percent of alumina, and the main process for isolating this substance has been used since the late nineteenth century.[30] Nepheline, in contrast, averaged 20–30 percent alumina content, less than Fersman's first assessment, and required a considerable amount of limestone to process. The Soviet Union, however, had a comparative dearth of bauxite that was especially stark before low-grade deposits of it just south of Lake Ladoga in the Leningrad region began to be exploited in 1931.[31]

While conducting research on the myriad uses for nepheline, Fersman and his colleagues devised an entire framework for waste management called the "complex utilization of natural resources," which they claimed could be applied to a variety of contexts in the mining industry. The process involved arranging multiple steps in order to separate out the different useful substances from the apatite-nepheline ore being extracted from the Khibiny Mountains. The compounds that made up apatite-nepheline ore possessed strong physical bonds with each other, which in turn gave rise to technical challenges for processing it and inspired creative thinking about how to economically treat nature. Complex utilization ultimately sought to use maximally all of the material removed from the earth and develop industry that would achieve regional self-sufficiency.[32]

At a conference of Gosplan (the state planning committee) in April 1932, Fersman claimed that the principle of complex utilization was "one of the vital nerves of the new forms of socialist industry." He defined a combined economy as resting "not in the external combining or the summation of diverse manufacture, united only by a determined territory or in the best case by a general power and energy economy, but in the deep interweaving of production processes, depending on the complete use of all mining mass extracted from the earth during the maximum constriction of the radius of the use of raw materials."[33] Such complex utilization of mineral material would rely on technological innovations in heavy industry processing and the recycling of tailings from enrichment in order to manufacture valuable subsidiary products. As his main example Fersman cited the apatite industry that he had helped establish in the Khibiny Mountains. Production of apatite concentrate would supply "an entire nepheline stream" of tailings that should be reused (see figure 9.1).[34]

Optimism about the potential of this model came through strongly when Fersman turned to its environmental implications. Complex utilization "is the idea of the protection of our natural resources from their predatory squandering, the idea of using raw material to the end, the idea of the possible preservation of our natural supplies for the future," exclaimed Fersman at the Gosplan meeting, "where not one gram of extracted mining mass is lost, where there is not one gram of waste, where nothing is emitted into the air and washed away by water."[35] The goal of complex utilization in the Khibiny, therefore, was not only to use mined ore efficiently but also to protect the local environment from pollution.

It is again important to stress here that research work with nepheline crucially influenced the development of this idea. By presenting a variety of potential uses—but ones that necessitated intricate processing—the material, nepheline, itself contributed to a scheme to reduce and reuse industrial wastes. To extract all value from this rock required such an extensive effort to reuse the raw materials that, in theory, it completely subsumed the ore and left no waste.

The belief that complex utilization would allow for environmentally beneficial outcomes of economic activities appeared in early plans for how to arrange

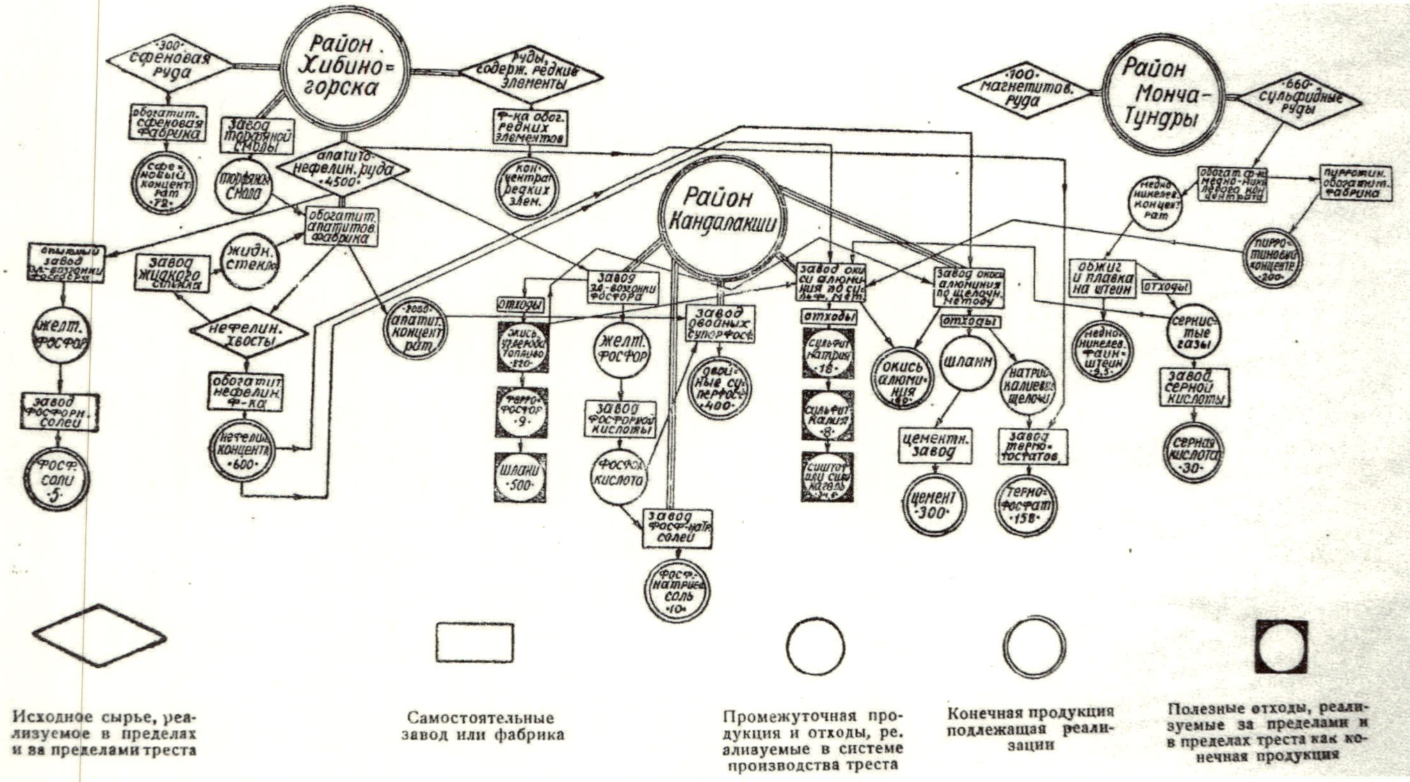

FIGURE 9.1. A proposed scheme for complex utilization in the Khibiny and surrounding regions from 1932. It shows three planned industrial centers in Khibinogorsk, Kandalaksha, and the Monche tundra located near each other and how different waste by-products would become inputs elsewhere. In the scheme, nepheline emissions from the Khibiny would go to alumina facilities in Kandalaksha. From V. I. Kondrikov, "Vybor ploshchadki Khimicheskogo kombinata na baze apatito-nefelinovogo syr'ia," in A. E. Fersman, ed., *Khibinskie Apatity i nefeliny: Nefelinovoi sbornik* (Goskhimtekhizdat Leningradskoe otdelenie, 1932), 4:140.

the various industrial and urban objects. Finding the ideal location for the city, the enrichment factory, and clean water sources for drinking and industrial uses concerned the designers of Khibinogorsk and Apatit. After considering a number of proposals in the first months of 1930, planners opted to place the city in a small valley nestled in the Khibiny Mountains and close to the mines. In this scheme Great Vud"iavr Lake provided water to the settlement and Apatit's enrichment factory sat at the point where Great Vud"iavr emptied into the White (Belaia) River.[36] Enterprise leaders viewed the use of the White River as an industrial sewer for dumping tailings as a temporary measure. The head of Apatit's enrichment factory, Nikolai Vorontsov, insisted that in the future the wastes from apatite processing would be reused to produce a variety of materials and this reprocessing would reduce pollution of the White River.[37]

Then from 1931 onward the Apatit trust became involved in a protracted campaign to turn nepheline wastes into a source material for aluminum production and to create a processing plant in nearby Kandalaksha. Near the end of the year trust leaders estimated that they could gradually increase their annual output of nepheline concentrate from tailings from 140,000 tons in 1933 to 750,000 tons in 1937, providing enough material for almost a half million tons of alumina over this period.[38] In February 1932 various agencies involved in the project protested a decision to place the alumina plant under the Aluminum Union (Soiuzaliuminiia) of the People's Commissariat of Heavy Industry (Narodnyi komissariat tiazheloi promyshlennosti [NKTP]), taking it away from Apatit.[39] In a complaint that would reach the Soviet leader Joseph Stalin, one chemist objected to this move because it would separate aluminum production from the "complex utilization of the minerals of the Khibiny."[40] A few months later, in May 1932, the issue was solved for the time being when the state transferred Apatit to the NKTP and allocated approximately seventeen million rubles for the construction of the new combine.[41] At the time called the Northern Mining-Chemical Trust (Severnyi gorno-khimicheskii trest), the facility was planned to have three main sections—cement, thermophosphate, and alumina—that would process materials extracted from the Khibiny.[42]

NEPHELINE AS POLLUTANT

At this point in the story the emphasis switches from the positive role played by nepheline in spawning a specific approach to conservation in heavy industry to the function of this material in disturbing the environment of the Khibiny region. As an industrial waste that possessed properties only partially controllable by the enterprise, nepheline undermined some of the more hopeful plans to avoid pollution. In this way it "bit back"—it exhibited unintended consequences related to its raw materiality instead of its social character (as was the case in its contribution to complex utilization).[43] This negative role of nepheline first degraded local waterways and the health of the new workforce. In later decades it became a major source of environmental destruction in the Khibiny region.

Vorontsov's assurances that reprocessing would soon address the problem of pollution from nepheline tailings proved premature. By the late 1930s wastes from apatite production had turned the White River literally white and killed off aquatic species in a number of local water bodies.[44] The impact of this environmental transformation fell disproportionately on a group of former peasants who had been de-kulakized and forcefully sent to help construct the apatite works in the Khibiny Mountains. Known by the government classification "special settlers," some of these forced migrants initially relied on fish from the White River to supplement their meager food rations from the state.[45] This deprivation combined with an array of other vulnerabilities to the natural hazards of the far north to contribute to the suffering of the "special settlers" (who, of course, had been deliberately targeted for oppression by the Soviet state).[46]

After the initial decision to build a chemical combine in Kandalaksha, the project faltered. State agencies refused further funding requests from the trust, allocating only one-fifth of the proposed amount for 1932–1935. During this period none of the main industrial objects were built and the surveying for local limestone sources had come to naught.[47] The discovery of new rich bauxite deposits in the Urals in the mid-1930s strengthened the position of the segments of the nonferrous metallurgy industry that opposed using nepheline.[48] As the project began moving again in the mid-1930s, researchers and enterprise leaders also faced new opposition to the alleged environmental benefits of reusing nepheline wastes. A controversy arose in 1935 as Apatit attempted to sell nepheline concentrate as a coagulant for water purification and the State Sanitation Inspectorate (Gosudarstvennaia sanitarnaia inspektsiia) prohibited this use. The agency cited concerns about the relationship between the purifying fluorine and the potentially poisonous phosphorous anhydride in Khibiny apatite and forced the enterprise to continue testing before giving its approval.[49] During a meeting the following January, Kondrikov vented his frustration with the obstacles that emerged in the enterprise's attempts to implement "complex utilization." He excoriated those who "want to receive refined alumina and to the 'holy mother' with everything else" and emphasized that the issue with waste was "not how to throw it away, but how to use it."[50]

In the second half of the 1930s some short-lived progress did occur in the efforts to reuse industrial wastes in the Khibiny. Engineers and consultants for Apatit continued to promote research on nepheline and in 1935 succeeded in getting an alumina plant in Volkhov, outside of Leningrad and near a limestone quarry, to convert to using nepheline concentrate.[51] A meeting of a bureau of the scientific research sector of the NKTP in 1936 devoted to "the questions of the complex utilization of Khibiny apatite-nepheline ore" passed a resolution endorsing the construction of a new factory in Leningrad, Red Chemist (Krasnyi khimik), that would process alumina from nepheline with sulfuric acid. G. F. Solov'ianov enthused around this time that "the only raw material source able to play a decisive role as a substitute of the world aluminum industry is Soviet/Kola/nepheline."[52] At this point the enterprise commenced construction of a nepheline enrichment facility to produce nepheline concentrate out of tailings.[53] A full capacity, combined apatite and nepheline enrichment factory, initially begun in 1933, went into operation in 1939, and that year construction began on a redesigned aluminum refinement plant in Kandalaksha.[54] Yet Apatit produced very little nepheline concentrate before the evacuation of the Kola Peninsula and the bombing of Kirovsk during the Second World War halted industrial activities.

Nepheline by-products from the Khibiny Mountains finally began to be reprocessed at the end of the 1940s. Two alumina plants south of Lake Ladoga, Volkhov and Pikalevo, started using Khibiny nepheline in 1949 and 1959, respectively. The Kandalaksha Aluminum Factory (Kandalakshskii aliuminievyi zavod) opened in 1951 and relied partially on alumina manufactured from Khibiny nepheline to produce refined aluminum.[55] Reflecting the long delay between the initial proposal for this reuse scheme and its implementation, Gosplan in 1952 somewhat disingenuously (given Fersman's speech two decades earlier) described the "complex utilization of raw material" from the Khibiny and in particular "the problem of reprocessing apatite and nepheline concentrates" to be "new and quite complicated."[56]

At this point the need for large quantities of limestone to process nepheline into alumina also altered the original conceptions of a localized economic geography.[57] Instead of complex utilization remaining in the Murmansk region, Khibiny nepheline traveled south to the locations of limestone deposits before returning to the Kola Peninsula for the final stage of production.[58] At the Volkhov plant one ton of alumina required from two to four tons of limestone per ton of nepheline concentrate and yielded from nine to ten tons of cement and one ton of soda ash (sodium carbonate for glass manufacturing) and potash (a form of potassium carbonate that can be applied as a fertilizer) as by-products.[59] Production also generated sludge and alkali wastes, but significantly less than from the bauxite process.[60] The value of soda, potash, and cement figured into the calculations that advocates of the reprocessing of nepheline tailings at Apatit had made to demonstrate its economic viability.[61]

This recycling of Khibiny nepheline, however, did not stop it from becoming a major pollutant in the region. In 1953 state sanitation inspectors reported: "The enrichment factory of the combine releases industrial discharges without purification into the White River, which therefore now has no value as a

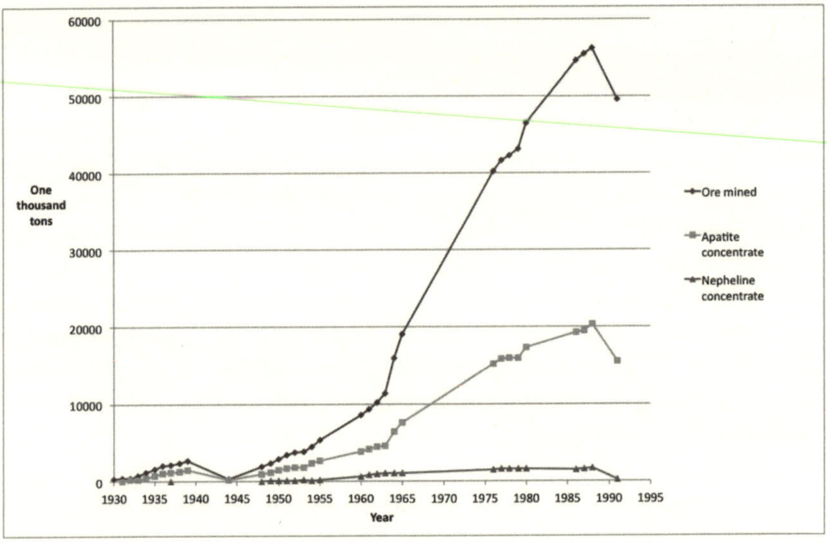

FIGURE 9.2. Reported Production Output at Apatit. The figures used to create this chart came from A. V. Barabanov et al., *Gigant v Khibinakh: Istoriia otkrytogo aktsionernogo obshchestva "Apatit" (1929–1999)* (Izdatel'skii dom "Ruda i metally," 1999), 44–66, 74, 78–82, 102, 173, 192, 193, 201, 213; P. V. Vladimirov and N. S. Morev, *Apatitovyi rudnik im. S. M. Kirova* (Seriia izdatel'stva po khibinskoi apatitovoi probleme, 1936), 3–4; M. D. Petrova, S. M. Salimova, and T. I. Podgorbunskaia, eds., *Kirovsk v dokumentakh i faktakh, 1920–1945 gg. Khrestomatiia* (Apatit-Media, 2006), 119. In some cases the numbers appeared as "more than" or "up to."

municipal drinking source. In this case the release of industrial sewage water causes an indirect impact on a large water source of the region (Lake Imandra, into which the White River flows)."[62] The inspectors also described how nepheline dusts from dry rooms, storehouses, loading facilities, and transport vehicles breached safe levels, causing respiratory illnesses in workers and damage to nearby flora. To counter this "severe pollution of the atmosphere," the firm installed better filters and purification equipment that reduced dust emissions from 370 kilograms per hour to 17–20 kilograms per hour. Further efforts additionally curbed the amount of dusts let out by Apatit's first enrichment factory in the second half of the 1950s before these emissions flattened out at between two and three tons a day during the 1960s.[63] The company also opened the first of three tailing dumps near the Pearl (Zhemchuzhnaia) River in 1957 to mitigate water pollution, but nepheline wastes continued to leak out of it. Even after a decade of expensive upgrades, which included reconstructing this dump, raising the level of a separating tank, repairing the tailing shoots, creating a new laboratory focused on effective work with tailings, and opening two new tailing dumps, Apatit still discharged liquid sewage with 90–180 milligrams of suspended matter per liter in the mid-1960s.[64]

Just as significantly, apatite production sprinted ahead of nepheline reprocessing. Industrial activities in the Khibiny grew astronomically in the Khrushchev and Brezhnev eras, with annual production figures exceeding forty million tons of mined apatite-nepheline ore and seven million tons of apatite concentrate in the 1980s.[65] While Apatit's production created an equivalently large amount of nepheline tailings, the amount of nepheline concentrate manufactured by the trust failed to keep pace (see figure 9.2). It only reached the level of 750,000 tons a year (which had been estimated for 1937) in the 1960s.[66] To the chagrin of Apatit, the country's aluminum industry remained more attracted to the idea of importing foreign bauxite than increasing processing with nepheline after Soviet bauxite sources began to dwindle in the 1970s and 1980s.[67] In this period only around 15 percent of the nepheline wastes created by the mining and processing of Kola apatite were used to make nepheline concentrate.[68] One scholar in the early twenty-first century estimated that over half of the three billion tons of ore (over a cubic kilometer in area) extracted from the Khibiny Mountains remained in rock heaps near the mines and close to 900 million tons of nepheline sat in tailing dumps.[69] This would mean that, at most, 20 percent of the material taken from the mountains had been transformed into an industrial product.

As massive amounts of nepheline went unused, the substance increasingly disturbed the local ecosystem. Decades of dumping led to the formation of a nepheline-rich layer of sediment up to eight meters deep in the White (Belyi) Bay of Lake Imandra. By the late 1970s the water in the bay was transparent for only one-half to one meter, in contrast to an average of six meters in cleaner sections of Lake Imandra (see figure 9.3). This pollution caused many of the native microorganisms in this part of the lake to disappear and almost all fish in the area to suffer from pathological sicknesses.[70] At the time Apatit reported that its first enrichment factory discharged "up to 14.4 tons of solid waste a day into the important—from an ecological standpoint—Lake Imandra." The enterprise also admitted that this dumping "does not align with active legislation on nature protection."[71] Yet, over time, measures taken by Apatit to reuse 80–90 percent of the water from enrichment, thereby reducing the dumping load, and the precipitous drop in production in the 1990s resulted in a gradual improvement in the water quality of Lake Imandra.[72]

But as the threats to Kola waterways moderated, new surface hazards and air pollution emerged. The size of Apatit's tailing dumps expanded rapidly with solid nepheline wastes, which primarily accounted for the surface area of industrial activities growing 2.17 times in Apatity between 1964 and 1996

FIGURE 9.3. Aerial image of the White Bay of Lake Imandra from June 1964. The white area at the bottom right of the image shows where nepheline tailings were entering the lake. From Olga Rigina, "Environmental Impact Assessment of the Mining and Concentration Activities in the Kola Peninsula, Russia by Multidate Remote Sensing," *Environmental Monitoring and Assessment* 75, no. 1 (April 2002): 19.

and by 6.6 times in Kirovsk over that period. In 1996 the 57.3 square kilometers of the Khibiny region were an industrial landscape. Furthermore, as Apatit's tailing dumps filled and dried out, they turned into sources of hazardous nepheline-saturated dusts that caused health problems for people in the area. As researcher Olga Rigina describes, "severe dusting episodes" and "polluted drinking water from Imandra" contributed to human impacts such as "lung and genitourinary diseases, cancer and allergies."[73]

A MATERIALIST STORY

From having contributed to hopeful schemes of waste management in the 1930s, nepheline became a recalcitrant material that repeatedly undermined planners' intentions and damaged the natural environment in the later Soviet era. But why, contra Fersman, were so many grams of extracted mining mass lost, so many grams of waste produced, and so much emitted into the air and washed away by water?

One could perhaps account for this episode in Soviet history as a case of bureaucratic infighting and the incompetence of communist central planning, as Ze'ev Wolfson did in the late 1970s:

> The matter is drawn out even longer when different ministries are involved in planning new multiple production processes. The end is not in sight for the "tug of war" over the multiple use of resources in Kar-Bogaz-Gol Bay on the Caspian Sea or the apatite-nepheline ores of the Kola Peninsula. What goes into the dumps at the Khibin mines contains large quantities of nepheline, a valuable raw material for the aluminum industry. But the Ministry of Nonferrous Metallurgy does not want to spend millions building a concentrating mill (this would lower the plan indices for the performance of the entire ministry for several years), nor does the Chemicals Ministry want to do so. Any combine they built to process tailings would not be reflected in any of the major plan indicators—why should the ministries make the effort? For a laudatory newspaper story? Hardly a powerful incentive when the issue is spending millions, and the risk is losing one's job if the plan is not fulfilled.[74]

Wolfson certainly captures part of the situation. The conflicting agendas of the chemical and nonferrous metallurgy ministries prevented nepheline reprocessing from becoming a higher priority.

However, such an explanation implies that a more rational approach to the natural environment could have solved the problem without sacrificing economic concerns. In theory, enterprise leaders and state planners could have taken bold steps to stop prioritizing economic growth and expanded production, but such moves were especially unlikely in this political, economic, and historical context of the late Soviet period. Without curbing output or investing scarce resources, Apatit did not find itself in the position to fully manage industrial wastes. Nepheline rock in the Khibiny therefore maintained the capacity to function as a pollutant that caused considerable environmental destruction. It was able to evade human mastery and forcefully interfere in Soviet history.

Thus, this case of Khibiny nepheline suggests that the materiality of the environment affected the emergence of an acute environmental situation in the Soviet Union. Without letting Soviet officials off the hook for not taking more radical actions to mitigate pollution, the incorporation of nonhuman actors into our explanation for the Soviet environmental legacy helps bring the leviathan state and its governing power down to earth. The Eurasian environment was not simply an object conquered and destroyed by twentieth-century communist modernizers (as much of the ecocide literature has implied). It also participated in this history, inspiring innovations, accommodating state schemes, and thwarting technological optimism with its raw physical recalcitrance. Even when considering the enormous changes that occurred on the Eurasian environment under Soviet rule (which compels us to acknowledge the unprecedented scale of human influences on the natural world in the recent past), the lurking power of nature to play a role in this process discredits illusions of total human control. The materiality of nepheline and other elements of nature show how political power extends beyond the human realm.

10 Encounters with Permafrost

THE RHETORIC OF CONQUEST AND PROCESSES OF ADAPTATION IN THE SOVIET UNION

PEY-YI CHU

IN 1938 the USSR Academy of Sciences published a work of popular nonfiction entitled *Zaveovanie severa (v oblasti vechnoi merzloty)* (Conquest of the North [in the region of permafrost]). The book was coauthored by a scientist, Mikhail Sumgin, and a writer, Boris Demchinskii. In it, the authors told their readers about the desirability and difficulties of economic development of "the North." For the authors, the North corresponded to "the region of permafrost," an area of ten million square kilometers within the Soviet Union with underlying perennially frozen earth. The vast space encompassed high-latitude regions above the Arctic Circle as well as sub-Arctic Siberia (see map 10.1).[1] The North, wrote Sumgin and Demchinskii, "intimidated with its severe mystique but also beckoned with its wealth." Its riches included gold deposits in the valleys of the Enisei, Lena, Aldan, Amur, and Kolyma Rivers; coal basins in the Trans-Baikal and Amur regions; and sources of silver, lead, zinc, tin, molybdenum, antimony, bismuth, and oil. But these resources remained beyond reach because "permafrost enfetters [*skovyvaet*] the earth's depths."[2] In their narrative of "conquering" the North, Sumgin and Demchinskii cast permafrost as the key enemy to be defeated.

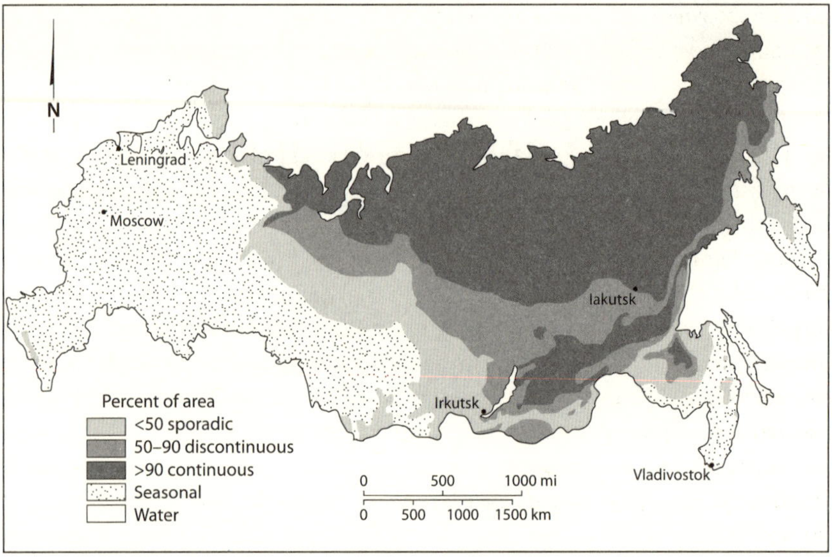

Map 10.1. The Extent of Permafrost in Russia and the Soviet Union. Data from V. Kotlyakov and T. Khromova, *Land Resources of Russia: Maps of Permafrost and Ground Ice, Version 1* (NSIDC: National Snow and Ice Data Center, 2002), available online at http://nsidc.org/data/ggd600/.

The visions of defeating permafrost presented in the book centered on its elimination. According to Sumgin and Demchinskii, thawing one cubic meter of frozen ground required three kilograms of coal. Thawing one square kilometer of the permafrost region, assuming an average permafrost thickness of fifty meters, therefore required 150,000 tons. For the entire region of permafrost, 1.5 trillion tons of coal were needed. However, humans cannot be "so reckless as to expend billions of tons of coal on thawing permafrost." Rather, they must "observe the most vulnerable [*uiazvymye*] areas of permafrost, and by directing their efforts there, they will even have the opportunity to cause the self-destruction [*samorazrushenie*] of frozen earth."[3]

In the context of global warming in the twenty-first century, the thawing of permafrost is considered a cause for concern. It may release more greenhouse gases into the atmosphere, creating a positive feedback loop that could trigger a tipping point for climate change.[4] That Sumgin and Demchinskii's book advocated deliberately degrading permafrost appears to confirm the callous and utilitarian impulse to destroy nature that writers have associated with the Soviet regime. Indeed, the theme of environmental destruction has featured prominently in studies of the Soviet Union.[5] But to what extent did Sumgin and Demchinskii's statements correspond to actual interactions with frozen earth in the USSR?

In this chapter I explore this question by focusing on the rhetoric and practices of "conquering" the region of permafrost during the Stalin era. In that period the ideological metaphor of "conquest" referred to the industrial transformation of undeveloped regions of the country. I analyze the rhetoric of conquest as it was applied to permafrost, the phenomenon of perennially frozen earth found in northern Eurasia, northern North America, and other high latitude and high altitude places of the world.[6] The rhetoric of conquest cast permafrost as an enemy to be struggled against and defeated. Permafrost was pictured as an animate and hostile agent, reinforcing a dualistic view that juxtaposed humans and nature. Notably, those who used the rhetoric of conquest included scientists invested in establishing permafrost research, an emerging field in the Soviet Union during the 1930s. For them, the rhetoric of conquest helped to draw attention to an unfamiliar phenomenon and to educate people about its importance for economic development. The rhetoric conveyed the need for specialized expertise about permafrost, thereby attracting resources necessary to institutionalize a new discipline. Finally, in a time of political terror, speaking the regime-approved rhetoric of conquest formed part of a strategy for survival. Instead of taking the rhetoric of conquest at face value, I argue that it served specific social purposes.

I also argue that the rhetoric of conquest was belied by the practices of conquest. The work of industrializing the region of permafrost entailed not degrading frozen earth but, rather, preserving, utilizing, and generating it. Soviet scientists and engineers developed such techniques while building infrastructure, including edifices, sanitation systems, and roads, in the region of permafrost. Despite the rhetoric, "conquest" relied upon adaptation, or learning to accommodate the presence of perennially frozen earth. I use "adaptation" to refer to a process, as the term is understood in evolutionary biology, rather than to any specific practice. To describe Soviet activities in terms of adaptation is therefore not to judge them as optimal or beneficial for the environment. Rather, the goal is to highlight the process of knowledge production necessitated by the encounter with unfamiliar environments.[7] While operating in the region of permafrost, scientists and engineers had to learn to negotiate the constraints it presented. The process of adaptation subverted the adversarial dualism trumpeted by the rhetoric of conquest. Seen through Soviet interactions with permafrost, humans did not transcend but became more embedded in the environment; their activities were not separate from but entangled with nature; their solutions were not conclusive but imperfect and ongoing.

By distinguishing between and contrasting the rhetoric and practice of conquest, I adopt a non-Hegelian approach that probes the disconnects between attitudes and behavior.[8] Literature on the environmental history of the Soviet Union has asserted the opposite: that a correspondence existed between rhetoric and practice, often with harmful consequences. Scholars have argued that the rhetoric of conquest promoted environmentally aggressive policies that led to ecological degradation.[9] However, this picture becomes more complicated when considering that Soviet actors strived to maintain permafrost for the sake of the stability of industrial infrastructure. By pointing out the disjuncture between the rhetoric and the practice of conquest, I argue that humans must inevitably respect environmental constraints. I avoid validating the rhetoric of conquest and its concomitant ideas of nature as separate, antagonistic, and subject to defeat by humans.[10] Human activities met with challenges under the particular conditions of permafrost, but the degree to which this was so depended upon the demands that people placed upon the land. Soviet propagandists were wont to assert claims about the hostility of permafrost. In fact, whether it was a source of difficulty depended upon the type of activity in which the population was engaged. The large-scale construction called for by state-driven industrialization created problems whose solution required a difficult process of adaptation.

Besides maintaining critical distance from the idea of conquest, pointing to discontinuities between rhetoric and practice can add nuance and specificity to the Soviet Union's environmental history. Scholars have been moving beyond straightforward narratives of declension and condemnation when analyzing Soviet interactions with the environment, drawing attention to the diversity of practices across time and space.[11] The purpose is not to exonerate the Soviet Union but, rather, to de-exceptionalize it in order to show that the Soviet experience is part of a broader pattern of interacting with the environment in which all societies are implicated. My focus on the Soviet encounter with permafrost in this chapter aims to contribute to that effort.

THE RHETORIC OF CONQUEST

Delving into the history of science yields clues about the context and motivations for using the rhetoric of conquest. Permafrost science as a field of research became institutionalized in the Soviet Union during the 1930s. Its leading spokesman was Mikhail Sumgin, the scientist author of *Conquest of the North* who became known as the founder of the new branch of science.[12] Sumgin worked simultaneously to build the discipline and to promote awareness of the subject of its study. The rhetoric of conquest served both goals. By portraying a heroic struggle with permafrost and permafrost-related phenomena, Sumgin emphasized that specialized investigations by experts were necessary to address the challenges presented by perennially frozen earth. The dualism of nature and culture embedded in the rhetoric of conquest helped to popularize an image of permafrost as a hostile adversary to be tamed and defeated. Not only did Sumgin borrow preexisting tropes in the rhetoric of conquest, but he also contributed notions of time that elevated the distinction between nature and socialist development. The Five-Year Plans mobilized the Soviet population for rapid industrialization in order to catch up and overtake the capitalist countries. By contrast, Sumgin established a scientific term for frozen earth, *vechnaia merzlota*, that included the adjective *vechnaia*, or "eternal." By embracing the concept of eternal, Sumgin identified *vechnaia merzlota* with the nonlinear time of nature, linking the conquest of nature to the conquest of time.

Conquering nature as a framework for describing humankind's relationship with the nonhuman world had precedents in the eighteenth and nineteenth centuries. When Europeans spoke of "conquering nature," they meant dramatically transforming the landscape to suit human needs. They imagined that building great machines and structures enabled them to emancipate themselves from historical limitations on human habitation, thereby defeating, controlling, or taming nonhuman elements.[13] In the USSR the familiar notion of conquering nature carried additional ideological significance. As scholars have shown, the struggle against nature comprised a central theme in Stalin-era culture and propaganda. Literature, film, and media portrayed Soviet society as engaged in battles with features of the environment, including rivers and oceans, ice and snow, the tundra and the taiga. They cast Soviet development projects, such as urban construction and resource extraction, as heroic endeavors akin to military campaigns.[14] The term *osvoenie*, the mastery or assimilation of territories, became conflated with the term *zavoevanie*, literally "conquest."[15] The rhetoric of conquest served to mobilize the Soviet population, amplifying Stalinism's emphasis on voluntarism, or the omnipotence of human will. It also resonated with the Soviet aspiration to remake humans encapsulated by Maksim Gorkii's famous statement that people, in the course of transforming nature, would also transform themselves.[16]

As a scientist Sumgin combined his interest in the nature of frozen earth with awareness of its relevance for economic development. His perspective emerged from experiences in the 1910s, when he spent six years in the Russian Far East employed by the tsarist Resettlement Administration (Pereselencheskoe upravlenie). As head of a meteorological station and then director of the newly created Meteorological Bureau (Meteorologicheskoe biuro) in Blagoveshchensk, Sumgin gathered observations about the soil and climate of the Amur region.[17] While recording temperatures of the soil, Sumgin became intrigued by the fact that, beginning at a certain depth, the earth maintained a temperature below zero degrees Celsius throughout the year. It remained below freezing even during summer, when atmospheric temperatures were high. What enabled such a condition, how widely did it prevail, and how deeply did it extend? How might it affect factors such as the groundwater regime that were relevant to agricultural colonization? Sumgin's sojourn in eastern Siberia also overlapped with the construction of the Amur section of the Trans-Siberian Railroad, which passed through areas with frozen earth.[18] He learned about the myriad difficulties associated with building roadbeds and edifices in such parts, where engineers contended with unstable landscapes that deformed their structures. Sumgin's intellectual curiosity and firsthand exposure to the practical significance of frozen earth convinced him of the need for systematic study of the phenomenon.[19]

Bolshevik rule after 1917 offered Sumgin new possibilities for organizing frozen earth research. During the 1920s the new regime sought cooperation with the prerevolutionary scientific and technical intelligentsia to produce knowledge that benefited the economy and society. It supported the efforts of a former tsarist organization, the Academy of Sciences Commission for the Study of Natural Productive Forces (Komissiia po izucheniiu estestvennykh proizvoditel'nykh sil), to establish institutes oriented toward applied science.[20] Accordingly, Sumgin reached out to the commission chairman, the geochemist Vladimir Vernadskii. With Vernadskii's backing, Sumgin presented to the Academy of Sciences (Akademiia nauk) a rationale for a commission that would "assume scientific leadership of all research on permafrost within the USSR." Among the goals of the new commission would be to ensure that "practical workers will have in their hands all the data needed for agriculture as well as for calculations when raising buildings and structures and constructing railroads and highways" in regions with frozen earth.[21] In December 1929, the Academy of Sciences approved the Commission for the Study of Permafrost (Komissiia po izucheniiu vechnoi merzloty [KIVM]). Sumgin not only succeeded in establishing an institutional framework for systematizing frozen earth research. He also advanced his career by being named academic secretary of KIVM, responsible for its day-to-day management.[22]

Sumgin was motivated to speak the language of the regime in order to attract increased resources. During the period of the Five-Year Plans, Sumgin and the emerging group of permafrost scientists at KIVM stressed their contributions to building socialism. "The closest coordination between sound theory and all branches of socialist construction," they noted in their work plan for 1933, "is the basic line that the Academy of Sciences' Commission for the Study of Permafrost carries out and will carry out." Such declarations framed repeated requests for more office space, laboratory equipment, personnel, and decision-making rights, including the right to organize expeditions, establish field stations, and approve building norms.[23] A common refrain consisted of the need to transform KIVM into a full-fledged institute. "Only with such an organization," scientists wrote to the governing body of the Academy of Sciences, "can KIVM conduct thorough research on permafrost, thereby facilitating the economic assimilation [*osvoenie*] of the North."[24] Over the course of the 1930s, their ambitions were realized. KIVM grew in size and stature, becoming the Committee on Permafrost (Komitet po vechnoi merzloty) in 1936 and the Institute for Frozen Earth Science (Institut merzlotovedeniia) in 1939. The upgrade in status was accompanied by additional facilities, including three field stations, and an expanded staff.[25]

Besides its importance for obtaining funding, the rhetoric of conquest also functioned as part of a strategy for institutional and personal survival. Sumgin's nascent group of permafrost scientists faced threats to its existence from within and without the Academy of Sciences. Given the competition for facilities, larger bodies of the Academy of Sciences made moves to absorb KIVM and its resources.[26] More ominously, amid the atmosphere of political terror in 1937, denunciations surfaced about the work of Sumgin and his colleagues. That year, complaints about permafrost scientists' flawed work reached the Academy's presidium, its executive body.[27] The journal *Sotsialisticheskii transport* (Socialist Transport) also published an article accusing permafrost scientists of "disorienting workers" with "speculative conclusions instead of tested designs." It hinted that their shoddy science amounted to collusion with "spies and wreckers" seeking "disruption of all work directed at the assimilation of the northern and eastern peripheries of the Soviet Union."[28] The attacks on the Committee on Permafrost took place as geologists were being arrested for past affiliations with opposition political groups. Sumgin was especially vulnerable given his previous membership in the Socialist Revolutionary Party (Partiia sotsialistov-revoliutsionerov), rivals of the Bolsheviks.[29] To defend themselves against charges of sabotage, permafrost scientists urged each other to "write more often in such organs as *Pravda* and journals about the activities of the Committee."[30] The tense political environment made it all the more important to communicate the value of their work using approved language.

Consistent with Soviet propaganda, permafrost scientists described economic development in terms of a struggle against frozen earth. In *Zavoevanie severa* (*The Conquest of the North*), the popular science book cowritten by Sumgin and Boris Demchinskii, the authors invoked metaphors of war. Civilization, embodied by "railroads, gold mines, coal shafts, settled cities, and industrial centers," had "in the first instance captured [*zakhvatyvali*] those regions where the conditions of life were less severe."[31] But "now, as civilization in such a systematic way is advancing north, the success of its advance depends on the outcome of the struggle [*bor'ba*] against permafrost." As an antagonist, permafrost exhibited unique aggressiveness, causing infrastructure to fall into ruin and equipment to break down. Show "the slightest disdain for its particularities," Sumgin and Demchinskii warned, and permafrost "would not be slow to mangle [*iskazit'*] those civilizing measures that would produce fruitful results under normal conditions."[32] Framing interactions with frozen earth as an adversarial confrontation in which nature and culture stood separate and opposing, scientists helped to sustain the rhetoric of conquest. Doing so enabled them to communicate in the idiom of their cultural environment, a prerequisite for drawing attention to their work. In the context of publicizing an emerging discipline, the rhetoric of conquering nature had instrumental value, educating audiences about an object of state and scientific importance.

When depicted with the rhetoric of conquest, permafrost appeared to have animate and hostile qualities. The motif of struggle pitted the agency of permafrost against the agency of human experts. In *Conquest of the North*, Sumgin and Demchinskii characterized permafrost as a live opponent, writing that "any work on the surface of the earth provokes a reaction on the part of permafrost."[33] Using anthropomorphic metaphors, they wrote that permafrost sometimes "submits [*podchiniaetsia*]," retreating in its extent, sometimes "dictates [*diktuet*]," shaping surrounding landforms. Permafrost seemed to possess a capricious will and exhibit unpredictable behavior. "The sensitivity [*chutkost'*] of permafrost is astonishing," Sumgin and Demchinskii asserted, stating that the phenomenon was "harmful" precisely in its "inconstancy [*nepostoianstvo*]." Against the arbitrariness of permafrost stood the "rational will [*razumnaia volia*]" of humans. Echoing the language of Stalinist voluntarism, Sumgin and Demchinskii stated that "rational will subdues [*podchiniaet sebe*] the forces of nature." Subduing nature meant making the environment amenable to economic development. But by its very existence, frozen earth presented an obstacle to certain activities such as agriculture and gold mining: it made tilling and grinding difficult. "Rational will" therefore had to be applied toward "the destruction [*razrushenie*] of permafrost." The struggle played out underground, where frozen earth either advanced closer to the surface or retreated further into the depths.[34] A picture of dramatic conflict with an active enemy helped to convey vividly the need for scientists' expertise.

Besides using the rhetoric of conquering nature, permafrost scientists also enriched it by contributing associations between nature and time. Sumgin himself popularized a perception of the temporality of frozen earth through the very name he chose for the phenomenon. It was he who advocated adopting the expression *vechnaia merzlota* to designate the phenomenon whereby the earth remained frozen throughout the year. He explained the term in a book published in 1927 that was later recognized as the first "comprehensive treatment" of *vechnaia merzlota*.[35] According to Sumgin, the noun *merzlota* referred to "frozen soil" or "frozen ground," defined as "soil (or ground) which has a negative temperature."[36] The adjective *vechnaia* designated the duration of existence of frozen soil or ground. By *vechnaia*, Sumgin aimed on the one hand to encapsulate the fact that the earth remained frozen across all seasons, without thawing during summer. On the other, he wished to indicate what he perceived to be the "geologically distant" origins of the phenomenon.[37] How long, then, was *vechnaia*? "A period of no less than two years I will consider the minimum boundary," Sumgin wrote. As for the maximum, he provided a guideline of "thousands or tens of thousands" of years, a span that encompassed the "time of the mammoths" to the present.[38] Thus, *vechnaia merzlota* was soil or ground that sustained a negative temperature continuously for a minimum of two years and a maximum of tens of thousands of years. Having advanced this term and definition, Sumgin reaffirmed them in decrees of the Academy of Sciences, the second edition of his monograph, and an official textbook.[39] The term *vechnaia merzlota* became institutionalized, forming part of his enduring legacy and giving rise to the English neologism "permafrost" coined in 1943 and accepted internationally today.[40]

Although *vechnaia merzlota* had not been invented by Sumgin, it gained wide acceptance as he defended and promulgated it in public lectures, radio broadcasts, and writings directed at general audiences. While promoting permafrost science in the 1930s, Sumgin simultaneously popularized his chosen Russian name for "permafrost."[41] Against detractors who criticized the use of *vechnaia* as wrongly suggesting something "everlasting, immeasurable, and inscrutable," he justified it on grounds of accessibility, vividness, and usage. The term, he admitted, contained more "artistic license than scientific precision," but it conveyed the phenomenon's essential characteristics: its continuity and long-lastingness.[42] It was telling, however, that Sumgin considered temporality a notable feature of frozen earth in the first place. Other scientists considered frozen earth the product of contemporary climatic conditions, not past geological eras. They also argued that no consistent difference existed between frozen earth of several months and several years vintage. If anything, frozen earth ought to be distinguished on the basis of its specific material composition, not its duration of existence.[43] But to Sumgin, a salient difference did exist between shorter-term and longer-term frozen earth: perennial frozen earth presented an obstacle throughout the lifetime of a construction project, whereas seasonal frozen earth did not. It also presented a particular set of engineering challenges because of the depths at which it was found and its effects on the surrounding landscape. The temporal perspective of economic development therefore informed Sumgin's understanding of *vechnaia merzlota*.

Vechnaia merzlota acquired new cultural significance during the Five-Year Plans by helping to link the struggle against nature to the struggle against time. As Stephen Hanson has shown, notions of time occupied a central position in the Soviet project. The ideology of Marxism-Leninism-Stalinism centered on what he calls a "charismatic-rational conception of time," which combined adherence to the linear time of modernity with the aspiration to transcend it.[44] Under Stalin, the goal of synthesizing the rational and the charismatic was manifested in the "planned heroism" of the Five-Year Plans. The party-state urged that production norms be overfulfilled in order to realize the Five-Year Plan in four years. Rapid economic development was necessary not only to catch up to the capitalist world but also to reach communism, the "end of human prehistory" according to Marx. Socialist industrialization therefore entailed applying labor discipline toward achieving the revolutionary compression of time.[45] In contrast to the charismatic-rational conception of time, the term *vechnaia merzlota* evoked the cyclical time of nature. As suggested by Sumgin's evocative expression, *vechnaia merzlota* was "eternal," set apart from the forward movement of socialism. By struggling against *vechnaia merzlota*, Soviet society also struggled against spontaneous, undisciplined time. The rhetoric of conquering nature reinforced the notion of conquering time embedded in the logic of the Five-Year Plans.

Ultimately, permafrost scientists embraced the rhetoric of conquest, albeit a technocratic version of it. Conquest demanded intelligent, methodical, organized, and disciplined labor. It consisted not simply of acquiring riches but of transforming wild spaces into industrial regions. In *Zavoevanie severa*, Sumgin and Demchinskii contrasted Soviet economic development with prerevolutionary resource extraction, which they lamented as "plunder [*raskhishchenie*] of the North's wealth, not its conquest [*zavoevanie*]." Their perspective resonated with those of late imperial and early Soviet bureaucrats, including the geochemist Aleksandr Fersman, who favored the maximum and efficient "complex utilization" of natural resources.[46] Sumgin and Demchinskii's vision was prescriptive as much as descriptive, conveying hope for policies that favored planning and expertise. They critiqued tsarist-era expeditions to the empire's peripheries as "isolated actions abounding in heroic forays, but not a single plan or coordinated system." Such undertakings, they claimed, "opened up the North but did not guarantee victory over it." Genuine, successful conquest required steady, comprehensive effort guided by the unified purpose of mastering the environment through knowledge. What was needed were "not great exploits, but great routines," that is, rational institutionalized practices.[47] While maintaining the adversarial framework of the struggle against nature, they crafted a less dramatic, but nevertheless inexorable, narrative of conquest.

PROCESSES OF ADAPTATION I: PRESERVING

While establishing and publicizing their emerging field, Soviet permafrost scientists participated in various construction projects called forth by the demands of socialist development. Given the goal of industrializing the entire country, urban, sanitary, and transportation infrastructure had to be created in regions with underlying frozen earth. Within the rhetoric of conquest, permafrost figured as an enemy to be defeated in order to advance industrial civilization. But a different picture emerged from scientists' and engineers' specific interactions with frozen earth on the ground. Building structures, wells, and roads required preserving, utilizing, and even generating frozen earth. "Conquest" in practice belied the rhetoric of conquest and instead required processes of adaptation. Adaptation meant learning how to coexist with and accommodate frozen earth, rather than eroding it. It entailed multiple approaches to producing knowledge, including studying older, prerevolutionary cultures of technology. Rather than nature and culture opposed, adaptation revealed that the land was shaped by both. Rather than "eternal" frozen earth, adaptation revealed that permafrost was all too prone to disappear.

When constructing buildings in places where permafrost prevailed, engineers faced challenges posed by varying patterns of freezing and thawing. Edifices, for example, suffered structural deformations from "heave [*puchenie*]," the swelling of the ground in winter and its sinking in summer.[48] The seasonal process manifested itself in the terrain's uppermost layers, where changes in the availability of heat in the atmosphere had the greatest effect. During the winter months, moisture in the earth froze and expanded. If the soil was well saturated with water, such expansion could be substantial. The ground bulged and exerted pressure on overlying structures. During warmer months, earth that had previously frozen subsequently thawed. It sank and lost some of its bearing capacity.[49] Although heave was possible in cold environments even where permafrost did not exist, its presence made the surface phenomenon more pronounced than otherwise. Soviet scientists attributed its amplifying role to its relative impermeability to water, which allowed moisture to accumulate above in larger quantities and the ground to heave more substantially.[50]

Heave was especially damaging to human constructions because it did not occur in a uniform fashion. Around buildings, land along the edges was the first to swell after the onset of cold weather, whereas ground directly beneath the building remained more insulated from atmospheric changes. Different parts of a structure therefore experienced pressure from below at different times, which strained the connectors between them. Not only was the upward action of heave uneven, but the terrain's sinking also happened asymmetrically. With the arrival of spring, land along the southern facade received greater exposure to the sun, thawing earlier than along the northern facade. As the ground at different places sank at different times, the edifice as a whole settled unevenly. "Thus the building rocks every year," a Soviet scientist explained, "first up, then down."[51] Taken together, these tensions could cause structures to fracture and adopt a southward slant. To complicate matters further, the ground's bearing capacity changed not only cyclically, because of annual heave at the surface, but also unidirectionally, because of irreversible thawing in the deeper permafrost layer.

The town of Iakutsk, capital of the Iakut Autonomous Soviet Socialist Republic, was a key site for learning to overcome challenges presented by frozen earth. Both older and newer buildings in the settlement of Iakutsk exhibited the deleterious effects of the ground's uneven freezing and thawing. The merchant court (*gostinyi dvor*), a low, one-story wood and brick structure built in the 1820s, appeared sunken and deformed by the twentieth century. An inspection in 1937 revealed that a trough filled with soft, moist earth had formed beneath the floor, which no longer rested on stable ground. Even edifices of more recent vintage, such as a tanning factory constructed during the Soviet Union's first Five-Year Plan, had become misshapen within a decade. By 1939 floor-to-ceiling cracks up to five centimeters wide had opened up along its walls, and engineers uncovered a deep pocket of thawed earth beneath the premises.[52] Because plans for urbanization entailed the construction of ever more installations in Iakutsk, including factories and power stations, it became necessary to find ways to counteract the forces of rapid decay.

In designing and systematizing norms for building in frozen earth conditions, Soviet scientists and engineers focused on preventive measures that preserved the ground's stability. They conducted careful study of local conditions and delved into historical archives in order to reconstruct patterns of interaction between earlier inhabitants and the environment. Their findings revealed, in part, that the land beneath Iakutsk consisted of a mixture of "natural [*estestvennyi*]" sandy and clayey soil, as well as "dung, construction debris, and garbage that has accumulated over the course of three hundred years of the city's existence." Scientists referred to the refuse left behind from previous settlement as the "cultivated layer [*kul'turnyi sloi*]." Its thickness was greater in parts of the town that had been settled longer and smaller in parts where human habitation was more recent.[53] Remnants of the past became embedded in the earth itself, attesting to the inextricability of nature and culture.

The "cultivated layer" introduced additional features to prevailing patterns of freezing and thawing that Soviet engineers had to consider when adapting construction to the local environment. It was a "characteristic peculiarity" of the land in Iakutsk that, where detritus from the past was plentiful, the associated accumulation of chloride and sulfate ions lowered the freezing point of water in the soil and increased its heat capacity. The subtle change in the uppermost stratum affected the underlying permafrost. It insulated the permafrost from heat in the atmosphere, and it also fed the permafrost with moisture. As water froze and became a constitutive part of the permafrost, "the surface of the permafrost rose [*podnimalas'*] by way of accretion [*nakopleniia*] of the 'cultivated' layer." Ultimately, scientists observed, the interaction between permafrost and the cultivated layer contributed to a "magnification of the heaving of the earth [*uvelichenie puchinistykh svoistv grunta*]," as evidenced by the "bulging up [*vypuchivanie*] of fences, gates, and porches" in areas of Iakutsk that had experienced a longer period of human settlement.[54] Traces of earlier human activities reverberated upon structures built by later generations.

Soviet scientists and engineers investigated not only local particularities of the soil but also preexisting building practices in Iakutsk. Prerevolutionary edifices that managed to sustain their solid condition shared a common feature: they helped to preserve frozen earth by preventing heat from reaching the underlying permafrost. By the first half of the nineteenth century, architects in Iakutsk had begun designing wooden houses with raised floors. Blueprints called for the lowest plane to rest, at various sections, on logs stacked lengthwise, which separated it from the ground by a distance of up to one meter. The gap between the floor and earth was important because buildings, especially those intended for human occupancy, discharged heat generated by indoor stoves. If underground permafrost gained exposure to the warmth, it would thaw and settle, causing overlying structures to buckle. Another example of a solution to the problem was the cellar space included in a stone building constructed in 1911. The cellar functioned as an insulating layer, protecting the earth from the building's heat in winter. In summer, inhabitants kept its doors open, allowing air to pass through and cool.[55] Examining past precedents thus provided the Soviets with techniques for counteracting permafrost disappearing, a key source of structural deformation.

In addition to methods for preventing the unidirectional thawing of permafrost, research into the local environment generated techniques for minimizing the effects of annual heave. Soviet scientists learned that, even before laying a single stone, selecting a propitious location for a building could increase its stability. They recommended avoiding sites where the cultivated layer was thick and the soil was silty and clayey because of the likelihood of pronounced heave and substantial subsiding. By selecting a place where the relief allowed surface water to drain, and where the earth was relatively sandy, engineers could reduce the extent that the ground's movement impacted their structures. Embedding foundations deeply was another protective measure. Doing so permitted the edifice to stand on the continually frozen permafrost itself, instead of the yearly heaving stratum above.[56] It complemented the strategy of maintaining permafrost.

By the 1940s Soviet engineers had begun constructing buildings in Iakutsk that incorporated multiple features to prevent and counteract heave and permafrost thawing. They recognized that measures oriented to preserving frozen earth were both more economical and more effective than working toward its elimination.[57] By building on "the principle of maintaining the permafrost," engineers found that they were able to accommodate higher structures that generated large amounts of heat required for industrial processes. One such four-story stone building in Iakutsk rested on pillars of reinforced concrete that lifted the structure one and a half meters off the earth, creating a ventilated cellar space. Its foundation was embedded four and a half meters into the ground, on the permafrost layer.[58] The techniques of raised floors and deep foundations represented adaptations that became norms for construction. As multistory structures proliferated, the newspaper *Sotsialisticheskaia Iakutiia* declared that Iakutsk had "become younger, has straightened its shoulders," and "unbent its back and become much 'taller.'"[59] Soviet construction practices embedded the visage of a city in the frozen earth landscape.

PROCESSES OF ADAPTATION II: UTILIZING

Besides raising buildings, permafrost scientists participated in establishing sanitation systems in regions with underlying frozen earth. Burgeoning urban centers such as Iakutsk required a plentiful and reliable supply of clean water for both industry and public health. Cities needed wells, but the presence of permafrost provoked doubt about the feasibility of creating them. To solve the problem, scientists sought to understand and utilize the permafrost stratum. Hydrogeological surveys investigated the relationship between frozen earth and groundwater, revealing that the permafrost stratum contained pockets of thawed earth that served as conduits for water. Water from above the permafrost layer, including atmospheric precipitation, traveled via the network of thawed earth into the earth's depths and fed into sub-permafrost aquifers. Scientists found that such subterranean reservoirs, nourished by passages within the permafrost stratum, could be tapped for use. Accessing groundwater from beneath the permafrost layer entailed drilling through frozen earth. But while burrowing underground, workers strived to preserve the integrity of the frozen earth that formed the walls of the borehole. Failure to do so led to accidents. Learning to drill a passage through frozen earth while maintaining the overall cohesion of the permafrost stratum was an iterative process. The accompanying delays and pressures belied the picture of planning, efficiency, and heroic transcendence presented by the rhetoric of conquest.

On the most basic level, some scientists and engineers suspected that groundwater simply did not exist in these environments.[60] In his attempt to dig a well in Iakutsk in the nineteenth century, Fedor Shergin had burrowed 116 meters deep and yet failed to breach the solidified stratum or to strike water. Yet with new industries and growing numbers of people in parts of northern and eastern Siberia during the 1920s and 1930s, the problem of water provision needed to be solved. In Iakutsk, for example, contaminated water contributed to annual outbreaks of typhoid fever and dysentery among the population.[61] Ensuring a reliable source of potable water became central to the Iakut regional government's plans to improve its capital.

Gathering information about local conditions was the first step. In 1932 the party-state allocated two million rubles to hydrogeological research in and around Iakutsk.[62] At the Third All-Union Conference on Permafrost, convened by the Academy of Sciences' KIVM in 1933, a working group of scientists formed around the question of supplying Iakutsk with running water.[63] Surveys were carried out by the People's Commissariat of Public Works and the Academy of Sciences.[64] The scientists found landforms in the region, such as powerful springs, that pointed to the existence of water-bearing strata below the permafrost.[65] Natural springs at isolated points along the right bank of the Lena spewed water at a sustained rate of between fifty and sixty liters per second.[66] Their robust activity even in winter suggested that establishing a regular year-round water supply was feasible. Analyses of soil and rock from boreholes revealed additional signs, including coal and other organic remains. Based on these clues Soviet scientists anticipated an aquifer encompassing the lower Jurassic and upper Cambrian series beginning some 220 or 350 meters beneath Iakutsk.[67] In light of their recommendations, the Iakut regional government authorized the drilling of an exploratory well down to 500 meters.[68] The work was carried out by scientists at the Iakutsk permafrost station, the Iakut ASSR People's Commissariat of Municipal Engineering, and the Iakut Geological Survey. Their hope was to locate an abundant source of groundwater underneath the permafrost layer.

Repeated investments and political pressure were brought to bear in order to overcome unplanned difficulties in the execution of the project. During the drilling of the exploratory well, hot drilling fluid thawed the frozen earth, causing the walls of the borehole to collapse. Having drilled down to eighty meters, workers were forced to abandon the initial well and start over.[69] After drilling through the permafrost layer and reaching the target depth of five hundred meters, test pumping began. However, the uncovered aquifer appeared initially to be unusable for practical purposes because of its small capacity. It provided from eight to ten thousand liters per day and, based on a rule of thumb of one hundred liters of water per day per individual, could support a population of just eighty to one hundred.[70] This discouraging development led the commissar for municipal engineering to declare that pumping water from such a well was impossible. The Iakut ASSR Sovnarkom, however, charged him with negligence and blamed him for allowing the borehole to freeze and congeal. Allocating twenty-one thousand rubles to clear the mess and resume efforts, the Sovnarkom made the commissar, as well as the head of the Iakutsk permafrost station, personally responsible for the project.[71]

With added funds as well as political pressure, the project resumed. After the presence of a viable aquifer was ascertained, the regional government gave the green light for a six-hundred-meter exploitation well near the city center.[72] Once again, the project met with stumbling blocks. At four hundred meters below the surface, the borehole began to silt up. Instead of stopping to assess the situation, workers continued drilling, which exacerbated the problem. Drilling fluid ceased to circulate, and a mud plug ten cubic meters in volume became an impenetrable barrier.[73] The head geologist made a perfunctory attempt to clear the well before pronouncing it waterless.[74] To the consternation of regional officials, it appeared that the millions of rubles sunk into the enterprise had gone for naught. Scientists were called from Moscow to review the situation. They determined that fissures in the stratum of water-bearing rock had become completely jammed with particulates from the drilling fluid. Instead of the clay-based drilling mud, which might have served well in a different setting, water should have been used. Under new oversight workers eventually evacuated the aquifer as well as the barrel and walls of the borehole. Blame for the accidents focused on human error, but the unfamiliar environment played a role in conditioning the mistakes.

Eventually, scientists and engineers found that the water-bearing stratum could produce up to 60,000 liters per hour, or 1.4 million liters per day, an abundant capacity. Construction began on a pumping station at the second well in 1945. In February 1947 the population of Iakutsk began to receive potable water drawn from beneath the stratum of frozen earth.[75]

PROCESSES OF ADAPTATION III: GENERATING

The appearance of new wells and urban structures in the permafrost regions of northern Eurasia accorded with the idea communicated by the metaphor of conquest of advancing industrial civilization. But despite the rhetoric of conquest, which emphasized struggling against and defeating seemingly hostile elements of nature, building infrastructure depended upon maintaining and utilizing permafrost, not degrading it. The practice of conquest necessitated a process of adaptation as Soviet scientists and engineers learned to accommodate and take advantage of the existence of frozen earth. As they studied past practices and conducted surveys, they accumulated and produced knowledge about permafrost, developing an appreciation of the value of perennially frozen earth for supporting industrialization. Road building provided an alternative illustration of the process of adaptation. In permafrost regions, a characteristic form of winter flooding known as *naled'* presented a particular challenge to road engineering. *Naled'* featured in permafrost scientists' rhetoric of conquest as "the most destructive manifestation of permafrost."[76] But the solution they found to mitigate its effects consisted of modifying the terrain to induce the formation of frozen earth. The frozen trenches carved into the land helped to divert the flow of water, thus circumventing *naled'*, but these required upkeep. In this case adaptation entailed generating frozen earth, as well as ongoing maintenance and repair, and highlighted the impermanence of frozen earth as well as its entanglements with human activities.

Managing the flow of water was an imperative for road building in frozen earth conditions. The *naled'*—or icings—that permafrost scientists depicted as an adversarial element of the environment became an engineering problem whenever they appeared near construction sites. Anticipating these occurrences was not always straightforward. *Naled'* derived in part from groundwater that traveled to the surface through interstices in the soil and rock. The pattern of these underground passages differed according to the structure and stratification of the terrain.[77] Where frozen earth was extensive, water became squeezed into narrower channels by its relatively impermeable walls. Pockets of thawed and permeable earth, which occasionally persisted within larger frozen masses, served as conduits. In winter the freezing and expansion of the uppermost layers of land further compressed the space where groundwater could flow. During the cold months, sufficient pressure built up as a result of the combination of high water volume and small outlets, causing massive cracks to appear on the earth's surface (see figure 10.1). Through such cracks, groundwater burst out into the open, pouring onto the surrounding landscape and forming successive layers of ice.[78]

Without a clear image of the intricate subterranean maze, predicting the location of *naled'* was difficult. More often, its source became evident only after it had appeared, by which time roadbeds, bridges, and any other object or structure in the vicinity would already be bent, flooded, or encased in ice.[79] Soviet engineers eventually devised strategies for managing *naled'*, but only after many frustrating experiences. An important occasion for research and development was the Amur-Iakutiia highway, a transport artery from the Trans-Siberian Railroad northward to Iakutsk whose route, over a thousand kilometers long, passed through regions with extensive frozen earth. After the Council for Labor and Defense (*sovet truda i oborony*) decreed its construction in 1925, surveying and building, in typical Soviet fashion, proceeded simultaneously.[80] The rushed circumstances meant that it was not possible to

Фиг. 21. Км 143. Наледь речная. Вершина наледного кургана с большой трещиной. (Снято 13 марта 1928 г.).

FIGURE 10.1. Photograph of naled', a characteristic and destructive form of winter flooding in permafrost regions, taken in 1928. Valerian Gavrilovich Petrov, *Naledi na Amursko-Iakutskoi magistrali: S al'bomom planov naledei* (AN SSSR i Nauch.-issled. automobil.'-dorozh. Institut NKPS SSSR, 1930). Published with permission from the National Library of Finland.

avoid areas where *naled'* occurred after performing a careful study of the terrain. In the winter of 1927 and 1928, over one hundred manifestations of *naled'* along its southern section rendered the would-be road impassable.[81]

In addition to reactive measures, such as deicing encrusted objects, Soviet engineers deployed modifications of the landscape in order to prevent *naled'* from occurring near construction sites. One strategy implemented on the Amur-Iakutiia highway entailed clearing swaths of land and digging extensive canals or trenches at calculated distances from the road. In order to be effective, the cutting needed to be properly situated. Ideally, it would be located between the roadbed and the source of a *naled'*, at a lower elevation than the source. The goal was to induce the formation of a layer of frozen earth that would become an impermeable barrier to the waters of a *naled'*. In winter the depression, denuded of vegetation and snow cover, froze more quickly and to greater depths from exposure to the atmosphere than did the surrounding land. Water flowing downhill from the source of the *naled'* would be sequestered away from the vulnerable structures lying on the trough's other side.[82] Along the Amur-Iakutiia highway, these "frozen earth belts [*merzlotnyi poias*]" measured up to fifteen meters wide, somewhat less than a meter deep, and two hundred meters in length.[83] Etching them in the terrain involved moving thousands of cubic meters of dirt.

The solution was relatively straightforward, but it was imperfect. Its soundness depended upon the trench freezing to the necessary depth each winter. Left to its own devices, the earthen canal would likely thaw in summer. As long as it froze to the same degree during future winters, it would still be able to function as an effective bulwark against *naled'*. If not, however, it would no longer be an impediment to water, which could subsequently flood the objects that Soviet engineers originally intended for it to protect. Indeed, the land was wont to change in structure and composition—and thereby thermal conductivity—over time. A scientist pointed out, for example, that contact with groundwater from *naled'* could increase the presence of minerals that lowered the freezing point of the earth contained within the human-made depression. This altered constitution would render the trough less susceptible to freezing than before, and more permeable.[84] Ensuring its integrity therefore required annual monitoring and maintenance.

As the effort to contain *naled'* demonstrates, adaptations sometimes generated further changes in the landscape that required additional fixes. In an ongoing effort to improve on the solution, Soviet engineers endeavored to conceive of more solid and heat-resistant earthworks using whatever was at hand. One design, for example, entailed excavating, mixing, and compacting locally available peat, sand, and clay, packing the different soils in layers, and covering the resulting embankment with insulating moss. This type of formation, engineers calculated, would maintain its frozen integrity better than the open trenches.[85] Despite these innovations, however, the behavior of *naled'* on the Amur-Iakutiia highway dogged Soviet scientists and engineers into the 1960s.[86] Although the road and its accompanying redoubts were far from perfect, they nevertheless changed the local landscape and became part of the environment, to the extent of being fashioned from the same material.

CONCLUSIONS

Framing Soviet interactions with permafrost as a process of adaptation provides an alternative to the rhetoric of conquest used by the historical actors themselves. It also connects Soviet experiences to other projects of industrial transformation. Studying and accommodating the environment similarly accompanied transport construction in northwestern Canada; gold mining in Alaska; offshore oil drilling in the Gulf of Mexico; and dam building in Egypt.[87] Like their counterparts elsewhere, Soviet scientists and engineers had no choice but to operate within constraints presented by the nonhuman world. Despite calls to defeat permafrost and force it to retreat, they learned to preserve, utilize, and generate frozen earth. The techniques they developed were later adopted in the permafrost regions of North America as well.[88]

Pointing out the tension between the rhetoric and practice of conquest also serves to highlight contemporary dilemmas surrounding climate change today. As Soviet scientists and engineers learned, building industrial infrastructure in the region of permafrost required maintaining frozen earth. Yet industrialization has contributed to global warming, which is causing permafrost to thaw. The thawing of permafrost threatens the stability of urban centers across the former Soviet North, an apt illustration of how anthropogenic climate change undermines humans' own creations.[89] People have become more aware than ever of permafrost's lack of permanence. Previously, the process of adaptation required accommodating the existence of frozen earth. In the all-too-near future, it will require accommodating its disappearance.

PART IV

Fruits of the Waters

11 The Christian Environmental Ethic of the Russian Pomor

Stephen Brain

Only a century ago Orthodox crosses dotted the frozen beaches of the White Sea. From Arkhangel'sk to the tip of Novaia Zemlia, and all along the shores of the Kola Peninsula, wooden monuments with the tented roofs and carved death's heads of funeral markers claimed the land as Christian territory, despite being many days' travel from the nearest church or permanent settlement (see figures 11.1 and 11.2).[1] In the second quarter of the twentieth century, the Bolsheviks went about tearing down these emblems of religious belief, but even before then English fishermen found reason to remove the crosses when they encountered them, for these crosses were not only expressions of Christian faith but also property claims.[2] The crosses had been put in place over the centuries by the Russian Pomors, seafaring Slavs who used the signposts to demarcate the boundaries of their ancestral fishing grounds.

The seagoing Slavs who erected these crosses stood apart from their Great Russian cousins for a number of reasons, but the most intriguing of these differences is the completely different daily lifestyle and economic strategy that they evolved after they migrated to the Far North.[3] The forbearers of the Pomor

FIGURE 11.1. Pomor cross on Kolguev Island in the White Sea. Photograph by author.

FIGURE 11.2. Detail of Pomor cross on Kolguev Island in the White Sea. Photograph by author.

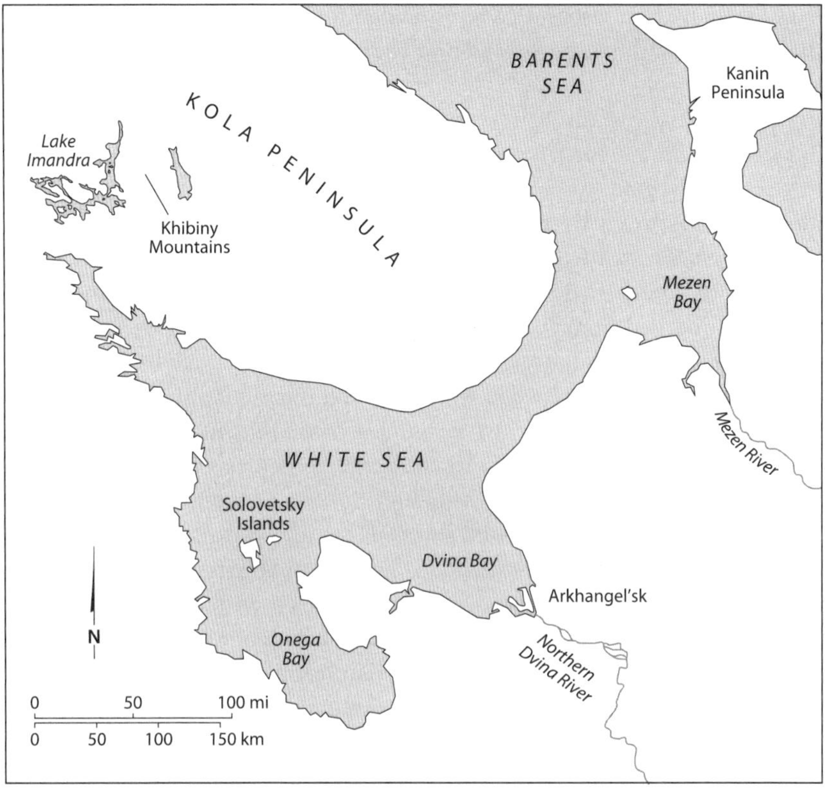

Map 11.1. The White Sea Region and Kola Peninsula.

first arrived on the shores of the White Sea (see map 11.1) in the twelfth century, drawn to the rich hunting and fishing grounds northeast of their home city of Novgorod, the most prosperous and influential town in northern medieval Russia. When the political power of Novgorod waned in the fourteenth century and its trade networks were disrupted by the Mongol invasion, the links between the center and the periphery became frayed. The temporary settlements of the hunters and fishermen became permanent, and the settlers themselves became more independent minded and self-reliant. By the time that the Muscovite princes reestablished external political control over the Far North in the seventeenth century, the settlers had taken on a name for themselves—the word *Pomor*, which is Russian for "people who live near (or go to) the sea"—and had developed an economy based not on agriculture, as was the case for the rest of Russia, but rather, on a form of maritime hunting and gathering.[4] Finding that traditional agriculture did not produce reliable harvests at their extreme northern latitude, the colonists instead came to rely upon seasonally available marine resources. Occasionally throughout the year, the men left their families and homesteads behind and ventured out to sea, following the herring runs from November to January and the seal migration from February to April, hunting walrus in June and July, and finally netting salmon and cod in the late summer. The rhythms and restrictions of agricultural life belonged to places left behind, and instead, as one observer put it, the Pomor "began to identify themselves with freedom of the sea."[5]

Relatively isolated from the Russian metropole and sources of central authority, the Pomor also developed local variants of the cultural institutions and practices inherited from their Great Russian ancestors, including a distinct dialect of the Russian language, a local customary law that allowed widows to inherit property at a time when women lacked this right elsewhere in Russia, and a unique architectural style for churches based on tent-shaped roofs.[6] But the most historically significant of these differences was the form of Christianity practiced by the Pomor. Although Russian Orthodoxy is customarily described as a syncretic faith, with survivals of Slavic paganism conspicuously coexisting with official Christian doctrine in the thoughts and practices of ordinary believers, Pomor Christianity provided an even more prominent and accepted place for animist beliefs.[7] Scholars of the Pomor have identified as central to Pomor culture a uniquely Christian sacral geography and argue that the Pomor, because of their exposure to the highly variable and unforgiving polar climate, invested the natural world with religious meaning. The sacral geography of the Pomor led them to see in the landscape signs of an active struggle between good and evil and thus had a significant impact on the way that individuals experienced personal religious matters. It also had another consequence: the mythopoeic quality of the Pomor worldview led them to perceive a land- and waterscape that was alive, which in turn helped them to create and abide by a moral economy that constituted a binding environmental ethic.[8] Accordingly, the Pomor represent a relatively rare example of a people who made conscious efforts to conserve resources precisely because of their Christian religious beliefs. In both the late tsarist period and the early Soviet period, the Pomor repeatedly resisted calls to intensify their economic activity, insisting that increased hunting and fishing would disturb the metaphysical and ecological balance of the White Sea ecosystem. They thereby provide evidence, although at variance with the Lynn White's claim in "The Historical Roots of Our Ecologic Crisis," that Christian belief can indeed give rise to an environmentalist consciousness provided that environmental conditions, like those found in the Russian Far North, encourage one to develop.

ENVIRONMENTALISM AND CHRISTIAN BELIEF

Although the article is very brief and more than forty years old, Lynn White's "The Historical Roots of Our Ecologic Crisis" continues to exert a remarkably strong influence over the scholarly understanding of the relationship between Christianity and environmentalism.[9] Since the article's publication in 1971, nearly every work touching upon the effect of religious belief on the natural world has discussed White's contention that the Judeo-Christian tradition, because of its anthropocentrism, its teleological component, and its emphasis upon humans' duty to subjugate creation, has bred into Western civilization a deep antipathy to the natural world.[10] Indeed, so great has been the impact of White's article that the prominent ecophilosopher J. Baird Callicott credited White with "implicitly set[ting] the theoretical agenda for future environmental philosophy."[11] White argues that the Judeo-Christian concept of an end to history, an idea that is utterly foreign to the ancients of East and West alike, encouraged in the minds of Westerners an expectation of progress that then translated into increased exploitation of natural resources. Yet these increased levels of exploitation would not have been possible were it not for the extremely central position of humanity in the Judeo-Christian universe—White claims that Western Christianity is the most anthropocentric religion the world has ever seen. Noting the tendency in the Book of Genesis to afford "no item in the physical creation [with] any purpose save to serve man's purposes," White suggests a direct connection between biblical attitudes and the later growth of science and technology, systems built on the assumption that physical matter is dead, devoid of lacking inherent value, and thus eligible for any manipulation humans deem fit.[12] In short, White links modern environmental destruction to Christianity's successful displacement of paganism, a cultural shift that "made it possible to exploit nature in a mood of indifference to the feelings of natural objects."[13]

The groundbreaking quality of White's article has not protected it from wide-ranging and sometimes vituperative criticism.[14] The critiques can be grouped into two basic categories: historically grounded examination and exegetical analyses. Members of the first group find fault with the highly schematic character of "Historical Roots," arguing that White's generalizations about East and West, about premodern and modern, are too simple or too vague to function as explanations. The ecologist and microbiologist Rene Dubos, for instance, pointed out that the ancient non-Christian cultures of Egypt, Mesopotamia, and Persia succeeded all too well in disrupting ecosystemic balances; Nebuchadnezzar and Hadrian exploited the cedars of Lebanon without compunction.[15] Likewise, Robin Attfield examined White's claims about the invention of the eight-oxen moldboard plough in northern Europe around the seventh century CE and expressed doubt about the metaphysical significance that White assigned to the innovation, meanwhile reasonably asking how a religion in existence for seven hundred years could possibly be to blame for the emergence of a technology almost a millennium after its foundation.[16] Other critics referencing the historical record point to the many influential Christian theologians and saints, including Basil the Great, Origen, Chrysostom, Augustine, Benedict, and Francis of Assisi, not to mention latter-day supporters of ecological thinking such as John Paul II, who wrote positively about the nonhuman world and offer proof that the Judeo-Christian ethic is wholly compatible with sympathy for nature.[17]

Critics in the second group adopt an exegetical approach and accuse White of distorting or misrepresenting the true meaning of biblical scripture and thus misconstruing the true meaning of Christianity. For instance, Jayapaul Azariah and Darryl Macer argue in "Historical Roots of Our Ecological Crisis in East and West" that the King James version of the Bible, upon which White relies, mistranslates the Hebrew word *radah* from Genesis 1:28 as "dominion over nature" whereas, they contend, "partnership" is a closer equivalent.[18] There has developed among exegetes a "stewardship school," scholars who either interpret Genesis differently or who acknowledge that the book of Genesis, Psalm 8, and the Pauline texts may contain an element of anthropocentrism, but who contend that the correct interpretation of other verses, such as many references to the Land of Israel as holy, or the obligation in Leviticus 25 for animals to rest on the Sabbath, yields a decidedly greener vision of Christianity.[19] In addition, there has emerged recently one additional challenge to White: empirical studies of the environmental opinions of believers who show that belief in Christianity does not appreciably influence one's environmental outlook or behavior, whereas other factors, including level of education, do.[20]

Conspicuously absent from critiques of White's argument, however, is any indication that Christianity, regardless of its potential compatibility with environmentalism, ever helped to encourage active and meaningful pro-environmental behaviors among its adherents, at least in its Western manifestations.[21] Numerous examples exist of Christianity serving to aid environmental destruction, perhaps foremost among them the way that medieval monasteries interpreted Genesis 1:28 to effect a "great age of forest clearance" across northern Europe.[22] However, rarely encountered are scholarly descriptions of Christian groups choosing to enact conservationist or preservationist policies because of their religious beliefs.[23] Such a lacuna matters because, at one level, the degree to which Christianity has led its followers to actual behaviors, rather than merely enabling a hypothetical compatibility of the set of beliefs with environmentalism, is the important consideration. As Roderick

Nash wrote: "White knew the relevant question was not, what does Christianity mean? but what did it mean to particular society at a given time and place? His approach, in other words, was pragmatic: How was the Judeo-Christian tradition used? . . . He agreed that there was a biblical basis for environmentalism. But his point was that for nearly two thousand years the Christian tradition had not been so construed."[24] White despaired that the history of Christianity, unlike the Zen Buddhist tradition, offered no instance, aside from the isolated case of St. Francis, of any believer who chose to moderate resource consumption because of a sense of moral obligation to nonhuman entities. Yet the historical record yields at least one such example: the Pomor of the Russian Far North, freed from the constraining influence of a religious central authority, developed precisely such an environmental ethic, one based upon empathy toward the natural world, and they defended it resolutely for decades until Stalin's plans for economic modernization finally destroyed that ethic as a functioning moral code of conduct.

POMOR SACRED GEOGRAPHY

The foundation for the Pomor protection of wildlife lay in their idiosyncratic perception of the landscape, an outlook that possessed pagan attributes but recognized no non-Christian deities, and their unswervingly passive methods of resource exploitation.[25] Scholars of Pomor culture emphasize the centrality of physical geography in creating the cultural landscapes that shape everyday life—that is, the way that specific places contain within them cultural values and meaning. A familiar illustration of this concept is the role that Broadway, the Brooklyn Bridge, the Statue of Liberty, and the former site of the World Trade Center play in the lived experience of New York City, not only as physical locations but as places with historical and cultural meaning.[26] However, for the Pomor, the key locations that helped produce the mythical understanding of the world are not human constructs such as bridges or buildings, or even conspicuous landforms such as mountains but, rather, natural phenomena in the landscape—the sun, the sea, the sky, and the weather—that can and must be read for signs of God's absence or presence.[27] According to the most accomplished expert on Pomor culture, Nikolai Terebikhin, this patient observation of the natural world for indications of God's favor created a "sacral geography" and a belief among the Pomor that the natural world was a living embodiment of God's will.[28]

The metaphysical and symbolic system that the Pomor created to interpret the natural world, as Terebikhin describes it, is complex and sometimes abstruse, but an easily explicated element of Pomor sacral geography, and one that helps demonstrate the unique blend of animism and Christianity, is the meaning of the winds.[29] The Pomor perceived the world as consisting of two spaces: a central, holy Russian space where the redeeming power of Orthodoxy predominated, and a surrounding peripheral zone populated by pagans. As the first ethnographer to study the Pomor, Peter Savvich Efimenko, noted, the Pomor saw "on the eastern side of their land a warm side, where lived Orthodox Christians, and beyond them Arabs, pygmies and one-legged people—and to the west, non-Christians, Germans, English, and French."[30] The winds were thought to carry spirits from their place of origin in addition to bringing the weather and thus were regarded in very personal terms, with personal names and stories for the winds that blew from each direction.[31] From the northwest, where dwelled the hostile shamans of Lapland, blew the *poberezhnik*, an ill-tempered wind that was said to bring illness and bad luck. The northeasterly *polunochnik*, blowing in from the Arctic, brought fog and ill will while taking boats; the Pomor said that once it had begun to blow, it never stopped. The western wind, wrote the Pomor folklorist Mikhail Prishvin, was the wind of the Antichrist.[32] Conversely, the winds that blew in from Christian lands were favorable omens. From the southeast, the direction of Muscovy, blew the *obednik*—a good wind, thought the Pomor, with a beautiful wife and the good manners to quiet down by dinner time.[33] This geography of the sacred also had a three-dimensional quality; when Pomors returned from the "German" lands to the west or the "islands of death" to the north, they said they were going "up to Russia."[34]

In addition to cardinal directions, Pomor sacral geography also ascribed transcendent meaning to the physical world itself, and as a result, according to Terebikhin, routine journeys to the sea possessed metaphysical as well as economic import. Although the sea, as noted by the famous Russian semiotician Boris Uspenskii, was equated with death in early Slavic culture, for the Pomor the meaning of the ocean was more complex. For the seagoing Pomor, the ocean was a manifestation of indifferent (but not hostile) fate, a field of chaos devoid of God's grace.[35] Their primary tool in their struggle with the anarchic force of death was the boat, which was regarded with a respect bordering on reverence: when a boat foundered at sea, its replacement was given the same name, emphasizing its immortal ability to mediate fate. Pomor folk poetry saw boats as microcosms of the world, sailing through danger on the way to safety, and often likened the pilot of a ship to Jesus, the mast to the Father, and the sails to the souls of believers, filled with the Holy Spirit.[36] If the land symbolized safety and the sea peril for the Pomor, then islands were outposts of death for the Pomor, places outside of the division of the lands and the waters. The Pomor used them only as graveyards, although the Orthodox Church famously chose islands in the White Sea as sites of monasteries. The Solovetskii Monastery is only the most famous of the numerous holy sites designed to overcome the death associated with islands in the culture of northern Russia.

In short, Pomor culture superimposed Christian soteriology onto the landscape around them, thereby enchanting the world. It is very important to note that this sacred geography of the Pomor was not a survival of pre-Christian paganism but, rather, a system of belief that developed following the Pomor's arrival in the Far North, centuries after the Baptism of Kiev and Russia's conversion to Orthodoxy.

Accordingly, the Pomor personification of the landscape, although animist, was not only non-pagan, but anti-pagan. The Pomor's sea was not ruled by a hostile god but, rather, represented a place where Providence did not dominate. Likewise, the winds themselves carried no will of their own but, rather, derived their power from the Christian divine essence (or lack thereof) in the places where they originated. The crucial quality of winds was their connection with the Holy Spirit and thus with the hope of overcoming death—the *polunochnik* and the *poberezhnik* arrived on the shores of the White Sea after blowing across land that God had forsaken, whereas the eastern wind (the *stok*), breezing in from familiar waters of the White Sea that fed the Pomor, was associated with life.[37] In sum, sacral geography mapped the Christian doctrine of everlasting life directly onto an active and alive landscape; one observer of the Pomor has claimed that their outlook placed them in a "dialogue with nature."[38] Accordingly, the natural world represented more for the Pomor than merely a source of sustenance; it provided indications of God's intentions for the fate of each individual, and it therefore deserved deference. Put another way, Pomor sacred geography represented an adaptation of Christianity to the White Sea environment; a unique expression of Christian thinking inspired by a particular landscape.

THE POMOR ENVIRONMENTAL ETHIC

The importance of sacral geography in Pomor cultural life might demonstrate nothing more than a hypothetical compatibility between Christianity and environmentalism were it not for the sustained record of the Pomor rejecting calls to intensify the exploitation of natural resources, defending instead their traditional methods that corresponded to their understanding of the moral economy. Scientists and local business leaders, as Julia Lajus mentions (this volume), urged the Pomor in the last decades of the tsarist era to adopt more modern methods so as to increase the economic productivity of the White Sea. However, the Pomor refused to adopt newer technologies and methods, frequently employing rationales that were implicitly or explicitly environmentalist.[39] The best example of this phenomenon is the resistance offered by Pomor fishermen to calls for more aggressive methods of fishing involving the pursuit of fish stocks. Local economic leaders, worried that "the trend of recent years has shown the Russian fishing industry in a sharp decline"—especially when

FIGURE 11.3. Pomor fishermen long-lining. Ministerstvo Gosudarstvennykh Imushchestv, *Risunki k izsledovaniiu rybnykh i zverinykh promyslov na belom i ledovitom moriakh* (Tip. V. Bezobrazova i ko., 1863), plate A.IV.a.1.

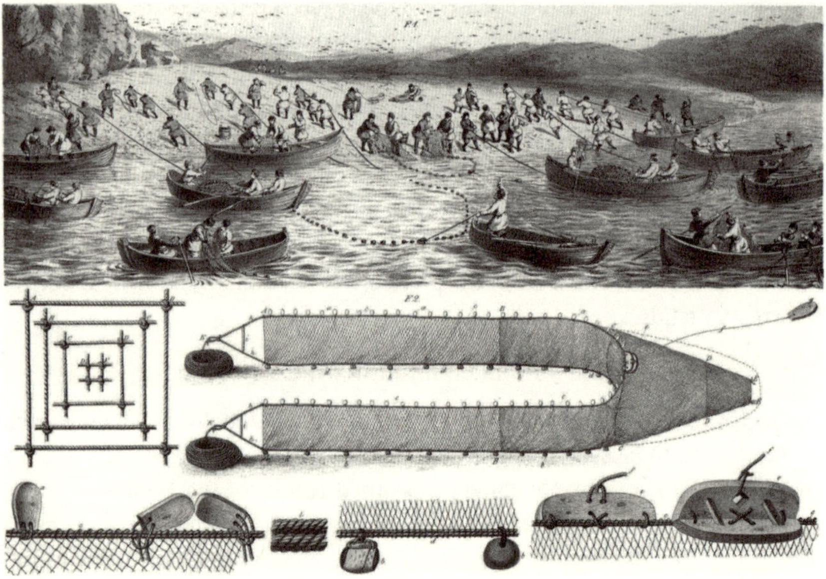

FIGURE 11.4. Pomor fishermen hauling in snares. Ministerstvo Gosudarstvennykh Imushchestv, *Risunki k izsledovaniiu rybnykh i zverinykh promyslov na Belom i Ledovitom moriakh* (Tip. V. Bezobrazova i ko., 1863), plate A.III.a.3.

compared with foreign countries—watched with mounting frustration as the native population adhered to more passive measures instead of maximizing output.[40]

The Pomor preferred fishing methods that brought the animals to their

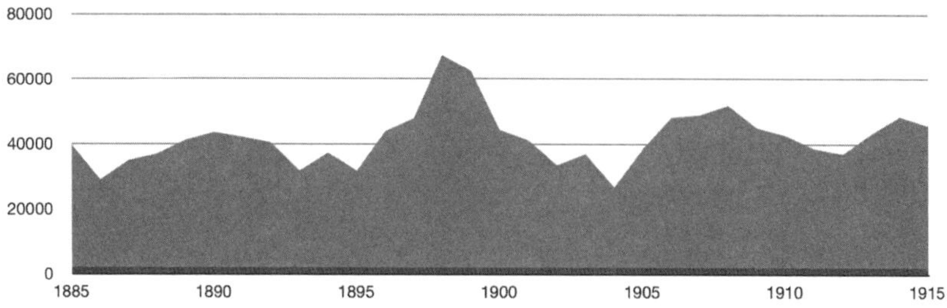

FIGURE 11.5. Annual Catch of Salmon from the White Sea, 1885–1915. Chart based on data presented in V. Aleev, "Gosudarstvennoe rybovodstvo v dele podderzhaniia semuzh'ego promysla," *Severnoe khoziaistvo* 1 (1923): 50. Year is on the x-axis. Annual catch is on the y-axis. All harvests are measured in *puds*, an old Russian unit equivalent to 16.38 kilograms, or 36.11 pounds.

boats rather than chasing the schools of fish; they favored long-lining—the placement in the water of a long rope with dozens or hundreds of hooks attached, which was then reeled in after a certain amount of time—as well as snares and traps (see figures 11.3 and 11.4). The result was a remarkably stable harvest of fish from year to year; according to Soviet accounting, the take of salmon remained in a relatively narrow range between 1875 and 1918, aside from a brief spike in 1898–1899, despite the fact that the Russian domestic population soared during the same period (see figure 11.5). These conservative practices, however, did force the fishermen to wait for fish to arrive and placed them, as Julia Lajus describes (this volume), at a distinct disadvantage to the Norwegian and English fishing trawlers that began to exploit the White Sea in the winter of 1906.[41] Once foreign craft appeared on the White Sea, there emerged a struggle that lasted for decades, between proponents of economic modernization on one side and Pomor resistance on the other.

The conflict came out into the open in the first decade of the twentieth century, when economic and intellectual leaders of Arkhangel'sk began to upbraid the Pomor publicly for sloth and lobbied for the adoption of more efficient fishing techniques. In 1909 the educated and well-connected stratum of Arkhangel'sk, including the vice governor of the province, formed a new public association, the Arkhangel'sk Society for the Study of the Russian North (Arkhangel'skoe obshchestvo izucheniia Russkogo severa [AOIRS]), dedicated to the maximization of the output of the region; the published proceedings of the society's meetings and commissioned reports testify to the wide gulf between the worldview of the Pomor and that of the economic elite. In one of the society's earliest publications a member of the society described the Pomor as "situated in a state of extreme backwardness," and hence as "enemies to any kind of progress whatsoever."[42] Specifically, the journal frequently criticized the Pomor for their lackadaisical fishing methods. One such report referred to the Pomor boats, their nets, and their longlines as "primitive" and "antediluvian" and urged the Pomor to adopt Norwegian methods instead.[43] Yet the Pomor, explained another article, could not be convinced to take the initiative: "In our waters, cod and other fish are so abundant that it might be possible to employ not hundreds of workers, but tens of thousands. Yet nothing is being done. Why is it that the Pomor are not catching cod in great quantities, but instead seem satisfied with catching small amounts of salmon for sale and smaller fish for their own use?"[44] In the second year of its existence, the AOIRS began to convene conferences of scientists, government experts, and industrial leaders so as to foster debate and issue resolutions but also to reach out directly to the Pomor, who might otherwise not hear about the society's activities. At the 1910 conference, held in January, the assembled experts prevailed upon the Pomor to "replace the outmoded tools of their trade with improved ones, such as have already been accepted abroad," urged the use of purse nets and the elimination of the proscription against *kil'not* nets, recommended the replacement of rowboats with motorboats, and offered to arrange favorable loans so as to allow this to happen.[45] Yet the Pomor showed no interest in the suggestions and indeed refused even to respond. The president of the section for fisheries, N. A. Smirnov, noted that in preparation for the conference, survey forms had been circulated among the Pomor in October 1909, but he lamented that not a single reply had been received from Arkhangel'sk and Mezen'sk districts, the most populous areas in the province, and as a result the society could reach no conclusion on this question.[46]

The society continued to hold public meetings and publish its bimonthly journal, and by 1913 the Pomor began to respond with their own forms of organization, still offering resistance but also providing reasons for their position. On December 1, 1913, there was convened a Pomor Fisheries Conference, with fifty attendees, including seven members of the AOIRS.[47] One speaker spoke out in opposition to the planned application of purse nets by claiming that "trawlers, sometimes lowering their purse nets to the very bottom of the sea, bring up not only fish, but mussels, mollusks and plants, and thereby destroy the food supply of the valuable fish. Where the trawlers have been, the fish disappear."[48] The congress passed a resolution calling for the banning of all trawlers within twelve miles of the coast. At other times during the conference, Pomor fishermen claimed that the trawlers working with purse nets tended to take all the fish from the water column, young and old alike, whereas the traditional methods took only the older fish and thus allowed the population to regenerate. These arguments did not persuade the members of the AOIRS to abandon their plans for economic modernization, but neither did the AOIRS succeed in convincing the Pomor to change their preferred methods. Instead, the debate remained at a rather distant, rhetorical level for the remainder of the tsarist era. There were moments when the conflict burst into violence, such as when a group of Pomor fishermen burned down two whale-processing factories on the Kola Peninsula, driven by the belief that the removal of whales from the marine ecosystem depressed the fish harvests, but in general the dispute remained peacefully unresolved.[49]

POMOR FISHING AFTER THE BOLSHEVIK REVOLUTION

The twin revolutions of 1917 and the ensuing civil war brought communism to Russia and hence an end to the pressure from business leaders, but after Soviet control over the Far North was reestablished in the mid-1920s, the communist leadership began to agitate for nearly the exact same changes recommended by the capitalists. However, the Soviets pushed for changes much more actively, and the direct interaction of Soviet fisheries experts with the Pomor made the incompatibility of sacral geography with economic modernization much more clear. Prior to the revolution, the Pomor—when they chose to respond to criticism at all—objected to economic modernization on ecological grounds, but during the Soviet period the influence of the Pomor religious outlook on their view of ecology found more explicit articulation.

The Soviets embarked upon their attempt to modernize the Pomor almost immediately after consolidating power in the Far North, but their first attempt resulted in almost complete failure. In 1924 notices began to appear in economic journals, seemingly verbatim copies of those published by the AOIRS, complaining about the Pomor approach to resource extraction. "The untouched riches of the White Sea are beginning to attract attention," announced the journal *Severnoe khoziaistvo* (Northern management), and the responsible party, it was claimed, was the Pomor, who preferred primitive tools to purse nets and motorboats, and who were "demonstrating no initiative whatsoever in their search for fish."[50] Even when provided with access to steam-powered icebreakers, the typical Pomor fisherman "equipped his boat like his grandfather and rejected those comforts afforded by the ships," since the Pomor, who preferred to wait for the fish to come to them, often argued that motorized boats scared away fish.[51] Scarcely helping matters were state policies (designed to help the poorer strata) that taxed fishermen using unmechanized methods at a lower rate than those using mechanized.[52] The Soviets believed that the solution to the problem was a thorough revision of state policy, and especially the application of collective labor, since the traditional "cabal-based" system favored the well-off who were not interested in expanding output.[53] Beginning in 1925 they sent political commissars from a new trade union agency, Promsoiuz, to the fishing villages to incite the fishermen to join *arteli*, or labor collectives.[54]

Although the Pomor willingly formed into arteli, and artel membership reached 80 percent within two years, this had frustratingly little effect upon the way that the Pomor went about their work. Pomor fishermen had always worked collectively, either in permanent groups based upon the family or in temporary groups based on acquaintanceship.[55] In most cases, the Pomor simply renamed their traditional working groups and carried on as before. One reporter noted that it was a common occurrence for "entire villages, and sometimes even entire districts," to register themselves as arteli, after which they continued to conduct their affairs autonomously and carried on with the political and management structure that had existed previously.[56] Accusations were levied in the press that the new collectives did not sell their goods to the state but to private speculators—"the cooperative works only as a trading place for the rich merchants," complained one report.[57] But most troublingly, fishing technology and fish harvests did not significantly increase after the first attempt at labor reform. In 1926 sailboats still outnumbered motorboats among the Pomor (965 to 30, a ratio unchanged from the years before the collectives), and by 1928 the number of motorboats had shrunk to only six.[58] A frustrated fishing inspector noted that the Pomor "still employed the same primitive measures, completely dependent upon the fortuitous appearance of fish near the shore."[59] The 1927 take for salmon was 614,250 kg—no higher than the year before, and well within the historical range (see figure 11.5).[60] The Soviets expressed repeated frustration with this state of affairs, but fearing that more thoroughgoing state control would be too expensive, they did not force the matter.[61]

The effort to force the Pomor into arteli did change matters, however, by increasing the level of communication between the villages and the government; the new arteli were obligated to meet with Promsoiuz officials occasionally, and the resulting meetings revealed in greater detail the objections of the Pomor based in sacral geography. Numerous government reports testify that the Pomor believed the use of purse nets would disturb the ecological balance of the White Sea, most specifically by adversely affecting the beluga whale—a small, dolphin-like toothed whale that inhabits the Arctic and North Atlantic Oceans. The Pomor told the government officials that they thought of the beluga as a "holy animal," and that it helped them by chasing fish into their nets.[62] Specifically, Pomor fishermen believed the belugas patrolled the mouths of gulfs and corralled the fish toward the Pomor nets; if adopted widely, circular nets would collect fish too efficiently and leave no food for the belugas.[63] This conception of fishing differed completely from the one held by the Soviets: in this arrangement, humans were supposed to wait patiently and take what nature provided, acting only as a part of larger system whose every component had value. Accordingly, the Pomor avoided hunting belugas. While Pomor hunters reported twenty-six thousand seals taken in 1909 and fifty thousand in 1915, they killed only eighteen belugas in 1909 and fifty-five in 1915—and they considered eating beluga meat to be taboo.[64] Ideas about the moral economy are not easily changed.

In 1929, as part of Stalin's nationwide push to place all food production under direct control of the state, the Soviet government dedicated itself to reorganizing Pomor society once more, this time much more comprehensively, so as to change the Pomor value system. Even in 1928, before the process of agricultural collectivization had begun in the rest of Russia, there were hints that the government intended to introduce a new policy; ominous reports began to appear in the Arkhangel'sk press, warning that class enemies had taken control of the arteli. An article in the newspaper *Volna* (The Wave) entitled "Under the Mark of the Artel" alleged that "under the screen of the artel system," the Soviet government had "made prosperous independent households, who kept the best for themselves and gave the worst to the poor."[65] The Pomor were again depicted as uncultured, "alienated individuals," employing "primitive, irrational, and unwise" methods.[66] Another article described their persistence in donating to the church what they called "God's share" of their catch, when the Soviet officials implored them to invest in motorboats instead.[67] Worse yet, also beginning in late 1928, the word *kulak*—Russian for "fist" and connoting in the Soviet context a class enemy—appeared in the popular press and in transcripts of Party meetings with greater frequency. The arteli were said to be riddled with kulaki, since the organizers in 1925 had merely tried to encourage maximum participation; one article claimed that "the cooperatives combine into one pile the poor peasants and the kulaki, since the only requirement for joining is the payment of a membership fee."[68] Similarly, the official government policy that assigned fishing parcels to those able to afford to pay an annual rental fee, it was claimed, barred entry to the indigent while allowing the kulaki to buy up extra parcels.[69] And the fishing methods remained far too passive for the Party's liking, with the Pomor laying out nets and waiting a number of days rather than chasing schools of fish and explaining poor catches by saying "that the fish had disappeared."[70]

There then commenced a more thorough and more effective alteration of Pomor society, focused on creating new state-centered labor units (a process called collectivization) and on removing the stratum of society that observed and enforced the Pomor environmental ethic (a process known as dekulakization). The former was accomplished by nationalizing all fishing equipment, while the latter was accomplished by assigning every owner of a boat kulak status, regardless of other economic indicators, and removing them from the Pomor village. Prior to 1929 the regime did not automatically assign kulak status to boat owners; individuals could own boats capable of carrying relatively large loads (as large as 5,000 or 10,000 *pudy*, or from 80,000 to 160,000 kilograms) and even hire one or two shipmates without arousing the suspicion of the government.[71] However, subsequent to the official dekulakization decree for Arkhangel'sk district that was issued on November 14, 1929, any owners or renters of "sailboats, motorboats, and other large boats" as well as anyone who had hired more than one worker was henceforth made subject to the criminal code.[72] Private ownership of the tools of the catch—including nets, snares, and rigging—was similarly criminalized. However, in practice, the dekulakization decree was interpreted much more loosely. As one memoir recounted, family composition weighed heavily in the application of the rules: "The local authorities from the party cell prepared lists of the poor, middle, and well-off farms. The families without a head-of-household were considered poor. Those families with heads of households and with fishing equipment were classified as middle peasants. Anyone owning a boat with a deck was considered a kulak."[73] A subsequent revision to the order, issued on January 20, 1930, indicated what would become of the dekulakized: they would be deprived of their possessions and resettled elsewhere.[74] The destination for the dispossessed was never specified and never needed specification because nearly all of the targeted families successfully interpreted the warnings and fled before resettlement, much to the consternation of the authorities.[75]

The displacement of the boat-owning class was accompanied by another policy that helped remake the structure of Pomor culture: the shuttering of the churches. Throughout the White Sea region, authorities visiting the villages to enforce the collectivization and dekulakization orders also chased away or shot the local clergy while converting the village churches into clubs and kindergartens, and the church bells into dinner bells. In one instance, the authorities forced the village priest to stand before the congregation, admit that he had intentionally deceived them, and then leave for the city.[76] In another, the authorities promised to pay a generous reward to anyone willing to chop down the crosses from the church steeples and deliver them to the village soviet. No one accepted the job.[77] In all cases, the collectivization effort was aimed at changing not only the economic system of the fishing villages but also the belief system that informed that economic system.

Collectivization, dekulakization, and the closure of the churches combined to dislocate Pomor society almost completely and severely disrupted the economy of the White Sea region—and the fishing industry in particular. With the boat-owning managerial stratum removed, the knowledge about how to organize daily work on the boats (and the value system that informed that knowledge) was lost. At the simplest level, a huge proportion of the labor force disappeared, either because of dekulakization or because people simply chose to leave the village. On the Murmansk coast, 450 fishermen reported for work in

1930 rather than the 2,000 of the year before.[78] At a deeper level, the absented people were precisely those who had previously built new boats and repaired those that had broken, and there then ensued a widespread boat shortage, leading to wasted workdays.[79] Furthermore, the workers suddenly thrust into supervisory positions more often than not followed the example set by the boat-owning class; according to the Party newspaper for Arkhangel'sk from early 1931, "the methods of the catch among [the fishermen] now are passive. The fishermen, not having scientific-based data about the migration of the fish, leave their nets, sit by the shore and wait for the catch. It is needed to end decisively this grandfatherly method, waiting on 'perhaps.'"[80] Fishermen attending a Party conference in November 1930 claimed that any failures were not their fault, since "the methods of our forefathers are still the accepted rule."[81] Again and again, the Party officials for fisheries management met and insisted on the need to find "practical measures for the replacement of passive catching of fish with active."[82] The widespread confusion resulted in years of economic disarray. The regional government called the fishing industry (having filled one-fifth of its quota for 1930 and one-quarter of its annual quota for the summer 1931) a "shameful failure" and threatened to arrest the members of cooperatives who failed to meet their quotas.[83]

Given how dire the situation appeared in 1930 and 1931, it is remarkable how quickly the White Sea fishing industry reached and surpassed its previous levels of production with the new Soviet environmental policies in place. One villager explained that after collectivization, the fishermen, newly arrived from other regions, "plundered everything, and took everything that they could haul away."[84] Motorboats began to appear on the White Sea; whereas in 1928 there had been six and in 1930 ten, by the end of 1930, there were seventy-nine small motorized boats on the water, and more every year, pursuing the fish with purse nets.[85] In 1933, for the first time, the White Sea fishery met and exceeded its quota, and in the years to come did so on an annual basis.[86] Furthermore, the new quotas were ten times larger than the harvests of the prerevolutionary era. By the mid-1930s the old Pomor environmental ethic had been successfully displaced, the "passive" methods of waiting for nature's gifts replaced with mechanization and industrialization.

CONCLUSIONS

The story of the Pomor suggests that Christianity and environmental protection can reinforce one another, and that the historical tension between them described by Lynn White may be because of the characteristics of the centralized form of Western Christianity that gained dominance in Europe and the geography of Christendom rather than the doctrines of Christianity themselves. The Pomor practiced a version of Christianity that, although self-consciously Christian and nonsyncretic in its hostility toward nearby pagan people, nevertheless resembled paganism because of its animistic regard for the natural world. Equipped with this "Christian animism" in the harsh climate of the Arctic, the Pomor were able to develop an alternative environmental ethic that not only comported with their Christian faith but, more precisely, grew directly out of their Christian understanding of the world.

However, it is important to note that the Pomor environmental ethic emerged and successfully evolved only in a place far removed from central authorities.[87] Perhaps, given the absence of similar ethical systems in Christendom, Pomor sacral geography would have been targeted by the church as dangerous heresy and rooted out had the Pomor lived in a less remote location. The determining factor in anti-environmental attitudes may not be the ideological content of Christian ethos itself, which after all is connected to a vast and diverse historical tradition, but rather, the presence or absence of entrenched institutional interests that stand to benefit from increased levels of resource exploitation. The Pomor were able to institute and observe an environmental protection scheme during that time when they lived in a distant corner of the Russian Empire. When the world trade system expanded to embrace the White Sea, the Pomor began to feel pressure to abandon the system, first from local business leaders promoting capitalism and then from Soviet officials promoting communism.

The persistence of Pomor sacral geography calls into question Lynn White's claim that the Judeo-Christian ethic carries within it an antipathy toward the natural world. It highlights the ability of Christianity to accommodate, and even encourage, environmentalist attitudes and practices, if external conditions are favorable. At the same time, the uniqueness of Pomor environmental protection, and its ultimate demise, demonstrate how fragile the combination of Christianity and environmentalism is—fragile, but not impossible.

12 Experts on Unknown Waters

Environmental Risk, Fisheries Science, and Local Knowledge in the Russian North

Julia Lajus

In the fall of 1894 a catastrophe took place in the north of the Russian Empire: twenty-five fishing boats were wrecked during a severe storm on their return to Arkhangel'sk from seasonal fisheries along the Murman coast of the Barents Sea (see map 12.1).[1] The summer cod and halibut fisheries were traditionally a male occupation among the local people, known as Pomors, who ventured out on a seasonal basis from their homes along the coast of the White Sea.[2] Because in some villages almost all of the male population participated in the fisheries, this catastrophic shipwreck left children orphaned and deprived families of their main sources of income. Those fishermen who survived the wreck lost their boats and gear.

High mortality at sea was a normal feature of Pomor life. We might consider it a "routine risk," but in 1894 the scale of the catastrophe was too large for the authorities and society not to notice, so it became understood as an "extreme event."[3] The event discussed here follows the pattern described by Kenneth Hewitt, that "rather being private and scattered, the tragedies are public and concentrated."[4] Thus, the catastrophe was discussed at the Imperial Society for the Assistance to Russian Trade Navigation (Imperatorskoe Obschestvo dlia

Sodeistviia Russkomu Torgovomu Morekhodstvu), which had existed since 1873. This society united shipowners, mariners, industrialists, merchants, and state officials interested in the development of water routes and marine trade. To manage the generous donations collected for victims of the catastrophe, the St. Petersburg branch of the Imperial Society formed the Committee for Aid to the Pomors of the Russian North (Komitet dlia Pomoschi Pomoram Russkogo Severa). The collection of donations was blessed by Tsar Nicholas II, who participated privately and also gave an order to the state treasury to contribute substantially. Grand Prince Alexei Mikhailovich became honorary head of the committee, while Mikhail Fedorovich Metz, chair of the St. Petersburg branch of the Imperial Society, was chosen as acting chair.[5]

The founding of this committee was one manifestation of the consolidation of Russian civil society in the 1890s, which was profoundly influenced by the Great Famine of 1891 and a major typhus epidemic the following year—and this civil society was located socially, as other historians have argued, "between tsar and people."[6] At first the new committee dealt with the payment of pensions and other social assistance to families of the fishermen. It also worked toward organizing a system of insurance for small fishing vessels and partly succeeded in that, although it is evident that most of the Pomors lived according to a traditional, religious understanding of the world in which the notion of risk made little sense.[7] Thus, the committee worked as a risk-management organization, a type of body that was becoming more common in Europe and the United States at that time and rapidly emerging in Russia.[8]

However, to fulfill these everyday practical tasks of dealing with risk was not enough for the committee. It wanted to use the money to substantially improve local communities, not just for ordinary social work. Because fish were at the core of the economic and social life of the Pomor, the leaders of the committee decided to invite natural scientists to discuss how to improve the lives of the Pomor by bringing them knowledge about fish resources. Such a proposal was very much in line with the activities of other organizations that promoted scientific research, such as the zemstva (local government organizations), which supported studies of local resources and the gathering of statistics.[9] Because there were no zemstva in the Russian North, the philanthropic committee served as a substitute organization with at least some of their functions, one of which was the promotion of applied science. And according to their professional interests and belief in natural science as a tool for solving all kinds of problems involving natural resources and environment, natural scientists proposed that a long-term research expedition was necessary to improve the northern fisheries and minimize their risks.

Using the work of the committee as an example, I will discuss attempts to connect the issue of disaster and risk with the development of scientific knowl-

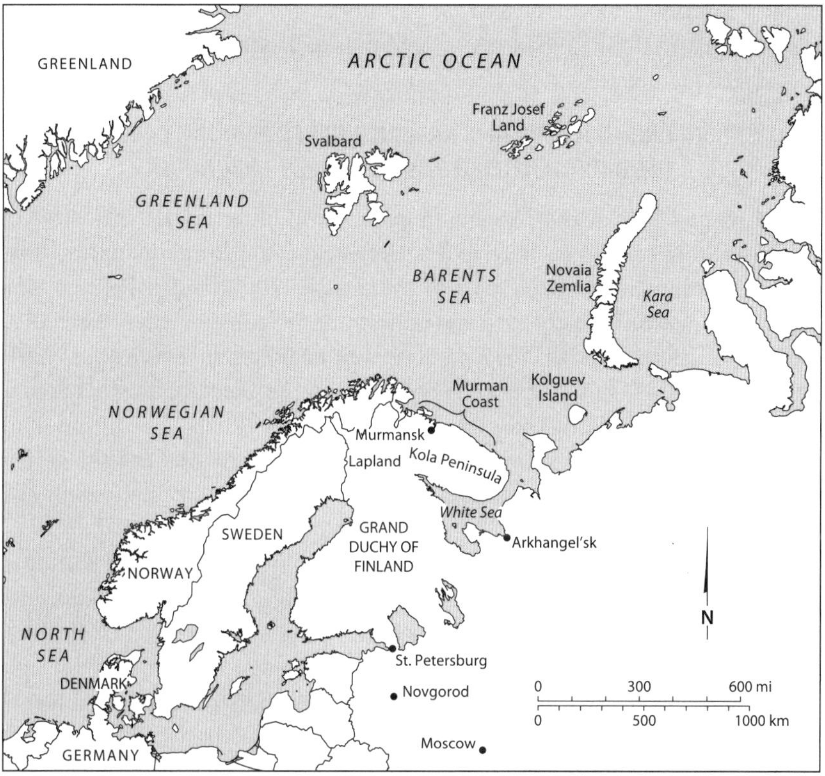

Map 12.1. Barents Sea Region and Murman Coast, Late Tsarist Period.

edge about the environment. I propose that risk and knowledge were closely connected in this case. I follow Ulrich Beck's notion that "the existence of and distribution of risks and hazards are *mediated in principle through argument*" and "qualified expert judgment is still required to determine [risk] 'objectively'" (original emphasis).[10] In this case, however, the experts did not focus on risk itself. Disaster was used as a trigger to develop scientific understandings of the environment and to envision social transformation.

The increasing involvement of natural scientists in the search for solid foundations of resource use and risk management occurred in an environment that they poorly understood. Nevertheless, they declared they could discover the general laws of the region's environment and resources and thereby articulate a reorganization of socioeconomic life based on rational scientific knowledge. Such a declaration and its acceptance by civil society and authorities is characteristic of the onset of what would come to be called the "knowledge society," a society that generates and shares knowledge that can be used to improve the human condition.[11] It is exactly that period of time when "science became a productive force in the 19th century" and thus "ceased to belong exclusively to the superstructure of society."[12]

RISK, CIVIL SOCIETY, AND NATURE'S EXPERTS IN THE CASE OF THE MURMAN SCIENTIFIC-FISHERIES EXPEDITION

In the second half of the nineteenth century, attitudes toward risk held by the government and especially civil society changed. Mortality at sea always was considered an inevitable part of the professional life of fishermen, and although it caused both concern and admiration in outsiders, there were no attempts to mitigate the risk. The idea that traditional Pomor vessels were unsafe had been widely discussed since the time of Peter the Great, well known for his strict policy of forbidding traditional northern shipbuilding while introducing foreign types of vessels, among other modernizing efforts. But in spite of new regulations, the Pomor themselves continued to build traditional vessels and found them suitable for the style of inshore fishing they conducted.[13]

Thus, it was first of all not the local citizens but the emerging civil society in the political centers whose attitude toward risk changed in the late nineteenth century. Civil society no longer was willing to tolerate the level of risk that several decades previously had been considered quite normal. For example, the Pomors' hunting expeditions to Spitsbergen (Svalbard) in the eighteenth century had been quite risky. The risk of death was between 1 and 10 percent, but the Pomors considered this risk relatively reasonable compared with their other activities.[14] However, in the middle of the nineteenth century, after several tragedies, hunting expeditions to Spitsbergen ended, at least in part because they came to be considered too risky.[15]

At the end of the nineteenth century, scientists and others in elite society ranked the improvement of vessels as the most important step to minimize the risks of the northern fisheries. Instead of the Netherlands, to whose technology Peter the Great looked as a model, Norway was then very rapidly developing near the northern coasts of Russia. Its fisheries flourished, and Norwegian fishermen, encouraged by large state subsidies, traded their traditional boats for modern decked vessels that could travel further from the shore. Authorities took their example as a model for Russian northern fishermen as well.[16] However, on the eve of the twentieth century it was no longer possible just to borrow technologies; rather, they had to be understood, rationalized, and approved by science.[17]

In that period, "science" was a magic word in Russia. According to Elizabeth Hachten, it was a period of "emergence of a newly vibrant public culture of science." Natural science (not social) was believed capable of solving most social and economic problems: "The natural sciences, theoretical and applied, contributed both ideologically and practically to the educated public's desire for purposive action to achieve the goals of renewal and reform of society."[18]

The Pomor case at first developed similarly to others wherein authorities and civil society advocated to scientists, and scientists convinced civil society of their power, in connection with challenges posed by environmental catastrophes or epidemics. Scientists were quite happy to propose a plan of research to provide clear reasons for these events and actions to prevent them. Well-known examples of the involvement of scientists in the public campaigns of that time range from confronting drought and famine with scientific studies of soil and vegetation to health campaigns in which bacteriologists and other medical scientists were highly involved. Changes in attitude toward risk appear in examples from bacteriology of the same time. For instance, although rabies in Russia was infrequently contracted and the risk of getting it was many times less than most other infectious diseases, the fatal character of rabies and the fear it aroused put anti-rabies vaccination at the forefront of public health concern in southern Russia. Struggle against this rare but terrible disease united the efforts of scientists and civil society for the benefit of both.[19] Like deaths from disease, deaths at sea might be considered barbaric and uncivilized, caused by the most terrifying forces of nature: wild animals and the depths of the ocean. In putting an end to these deaths, modern science could secure for humans freedom from malevolent aspects of the natural world.

Some general assumptions about science circulated in Russian society at that time: the usefulness of applied science to improve human interactions with nature—in agriculture, forestry, fisheries, and mining—and an understanding of the importance of science for better risk management in all sides of human life. These two approaches coexisted with the notion of science for its own sake and for the improvement of society in general. All these assumptions led to broad support of scientific education for improvement of economic life and better social order.

Improvement of maritime education in the Russian Empire, including that of local coastal dwellers, was a goal of the Imperial Society for the Assistance to Russian Trade Navigation. Accordingly, the meeting of the society in November 1896 was devoted to the lecture "On Fishermen's Schools for Our North," presented by the professional ichthyologist and fisheries inspector Nikolai Arkadievich Varpakhovskii. The lecture provoked a vibrant discussion, largely induced by previous attention to the fate of the northern fishermen in connection with the 1894 catastrophe and activities of the Committee for Aid to the Pomors. Discussion led to the organization of one more institution, the Northern Commission (Severnaia Komissiia), intended to help the committee define the problems and develop a plan for their solution.[20] The main items on the commission's agenda included organizing scientific-fisheries research of the northern seas, fishermen's schools, and fisheries inspection in Arkhangel'sk region.[21] Mikhail Metz led the twenty-five-member commission. Members included three marine officers (including one hydrographer), five engineers, two diplomats working in Norway, and thirteen scientists (eight zoologists, one botanist, one geologist, and three meteorologists). The composition of the commission clearly shows that the Imperial Society considered the problem of improving the education and economic life of the northern fishermen to be primarily a zoological one.[22]

Zoologist Nikolai Knipovich, who already had some experience in surveying northern fisheries, was the most active proponent of the idea of a scientific expedition funded by the committee.[23] Knipovich relied heavily on the emerging ideas and methods of international—and above all, Scandinavian—fishery science and oceanography. He was appointed a head of the large, permanent Murman Scientific-Fisheries Expedition (Murmanskaia Nauchno-Promyslovaia Ekspeditsiia), well funded by the Committee for Aid to the Pomors.[24] The expedition, in turn, became the major Russian institution to participate in international oceanographic research under the umbrella of the International Council for the Exploration of the Sea (ICES), founded in 1902. Thus, Knipovich was trying to steer the program of the expedition toward two rather contradictory objectives: modern oceanographic research, on one hand, and the practical tasks of improving local fisheries that were demanded by the philanthropic committee, on the other.[25] He was in favor of modernizing fisheries and wanted science to go beyond purely practical applications, which he largely left to other institutions and people.

Members of the commission justified undertaking comprehensive research of the northern seas on the basis of two interrelated points: the voices of local fishermen, complaining about the lack of knowledge of fish behavior, and references to successful foreign examples where scientific research helped improve fisheries. The meetings of all these societies, committees, and commissions took place in St. Petersburg. The Pomors lived and fished in the north, one thousand and fifteen hundred kilometers from the capital. Most had never traveled to St. Petersburg at all. However, several people who are named in the protocols as "representatives of Pomors" attended some meetings, and several others sent letters that were published in the protocols. In contrast to the opinions of scientists and committee members, which were published in the first person, the voices of representatives of local fishermen are articulated by others—the same scientists and committee members. For instance, Metz opened his speech at the meeting on February 7, 1897, later published under the title "The Necessity for Scientific-Fisheries Marine Research near the Murman Coast," with a reference to the pro-research opinion of a Pomor representative, S. V. Postnikov. Metz told the audience that Postnikov, while participating in many voyages on the northern seas over almost half a century, had witnessed many expeditions to study coastal areas, but never the sea itself. Postnikov had observed no practical results from these expeditions and doubted that a single need of the fisheries had been satisfied by them. Postnikov stressed the usefulness of scientific research for the local fishermen and called on scientists to study the sea and marine fish; he emphasized that they should hurry to assist the Pomors, who urgently needed enlightenment.[26]

It is not clear from the text whether Metz continued to paraphrase Postnikov's opinion or added his own thoughts when he pointed to the example of Norway:

> Where in order to assist the "dark" working people, enlightened people equipped with knowledge and science appeared . . . they learned . . . where, how and when the fish came . . . and after that the "dark" fishermen needed only to follow their exact instructions, without wasting their time and labor on ineffective efforts. It is completely different here. Nobody knows so far from where and how cod was coming to the Murman coast, and whether it is right that there are no fish nearby when fishermen wait for a long time for fish. It is also not known whether there are any fish near the coasts in winter time.[27]

Details in Metz's speech indicate that Postnikov was not an ordinary fisherman (which is clear also from his ability to attend the meeting) but a rich industrialist who had a larger vessel designed for the high seas that allowed him to search for fish offshore where ordinary fishermen on their small, unsafe boats could not go at all. The results of his successful fishing offshore were used to legitimize a scientific program of studies of fish distribution and ocean currents on a large scale. From this example it can clearly be seen that Postnikov, Metz, and other members of the commission claimed to represent local voices in justifying the development of scientific research. Because the Committee for Aid to the Pomors was not a governmental but a philanthropic organization, funded by both charity and state money, it probably required this kind of justification.

Pomor Barents Sea fisheries were seasonal inshore fisheries, dependent on unpredictable patterns of fish migration. The most valuable fish species in the Barents Sea, like cod and herring, spawned in the west near Norwegian shores and migrated to the Barents Sea for feeding only.[28] Too often fisheries were interrupted not only by bad weather but by inexplicable absences of fish near the shore. Sometimes fish appeared around western Murman but did not go further along the coast to eastern Murman, where most of the seasonal fishing stations were located. Sometimes the pattern was the opposite—the number of fish seemed to drop rapidly in the waters near western Murman but appeared for a longer time and in larger quantities near eastern Murman. Scientists believed that, because of unknown reasons, even when fish were not coming close enough to the shore to be caught from small artisanal boats they nevertheless occurred in large quantities offshore. At the meeting of the society, Metz reported some cases such as these, among them the successful attempts made by Postnikov. The same evidence, also with reference to Postnikov, was provided by Knipovich who, at a meeting in February 1897, introduced into the record an 1895 letter from Postnikov he had found in his papers. It described successful offshore fishing in early spring when fish did not yet come close to the shore. Postnikov wrote that some Pomor followed him and also got good catches. Thereafter, one of the main tasks of the scientific expedition was "to point out to fishermen where they could find fish." Almost everyone agreed the information would be useful to the Pomor. The head of the committee even alleged that local fishermen were so "at one" with the sea that there was no reason to doubt their ability to use the new knowledge that they would be given. But several members of the committee more familiar with local conditions, for instance the Russian consul in Finnmark, V. A. Bereznikov, doubted the ability of fishermen in an economically underdeveloped area to reap immediate benefits from the scientific knowledge. Bereznikov asked how the Pomor would find these places in the absence of good charts of the offshore areas and navigation instruments.[29]

These discussions also show the juxtaposition of egalitarian, objective knowledge to exclusive, traditional knowledge. N. I. Gunin, another "representative of the Pomor" who attended the meetings, emphasized that the Pomor needed scientific research to provide them with equal knowledge about where the fish occur. "Nowadays," he continued, "there are some Pomor who know more than others and use this knowledge for their own profit only."[30]

Committee members assumed that once the Pomor understood how to find fish in the open sea and developed offshore fisheries, they would see the necessity of exchanging their poor vessels for better Norwegian ones. The committee even began to set up a program for assisting them to do that. Again, only Bereznikov was skeptical about the need to advise Russian fishermen to use Norwegian fishing vessels and advocated for the ancient Russian *shniaka*, a boat commonly used by the Pomors. The local boat was better adapted to the coasts of the eastern Murman, where there were not enough sheltered places to put boats at anchor, because shniaka could easily be pulled out of rough waters onto shore. Thus the problems and controversies that might be caused by the proposed rapid changes of long-term traditional practices and technologies were significant. After long discussions during the winter and spring of 1897, the society decided to apply to the minister of agriculture and state domains, the minister of finance, and the navy to convince them of the need for scientific fisheries studies near the Murman coast. Minister of Finance Sergei Witte, "having a steadfast sympathy with the activities of the committee," was willing to finance scientific studies.[31] To be able to fulfill a very ambitious program and to build a specially designed research vessel (the very first not only in Russia, but in Europe), the committee again announced a collection of charity money. Later it also received generous state funding.

Thus, we see that the initial goal of the discussions, to minimize the risks of marine fisheries in the wake of the 1894 catastrophe, was soon almost forgotten. It was replaced by a general aim of economic and technological progress based on comprehensive scientific studies. The practical work of establishing regulations and safety measures for these fisheries was mostly postponed to a later time when, because of the work of natural scientists, knowledge of the marine environment and fish resources would increase.

RESOURCE CONSTRUCTION IN THE ABSENCE OF STAKEHOLDERS

What was the knowledge that the committee and its scientists wanted to obtain? The Barents Sea remained almost unstudied until the end of the nineteenth century. Maps of the coast itself were suitable, but maps of the sea bed were absent and there was little understanding of the system of currents. Currents were important because fisheries were based on stocks of migratory fish: cod, halibut, haddock, and pollock appeared near Russian shores following food—smaller fish such as herring and capelin—and then disappeared. High fluctuations in the time of migrations and quantity of fish led to high fluctuations in catches, making fisheries an inconsistent economic activity, too dependent on unpredictable natural forces. The need to predict and rationalize catch fluctuations everywhere at that time was a driving force behind the development of fisheries science.[32] As in agriculture, this work was presumed essential to rendering fisheries a reliable economic enterprise, in an inevitable shift from intuitive artisanal activity to part of the modern economy based on rational knowledge.

As emphasized above, Russian civil institutions that supported scientific research and the scientists themselves looked to Western countries for models of the application of science to economic life. Thus, it is important to discuss briefly the kinds of science these countries could offer for studying fish resources. Concerns about the fate of fishermen in rapidly changing societies across Europe led to the organization of research on a wide scale. The first method approved in many countries was the questionnaire. For instance, the usual approach in the United Kingdom was to create temporary royal commissions to analyze fishermen's more strident complaints.[33] In Germany the biologist Victor Hensen, hired by the Kiel Commission in the 1870s, also started his studies of fisheries with questionnaires, though he regarded them as complementary to the results from observation stations, since the total catch from an area was a necessary factor in any attempt to evaluate what he called the general metabolism of the sea.[34]

Dissatisfied with information gathered from fisherfolk, many researchers shifted their focus from fishermen and fisheries to the fish themselves. This was a time when the leading US expert on fisheries, Spencer Baird, had "realized that the usual method of interviewing fishermen and other interested and knowledgeable parties would not be sufficient, and had initiated a program of scientific study to answer the many questions that had to be addressed."[35] In Norway, also from the middle of the nineteenth century, the state became highly involved in research into the biological resources of the fishing industry, where previously it had been content with only "regulating the trade and the way in which fishing gear was used."[36] This shift is connected with the name of Georg Ossian Sars and his studies of cod. Some methods developed for studies of humans were extended to fish: Friedrich Heincke in Germany adopted the concept of the average man for studies of fish populations, while Johan Hjort in Norway later transferred the approach of determining generations from the censuses of human populations to those of fish.[37]

The Scandinavian countries led the field in the twentieth-century modernization of the use of marine resources. The development of fisheries oceanography and biology was one of the expressions of the modernist thrust of Scandinavian countries that soon spread rapidly to other areas. Scientists from these countries authored a new scientific program that linked fish to their ocean environment. They disseminated it on an international scale, leading to the foundation of ICES.[38] The goal of ICES—to link fish distribution to the properties of the waters the fish inhabit—at least partly redirected the attention of fishery scientists from studies of fishing as a socioeconomic activity toward what would later be known as biological oceanography.

Knipovich hoped he would be able to discover the general laws of fish distribution according to environmental factors. This aim was in line with contemporary thinking in other studies of nature in Russia at that time, especially geobotany, in which practitioners tried to find a general law of the distribution of plants in connection with soil and climatic conditions.[39] Having graduated from St. Petersburg University with its strong tradition of environmental thinking, represented by soil scientist Vasilii Dokuchaev, botanist Andrei Beketov, and others, Knipovich was engrossed in searching for clear links between the distribution of marine fauna, including fish, and environmental factors.[40]

Knipovich was well aware of the socioeconomic problems of the region: an outmoded system of forming fishing teams wherein fishermen, lacking credit to invest in their own endeavors, relied on the owners of boats and gear; primitive travel and communication between their residences on the White

Sea coast and the Murman coast; poor infrastructure and trade; and so on. The lack of fishermen was also acknowledged by the committee, one of the tasks of which was also to facilitate the colonization of the coast, a program started in the 1860s and heavily supported by the state.[41] In 1893 and 1895, Knipovich led two surveys of the northern fisheries organized by the Ministry of Agriculture and State Domains. In his reports, he supported the development of infrastructure in the region and, first of all, the construction of the railroad to Murman. Witte approved the plan in 1894 but it was not fulfilled until the First World War.

But Knipovich was a zoologist, not a state bureaucrat or a rich industrialist, so he could not help much with solving the large infrastructural and social problems of the region. What could he promise when he was lucky enough to be appointed as head of a long-term, well-funded expedition—the first one of that scale in Russia? Very logically, he promised the committee that funded his expedition that he would find plenty of fish offshore, which should drive more fishermen to the Murman coast. Enchanted, the committee did not seriously entertain doubts about the practical implementations of the proposed scientific studies, nor did it support Bereznikov's proposal to separate long-term biological studies from more urgent and more practically oriented tasks. The decision was, instead, to support the proposal for a large-scale expedition based on the program presented by Knipovich.

Thus, it is possible to argue that Knipovich and other experts who worked in the expedition used scientific knowledge, methodology, and instruments (research vessels, hydrographical instruments, fishing gear, and so on) in order to construct a new resource—fish that would be suitable for the development of offshore fisheries. Theoretically, fish in the ocean are not considered a resource when they are not caught and their distribution is not more or less understood and predicted. To become a resource, the natural objects must be "enrolled in networks that code those objects with the status of resources, and embed them within different visions of future prosperity."[42]

It is evident that fish in unknown waters with unknown migration patterns could not be a manageable resource.[43] Knipovich hoped, however, that this resource would be larger and more stable and, although not easily accessible in his times, would draw attention and stimulate the technological development of Russian fisheries. Thus he clearly worked for the future. In his program he argued that the studies he proposed needed a long period of time. The well-known polar geologist Feodosii Chernyshev, a member of the Northern Commission, compared Knipovich's scientific program with the program of surveying for mineral resources when the state was investing money in the hope that mining would someday generate a much larger profit.[44] So the scientists did not see much difference between fisheries and mines, concentrating on the resources themselves and not on the local resource users, the Pomor, whom they were called on to assist.

The word "stakeholder" did not exist at that time, and scientists did not think too much about who would use the knowledge they created, or how. The problem of how to proceed "from knowledge to action"—much discussed during the past several decades in application to environmental knowledge and actions to protect natural resources and local people—was not even understood a century ago. Knowledge was considered to be a power in itself. Additionally, the attitude toward the future was very different. Whereas we now see that knowledge and action are both necessary for change, on the eve of the twentieth century scientists believed that their main task was to gain new knowledge on behalf of later generations who would be better prepared to use it. When the future was seen in the light of progress, and even more when it was seen in the projections of future social revolution that would completely change the existing social order, the problem of the practical application of knowledge moved into second place.

Knipovich's political views were radical: as a student at St. Petersburg University, he was a member of one of the first illegal social-democrat groups in Russia, founded by the Bulgarian Marxist Dimitar Blagoev. Knipovich was arrested twice: in 1887 and then in 1896, just before he began to work for the committee. He was not allowed to teach at the university and in 1887–1893 was also not allowed to have any job at governmental or civil institutions. However, he did not become a professional revolutionary like his sister, Lidiya Knipovich, and in 1892 he defended his master's dissertation. In 1893, when the restrictions on work in state institutions ended, he was hired by the Ministry of Agriculture and State Domains to conduct a survey of northern fisheries.[45] His assistant Leonid Breitfuss was a "politically unreliable person" as well. This did not prevent the committee from giving them responsibility for the generously funded expedition. It seems that it was not possible to find established and politically moderate people to go and work on the remote and severe Murman coast.[46] Liberal members of the committee, the official head of which had been the Prince Alexander Mikhailovich, and the radical Knipovich were all very critical of the contemporary economic and cultural lives of the northern fishermen and saw their traditional pattern of resource use as terribly backward. Thus they were not interested in improving the fishermen's lives but, rather, wanted to change them completely. They wanted to provide the foundations for a better organized and more reasonable life based on scientific knowledge instead of on traditional practices.

Thus, a rather peculiar situation occurred whereby civil society (in this case the Committee for Aid to the Pomors) supported activities to advance "pure" science, not to help local communities. This direction was taken very much on the basis of the successes of scientists involved in solving fisheries problems in foreign countries such as Norway and Denmark. Scientists emphasized that Russia was behind other countries in the level of marine scientific knowledge, and they paid considerable attention to the necessity of reducing this difference. As was so common in Russia during this period, the practical applications of research were not clearly formulated and the results were considered as self-evidently useful.[47]

Local authorities could not mediate between scientists and fishing communities: over such vast territory there were simply not enough officials. Private capital was not interested in investing in fisheries in this remote northern region, because there were already enough profitable fisheries in the south, where the whole infrastructure was far better developed. It is interesting that the committee and the scientists at that time did not expect private capital to try to use the fish resources in the north. They did not see the future of the region in the development of capitalism, as some scientists started to think later. They either supported a communal way of life by the Pomor that, from their point of view, could be improved by science and technology, or they dreamed of future revolution that would radically transform the region.

However, the process of colonization of the Russian North offers us the best window on the construction in uncharted waters of new fishing "resources" led by scientists. As everywhere at the periphery of the Russian Empire, colonization went in two directions—external and internal. Russia was at the same time a colonial empire and a colonized territory.[48] On the one hand, the Russian North was a part of an old heartland realm that began to be colonized under Novgorodian rule, with Russian fishermen establishing their seasonal fisheries on the Murman coast by the mid-sixteenth century. On the other hand, it remained a newly colonized territory, both the sparsely inhabited coastline but especially the sea itself. Many debated the need to establish a border of territorial waters and organize effective defense against foreign vessels, since the area had proved highly vulnerable during the Crimean War. An attempt to "colonize the open sea" by the expedition for the sake of local fishing communities was closely connected to activities that were paving the way for the intensification of internal colonization of the area. Authorities hoped that obtaining a new resource would bring new settlers to the coast and thus facilitate the governmental colonization program. The same colonization policy idea was implemented slightly later in the Russian Far East, as Mark Sokolsky describes (this volume).

Tsarist leaders expected either that new open-sea fishermen would come from other parts of Russia or that the Pomor themselves would miraculously and rapidly transform their mode of life and technology. However, when the results of experimental fishing conducted by the expedition were obtained and then published, none of their predictions came true. The expedition's exploratory offshore fishing did not bring any immediate profit for Russian fishermen, who had no suitable vessels. Instead, the publication of the data in fact led to considerable earnings for foreign trawlers. The Murman expedition was then accused of facilitating the arrival of British trawlers at the shores of the Barents Sea. Thus, an anonymous author wrote in 1907 about the enormous catches of foreign trawlers that exploited fishing grounds founded by the expedition, and "thus about one million rubles that were spent on the Committee brought benefit not to Russian Pomor but to German and British businessmen."[49] As early as 1900 the expedition was accused of "forgetting the needs of local fishermen in favor of foreign science."[50] During the following five years, after Knipovich lost his position as head of the expedition because of conflicts with committee members who wanted to shift the expedition's activities to more practical tasks, Breitfuss managed amazingly well to maintain a high level of scientific research despite being forced to fulfill numerous practical tasks, such as building baths for seasonal fishermen and maintaining a school for their orphans.[51]

In 1907–1908, in the new circumstances that followed the 1905 Revolution, several attempts were made to revive the activity of the expedition. There were efforts to remove the expedition from the jurisdiction of the committee—which, in the words of Breitfuss, "was a closed institution that did not experience any control, either from the Society for Navigation, or the Ministry which subsidized it, and functioned as a dilettante without a clear plan"—and hand over supervision to local authorities and local civil society, then developing rapidly.[52] For instance, when in 1908 the local Arkhangel'sk Society for the Study of the Russian North was founded, this immediately provoked a discussion of the fate of the expedition. However, the local authorities were not able to allocate the large amount of money needed for continuation of the research. Thus an application to the central authorities was inevitable. The proposal for a new stage of research was submitted to the State Duma but failed because the authorities did not see any reliable institution that could lead this work.

CONCLUSIONS

After several years it became obvious that in spite of the expedition's success in obtaining high-quality scientific knowledge on oceanographic patterns in the Barents Sea, as well as fish distribution discoveries and finding new grounds for offshore fisheries, local fishing communities were proving unable to put this new knowledge into practice. The expedition attempted to institute new types of fishing vessels, mostly borrowed from the Norwegians, but they were too expensive for local Russian fishermen. Knipovich complained that the expedition was blamed by the committee for fishing too few fish.[53] But for him, the primary goal of the expedition was not actual fishing or testing and improving fishing technology but, rather, developing marine science. Thus, to free the expedition from these tasks, he proposed giving the small fishing vessels bought for the expedition to fishermen. By 1900 the expedition had already sold most of its small vessels, which much reduced its capacity to develop other practical fisheries research. He wanted to encourage the fishermen themselves to experiment with these boats and fishing gear. "Otherwise it would be too hard to struggle with the laziness of fishermen, although these features of their character harm them so much," he explained.[54]

Basic knowledge about fishing grounds gathered by the Murman expedition needed to be transformed into practical recommendations. This task was taken up by several private industrialists who bought English trawlers and, in addition to their commercial fishing, systematically surveyed the waters. As in Norway, most local citizens in the Russian North opposed the idea of the development of trawling fisheries. They declared the trawl a "rapacious foreign device" that threatened traditional artisanal fisheries. During many years of research on the history of Pomor fisheries, I have not seen in the documents any clear sign of environmental consciousness among the Pomors in our modern understanding of the term that would explain their opposition to industrial fishing. As a result, I have seen no evidence to support Stephen Brain's arguments (this volume) that Orthodox religious beliefs of the Pomor led to any conscious protection of resources. Rather than the religion itself, it was the scant population density, traditional social life, and poor technology that were responsible for relatively small human pressure on marine environment.[55]

The invasion of English trawlers, along with Norwegian fishing vessels and marine hunting expeditions, in a region where no formal borders of territorial waters had yet been established, redirected the attention of central and local authorities from the modernization of fisheries to guarding the region from foreigners—a concern about foreign resource exploitation that Mark Sokolsky and Ryan Jones underscore in their chapters on the Pacific Ocean (this volume). It also became evident that knowledge itself could not help without significant changes in technology. In 1908 Breitfuss correctly diagnosed the problem of Russian northern fisheries (which might be generalized to Russian fisheries at that time as a whole): "by contrast with western European countries, which are strong in technology and economic power and which rush for fish to our North, to the shores of Iceland and even to the coasts of northern and western Africa, because their fishermen do not know 'where' to get fish, we, living so close to the best fishing place, do not know 'how' to get fish."[56] Two decades later, however, when industrialization became a definite goal of the Soviet government, the scientific legacy of Knipovich's program led to the foundation of Soviet industrial fishing, which over the subsequent twenty years almost eliminated local users of marine resources in the Russian North. This process proves Stephen Bocking's contention that "Conventional resource science can also exclude other forms of knowledge, including that of resource users."[57] Thus, the resources the expedition "constructed" in the absence of stakeholders became useful in a future when the Soviet state became the ultimate stakeholder and user of these resources.

The scientific enterprise brought into being by concerns of civil society about risks for the local users of fish resources provided knowledge that was used for the development of industrial fisheries, which eliminated traditional technologies and local users of marine resources with their risky marine ways of life. This chain of causation illustrates the often realized politics of risk reduction or even elimination through a full transformation of traditional lifeways. The notion of risk is often used and even abused by experts, industrialists, and authorities to justify the radical shift from traditional to modern economies. Risk mitigation serves as a trigger for large transformations that, with the success of modernization, could then be more or less forgotten.

13 Fishing, Settlement, and Conservation in the Russian Far East, 1860–1940

Mark Sokolsky

The Russian Far Eastern province of Primor'e, also known as the Maritime Territory, is, as its name suggests, defined by water. Bordering the Sea of Japan, its rugged coast stretches from the Tatar Strait, opposite Sakhalin Island, to Russia's border with North Korea at the Tiumen' River. The Ussuri River, a tributary of the Amur, forms most of its western edge, separating the province from Manchuria, while broad, shallow Lake Khanka, the largest freshwater lake in northeast Asia, lies astride the Russo-Chinese border. Since 1938 the line between Primorskii *krai* (territory) and its neighbor, Khabarovskii krai, has passed through the Sikhote-Alin Mountains, but a bit further north the mighty Amur forms another riverine boundary, running north of the mountains on its way to the sea (see map 13.1).

Despite the maritime character of Primor'e and its rich aquatic environments, the development of a Russian fishing industry in the territory emerged decades after Russian colonization of the territory began in the late 1850s. Indigenous peoples had long harvested the bounty of its coast and rivers. Chinese and Koreans historically traveled to the territory to fish and gather seaweed, sea cucumbers, and other marine species, and they continued to do so

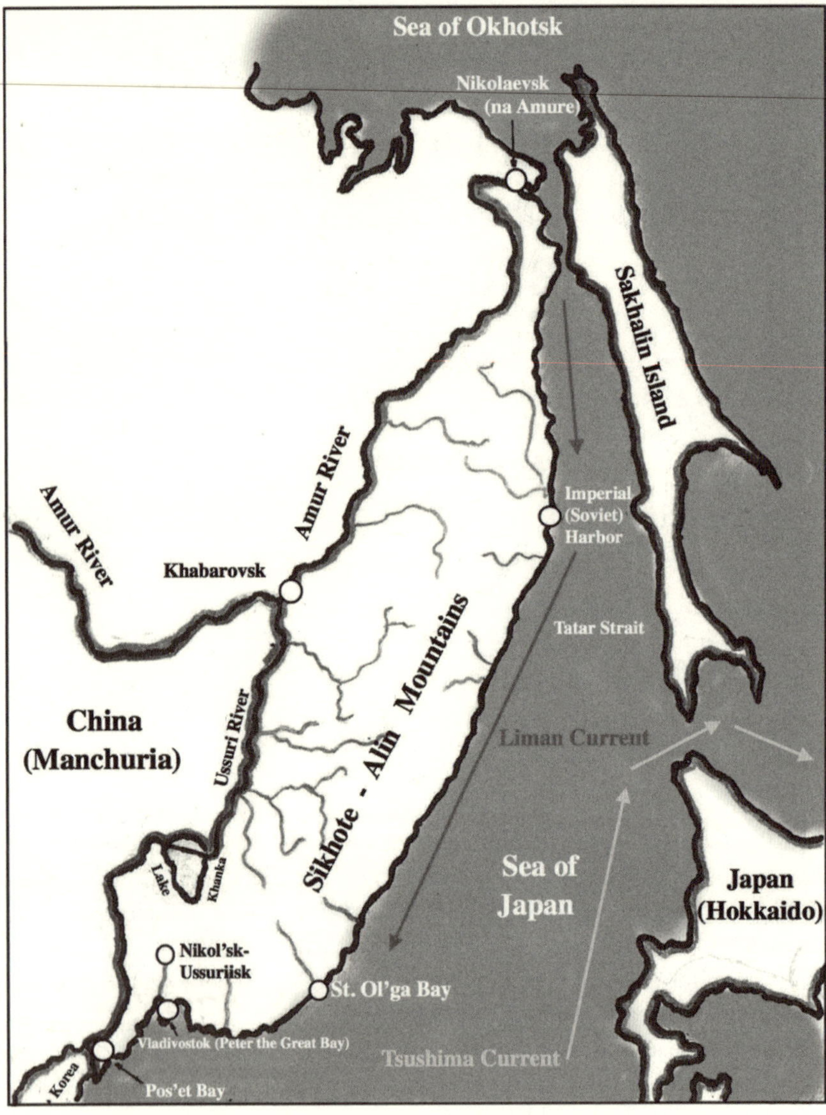

Map 13.1. Primor'e and Neighboring Seas. Author's rendition, adapted from *Atlas SSSR*, 2nd ed. (Glavnoe upravlenie geodezii i kartografii, 1969), 56.

after Russia acquired the territory in 1860. The Russians, Ukrainians, and other tsarist subjects who migrated to the territory in Primor'e, in contrast, were much more terrestrially oriented.

Beginning in the late nineteenth century, however, and especially in the early Soviet period Primor'e became a major center of the Russian/Soviet Pacific fishing industry. The manner in which this industry emerged speaks to a number of questions in Russian environmental history. First, it underlines the importance of international competition in motivating changes in Russians' use and management of their environments. As in the case of whaling, as Ryan Jones shows, the Russian entrée into fishing around Primor'e owed much to resentment and fear of foreigners operating in home waters.[1] Both tsarist and Soviet authorities believed that the exploitation of aquatic life by foreign (primarily Chinese and Japanese) fisher-gatherers represented a strategic threat, a loss of income, and an affront to national pride. In response, tsarist leaders attempted to attract able fishermen to the shores of Primor'e in order to assert the empire's territorial claims and to displace the foreigners from Russian waters, a policy that continued with few changes during the Soviet era. Unlike fishing disputes elsewhere in the world, where competing commercial interests drew states into conflict (as in the Grand Banks, for instance), contestation over Primor'e marine resources was the result, not the cause, of conflicting imperial ambitions.

Second, the history of fishing in Primor'e also shows that, early on, there was sensitivity to ecological questions among Russian and Soviet leaders. In both pre- and postrevolutionary periods, officials were anxious about the impact of human activity on fish (primarily salmon) and other marine organisms. At the same time, as in the case of whaling, concern with the state of aquatic life tended to be inflected with nationalist *ressentiment*, focusing on the actions of foreign fisher-gatherers. Conservationism emerged in tandem with attempts to take firmer control of the Primor'e coast and drew strength from the same national-imperial sentiments.

Third, Primor'e's experience also challenges the idea that there was a particularly Soviet—or for that matter, Russian—approach to the exploitation and management of natural resources. Many scholars have commented upon the Soviets' faith in the ability of science and technology to furnish human prosperity without causing the ecological problems characteristic (in the Marxist-Leninist view) of capitalism, and certainly there were elements of Soviet ideology and practice that reflected such naïveté.[2] However, discussions and policies concerning the Primor'e fisheries suggest that such a view predated the Soviet period and was not confined to Russia's corner of the Pacific. This trust in science emerged from the prerevolutionary belief that technologically advanced development could facilitate the growth of the Russian/Soviet fishing industry (which would help oust foreigners) enabling the conservation of fish stocks. When directed resettlement failed to displace Chinese and Japanese fishermen from Russian waters, Primor'e officials began to advocate capital-intensive methods of fish extraction, arguing that large ships and new technologies would enable Russian fishermen to better compete with their Japanese counterparts and, at the same time, would be more conducive to the conservation of fish stocks. After the civil war, Soviet authorities followed the same line, believing that industrial fishing and conservation were perfectly compatible. In this regard, the Russian and Soviet approach to fisheries resonates with the Chinese experience. As Micah Muscolino has shown, Chinese authorities viewed fishing as "a promising industry that needed to be managed by proper scientific research and the improvement of technology," a viewpoint that developed in response to imperialist threats—particularly from the Japanese.[3] Although Russia was itself an imperial power in the North Pacific, Primor'e administrators also tended to view fisheries through a nationalist lens and to see "modern" fishing methods as a solution to underdevelopment and ecological decline.

Not surprisingly, in practice, the expansion of Russian/Soviet fisheries served strategic and economic goals more effectively than it did conservation-minded aspirations. Primor'e's fish and other marine organisms were not inert objects to be annexed, and although the territory's fisheries proved fairly resilient in the long term, they suffered a mid-century decline in the face of combined Soviet and Japanese fishing. Ultimately, conservation of fish stocks and coastal ecology proved more complex and difficult than the nationalization of maritime spaces.

PLACE, PEOPLE, AND ECOSYSTEMS

The area known as Primor'e has undergone many permutations since 1860, but for the purposes of this paper, it will refer to the area east of the Ussuri River and south of the Amur, including the coastal region running from the Korean border to the Tatar Strait, a distance of roughly two thousand kilometers. From the Korean border to the area around the Murav'ev-Amurskii Peninsula on which Vladivostok is located, the coast is marked by broad bays, the outlets of several rivers, and many offshore islands. As one moves southeast and then northeast away from Vladivostok, the coastline becomes more uniform but also more rugged, rising sharply into the Sikhote-Alin Mountains. In the far north, in what is now Khabarovskii krai, the Amur River meets the Tatar Strait, having swung northward after colliding with the Sikhote-Alin (see Map 13.1).

The Primor'e coast, nourished by ocean currents and nutrients provided by the region's rivers, provides habitat for a great variety of aquatic life. Cold water from the Sea of Okhotsk flows southward along the shoreline after passing through the Tatar Strait. To the south, it combines with the waters of the Tsushima current bringing warmer water northward, creating a counterclockwise circulation in the Sea of Japan.[4] Many of the region's waterways flow directly into the sea from the mountains. Others (including the Ussuri) form part of the Amur River system, which drains a watershed stretching into Mongolia and eastern Siberia.

Where it meets the sea, the Amur creates an enormous estuary marked by shifting sandbars and notoriously fickle conditions. Among the principal fish species in the rivers and coastal waters are chum and humpback salmon (*Oncorhynchus keta* and *Oncorhynchus gorbuscha*), two types of sturgeon (including the enormous kaluga, *Huso dauricus*), cod, herring, and flounder. The estuary of the Amur River is very productive, particularly for salmon, and in the nineteenth century for sturgeon as well. Primor'e's rocky coast is very rich in seaweed, crustaceans, and mollusks. The region has also long been renowned for its sea cucumbers (also known as trepang). Indeed, the original Chinese name for the area that became Vladivostok was Haishenwei, or "Sea Cucumber Cliffs."[5]

Primor'e's rich and diverse aquatic environments played an important role in the economic life of the region's indigenous groups: the Nivkhi along the coast and around the mouth of the Amur, the Ul'chi on the lower Amur, and the Nanai, Orochi, and Udeghe along the Amur and Ussuri Rivers and in the interior. To varying degrees, all of these peoples relied on fish, particularly salmon, for nourishment. The Nivkhi, Ul'chi, and Nanai were especially dependent on the productivity of coastal and inland waterways. One observer estimated that the Nanai and Udeghe may have consumed one and a half tons of fish per person per year.[6] A recent estimate suggests that, together, the native groups of the Amur basin caught roughly eleven thousand to thirteen thousand tons of salmon annually, about 4 percent of the total run.[7] Fish provided more than food; the peoples of the Amur and Ussuri Rivers were known to European missionaries as Fish-Skin Tatars, because some made waterproof shoes and clothing out of fish skins.[8]

When Russian and European visitors began arriving in the 1850s, they found an abundance of aquatic life and intensive indigenous fishing practices. The explorer Nikolai Przheval'skii wrote that the Sungacha River, a tributary of the Ussuri, "in the literal sense teems with fish. The abundance of [fish] is so great, that they are often killed by the wheels of steamboats. Moreover, fish jumping out of the water often drop into boats and sometimes even onto the decks of steamships." Przheval'skii wrote that he himself witnessed, on one occasion, a large fish land on a steamship's deck beneath a table where people were having tea.[9] Indigenous fishermen also used detachable harpoons attached to floats to catch larger prey, along with nets up to 630 feet wide that could net thousands of fish at a time.[10]

When Russians began arriving in the 1860s, Chinese and Korean fisher-gatherers were also making use of Primor'e's aquatic riches, traveling to the region on a seasonal basis in search of seaweed and sea cucumbers, as well as salmon and other fishes. Some Chinese brought fish from Lake Khanka and seaweed to Beijing for the emperor.[11] A Cossack commander reconnoitering southern Primor'e in July 1869 found that a small local population of Chinese and Udeghe-Chinese métis swelled to one thousand during the seaweed harvest, as migrants arrived from China. He estimated that they removed an estimated 8,000 tons of (dried) seaweed annually, along with herring and salmon.[12] Similarly, Przheval'skii wrote that each spring between three and four thousand migrant workers descended on the shores of Primor'e for fishing and gathering, where they harvested some 6,650 tons of seaweed.[13] In addition, Japanese fishermen began to operate in the vicinity of Primor'e (as well as off Sakhalin and in the Sea of Okhotsk) on a large scale beginning in the late 1870s.[14]

RUSSIAN COASTAL SETTLEMENT

In comparison to the Chinese and Korean fisher-gatherers (let alone the indigenous groups), the Russian presence along Primor'e's coast and waterways and Russians' involvement in exploiting the region's various aquatic environments appeared quite suddenly and as a result of deliberate policy decisions. Primor'e and the neighboring Amur Province were relative latecomers to the Russian Empire. As the Qing Empire of China faced internal turmoil and European aggression in the mid-nineteenth century, Russian diplomats were able to acquire large tracts of territory in what had formerly been part of greater Manchuria. Through the treaties of Aigun (1858) and Beijing (1860), the Qing ceded the left (north) bank of the Amur River, as well as the region south of the Amur and east of the Ussuri River. The governor-general of eastern Siberia, Nikolai Murav'ev, divided the new territories into the Amur oblast (province), situated along the middle and upper Amur, and the Primorskaia (Maritime) oblast, which encompassed the coastal regions (see map 13.2).

To secure the newly won territory, tsarist officials directed Russian subjects to Primor'e and the Amur oblast. They hoped that settlers would "make the land Russian," by their physical presence, cultural influence, and their exploitation of natural riches. As Murav'ev explained to the tsar's brother, Grand Duke Constantine, in the autumn of 1858, "the reinforcement of our maritime positions in the Priamur [the Amur Valley and Primor'e] urgently demands the rapid settlement of the country."[15] On a more ideological level, explorers and publicists enthused that "even here we will make Russia [*my i zdes' Rossiiu sdelaem*]," carving a homeland and livelihood from an unfamiliar landscape.[16]

Coastal settlement proved particularly tricky, and from the beginning it reflected a desire to bring the fruits of nature into the imperial fold—and out of foreign hands. Tsarist officials believed that use of natural resources and territorial claims were closely linked, either in a symbolic sense or for more direct,

Map 13.2. The Russian Far East (including Primor'e) in 1860. From Kimitaka Matsuzato, "The Creation of the Priamur Governor-Generalship in 1884 and the Reconfiguration of Asiatic Russia," *Russian Review* 71, no. 3 (July 2012): 367. Reprinted with the author's permission.

practical reasons (demographic, economic, and/or strategic). The first military governor of Primor'e, Rear Admiral P. V. Kazakevich, called in 1866 for the "rapid formation of maritime settlement" through the directed migration from central Russia. Kazakevich argued that Russians in Primor'e "derive[d] almost no benefit from the riches with which nature has endowed the local rivers and seacoast." Without people, Russia would have only a province "for Chinese and Koreans."[17] In the late 1860s Kazakevich attracted a small group of fishermen from the Caspian Sea to Nikolaevsk, believing that they could transfer their sturgeon-fishing skills to the lower Amur. When this effort fizzled out (for lack of markets, primarily), Kazakevich's successor, I. V. Furugel'm, also sought out settlers accustomed to marine environments. He attracted a small group of Finns, including the sea captain Fridol'f Gek, to southern Primor'e. The Finns attempted unsuccessfully to break into the coastal cargo trade, but they stayed in the territory, forming settlements on the Ussuri Gulf near Vladivostok, and Gek went on to command the first coastal patrol ship in the Sea of Japan, the *Storozh*.[18]

Beginning in the 1890s officials undertook more concerted efforts to move fishermen to the coast. In 1891, at the request of Governor-General A. N. Korf, the South-Ussuri Resettlement Administration (founded in 1883) initiated a program designed to bring capable seafarers from European Russia. The chief of the Resettlement Administration at the time, Fedor Busse, was aware that Russians had little presence on the Primor'e shores and remained a minority in maritime trades. Busse warned that the Chinese were "masters of a Russian sea," harmful to the economic development of Russian settlements and also to the "prestige of the Russian government among the indigenous population." He argued that they exhibited the "well known malevolence of the yellow race toward Europeans" and a "proclivity for treachery." He argued for the resettlement of pomory (seafarers) on Primor'e's southern coast in order to displace the Chinese.[19]

To do so, Busse and his successors embarked on an ambitious and expensive program designed to resettle able fishermen from elsewhere in the empire to Primor'e. Beginning in 1890, the Resettlement Administration began to offer fisher-settlers loans of up to 850 rubles (roughly 400 dollars) per family, along with free logging rights in coastal forests and other benefits, to settle along the coast and take up fishing.[20] Busse found a willing partner in the governor of Astrakhan province, who arranged for the transportation of fifteen families to Vladivostok in 1891. The Resettlement Administration also provided these first families with an additional 3,000 rubles' worth of fishing equipment. Much to his disappointment, Busse and his assistants soon discovered that the "seafarers" were simply poor dock workers with neither interest nor ability in fishing.[21]

Undeterred, Busse turned to settlers from the Baltic who were, in his view, "excellent sailors" able to "secure for Russia the adjacent sea, which the yellow race now dominates."[22] This idea soon gained the support of Alexander Krivoshein, at the time head of the land section of the Ministry of Internal Affairs. As for the fact that they were not ethnic Russians, Krivoshein wrote, "the Ministry believes that when placed face-to-face with Chinese and Japanese, they will naturally join with the Russian population and, with time, will merge with it completely [*vpolne s nim sol'iutsia*]."[23] So, in 1899 the Resettlement Administration offered willing parties from the Baltic regions of Estland, Livland, and Courland up to 1,000 rubles toward settlement along the Sea of Japan, much more than the 135 rubles available to agricultural migrants, along with land grants and other incentives. Within the year, some seventy-five Baltic fishermen arrived to settle on the coast, and their numbers swelled during the following years.[24]

Coastal resettlement continued into the early 1900s, but foreign fishermen still predominated along the coast of Primor'e. Writing in 1903 one official lamented that "some kind of evil fate hangs over colonization of this coast.

Hitherto it has occurred without any success."[25] Busse's first settlers were more interested in farming than fishing, while the Estonians and Latvians who settled in the Far East were in fact not "natural sailors." Agriculture, not fishing, had been their main occupation in the Baltic region, and it was land hunger and state benefits—not the prospect of rich fisheries—that drew them to the Far East.[26] By 1917 about one thousand maritime settlers of various origins were living along Primor'e's shores, and some engaged in fishing and the gathering of seaweed and sea cucumbers, but they were still the minority in these trades.[27] Nevertheless, the effort to replace East Asian fishermen and gatherers with "Russians" (broadly defined) would continue in almost identical form—though with somewhat more success—after the revolution.

PROTECTING NATURE

While resettlement officials were clamoring to assert Russian control over fisheries and Primor'e's coast more generally, some began to express concerns about the state of the region's aquatic life. They tended to focus on foreigners' culpability, implying that Russian (state) control would be beneficial for nature. The publicist Konstantin Krestovskii, for instance (among the first to voice such views), claimed in 1881 that Chinese and Korean harvesters had exhausted wild seaweed supplies by not allowing existing stands to regrow. "It is said," he wrote, "that in former years the Korean coast was, no less than ours, rich in seaweed, but now there is none: all was destroyed as a result of the incorrectly conducted industry, which therefore was forced to turn further north, to our shores." Only with the introduction of "pure Russian commercial fishermen [*promyshlenniki*] and Russian coastal settlers" and enforcement of fallow periods for seaweed could the industry thrive.[28] Similarly, in 1899, resettlement officer A. A. Rittikh reported that, according to local inhabitants, fishing grounds had been much richer, attracting "whole flotillas of foreign fishermen," but "[the sea] is now exhausted, and the hunt has moved north." Rittikh asserted that salmon fisheries were the hardest hit; whereas previously Peter the Great Bay had teemed with fish, foreigners had exhausted it almost entirely.[29] D. I. Shreider, who published a popular account of his travels in the Far East, claimed that in 1894, the Japanese had extracted some 60,000 rubles' worth of fish from Russian shores, while the Chinese had taken 126,000 rubles' worth of seaweed. No traveler in the territory, he wrote, would get used to "the picture of reckless embezzlement of those gifts which nature has so generously provided. . . . [I]t would be better to change the word 'exploitation' [*eksploatatsiia*] (of natural wealth) for the term 'extermination' [*istreblenie*]."[30] Ferdinand Ossendowski, a Polish adventurer who lived in Primor'e in the early 1900s, observed that Japanese fishermen had completely denuded a bay on

Primor'e's south coast. "I saw no traces of life," he wrote, "no seaweed, no fish, no sea anemones or medusas or even shells. It was truly a waste, abandoned by live beings as though it were a plague-stricken country."[31]

By the early 1900s state officials regularly accused foreigners (particularly the Japanese) of overfishing, harming fish stocks and the indigenous peoples who depended on them. In 1908 the regional division of the Department of Agriculture and Land Management (Gosudarstvennoe upravelnie zemledeliia i zemleustroistva [GUZZ]) argued that the Japanese were practicing "predatory, unlawful" kinds of fishing, since they were operating illegally within river mouths and preventing fish from ascending to their spawning grounds upstream. The result, according to GUZZ, was not only a reduction in the amount of available fish but also the threat of depriving "our indigenous peoples [*inorodtsy*], living upstream, of their food reserves." GUZZ compared Japanese actions around Primor'e to those of American whalers and cod fishermen in the Far North, whose approach to natural resources and native populations showed "a predatory character and which [was] aimed at the destruction of natural resources and the exploitation . . . of the locals."[32]

Pavel Unterberger, who served as Priamur Governor-General from 1906 to 1910, also singled out the Japanese for overfishing. He argued that the extraction of salmon from Russian waters by the Japanese had "taken on such scale and character that it must inspire apprehension for the future." He too spoke of the "predatory nature" of Japanese fishing, citing wastage and overuse; he alleged that the Japanese caught more fish than they could process and left much of the catch to spoil. Unterberger pointed to the example of Peter the Great Gulf (near Vladivostok), where during the previous twenty or thirty years "abundant reserves of chum salmon were exterminated almost completely. . . . We see the same thing," he wrote "on the coasts of Japan and Korea." Unterberger also cited the Chinese practice of gathering sea cucumbers using diving bells, which he also deemed predatory, as "in this way even the small [sea cucumbers] are caught, without having reached full size."[33]

Fishing regulations, though generally ineffective, were also indicative of concern for protecting marine organisms and Russian interests. In 1892 the governor-general promulgated laws allowing Russians and non-Russians access to fisheries, seaweed, and sea cucumbers but banned fishing that impeded the "free movement of fish" upriver.[34] Subsequent revisions limited the size of gill nets and the distance between them to facilitate the movement of spawning salmon.[35] In addition, in 1903 the governor-general's office outlawed salmon fishing at points where tributaries of the Amur flowed into the river and established a fishing season for sturgeon that extended to the Amur, the Ussuri, and Lake Khanka. Lake Kizi, near the lower Amur, was declared a protected zone (*zapovednik*) for spawning sturgeon, and the catching of undersize sturgeon was banned year-round in light of the threat of "rapid destruction of sturgeon-type fishes."[36]

Enforcement of such regulations was evidently quite rare. At first, there was only one vessel (the *Storozh*) tasked with patrolling Primor'e's long coastline.[37] In 1899 a newly created division of the Ministry of State Domains and Agriculture (MGIZ) began to oversee fisheries in the region, but this too suffered from a shortage of patrol boats and had difficulty keeping foreign fishermen out of restricted waters.[38] Still, laws aimed at regulating fisheries reflected rising concerns among Far Eastern officials about the condition and exploitation of fisheries.

Because oversight was wanting and catch data were scarce, the impact of human activity on coastal ecosystems during the late tsarist era is difficult to judge. It seems unlikely that the anecdotal observations about overfishing and gathering cited here were simply invented, but they may have reflected localized or short-lived changes. The only instance of evident decline was that of sturgeon (both *kaluga* and Amur sturgeon). According to the MGIZ, a total of 73,415 sturgeon were caught on the Amur and Ussuri Rivers in 1891. By 1909 the catch had fallen to 17,626, and then leveled off before dropping to 9,012 in 1914. The average size of fish caught also decreased. By 1927 one fisheries inspector wrote, sturgeon in the Amur basin had become "a thing of the past."[39] On the other hand, duties collected from seaweed harvesters increased through the 1890s and early 1900s, suggesting that Krestovskii's warnings about exhaustion were probably premature.[40] Total fish catches for the Far East as a whole increased seven times between 1893 and 1913, while the salmon catch on the lower Amur increased until around 1912.[41]

Nevertheless, the dominant impression among officials and other contemporary Russian observers around the turn of the century was that aquatic life was under threat of exhaustion, primarily because of foreign activities. Taking over coastal fisheries had a strong ideological and strategic rationale. At the same time, ecological concerns provided an additional argument in favor of greater control.

NATURAL CAPITAL

Despite—or rather because of—environmental concerns, tsarist officials continued to press for greater investment in coastal fisheries during the regime's last decade. After the Russo-Japanese War (1904–1905), however, they began to emphasize large-scale, "modern" fishing operations, which many believed would help reduce foreign influence while also facilitating conservation. Such visions presaged the creation of an industrial fishing fleet in the Far East during the Soviet period.

The war, which had been disastrous for Russia, reinforced strategic fears about the Japanese and defensiveness about resource use. The Russo-Japanese Fishing Convention of 1907, part of the Treaty of Portsmouth that ended the war, permitted Japanese fishermen to operate in Russian territorial waters, excluding only inland waterways, river mouths, and inlets.[42] The Japanese could purchase the rights to fishing parcels at annual auctions held in Vladivostok on equal terms with Russian subjects. In the following years, "small armadas" of Japanese fishing and crabbing fleets operated in Russian waters off Primor'e, Sakhalin, and Kamchatka, and in the Sea of Okhotsk, contributing to the rapid growth in Japan's fishing industry.[43]

In this context, resettlement officials became less concerned with relocating individual households of fishermen and more with fishing operations that operated on a large scale, employed capital investment and new technologies, and were oriented toward global markets. In their view, a "modern" fishing industry would be better able to compete with the Japanese and would be a more effective way of populating coastal regions. Several Russian entrepreneurs had been engaged in commercial fishing and seaweed gathering on the lower Amur and elsewhere since the 1890s (and in some cases earlier), but Japanese fishermen were much greater in number.[44] Assessing the failures of directed coastal settlement, Resettlement Administration officer S. I. Shlikevich wrote in 1908 that Russian fishermen could not be expected to lead the life of "a Giliak [Nivkhi] or a Tungus [Evenk]," living solely off the fruits of nature. They needed opportunities to sell their fish, which in turn required better communication routes, larger local markets, and equipment, from salt to icebreakers. The state, he argued, could offer incentives and direct investment in infrastructure, like large refrigerated steamships, with the goal of exporting fish to European markets. This would, in turn, give a strong stimulus to the development of industry "and its liberation from Japanese influence."[45]

However, his was not an argument for all-out exploitation of fish stocks. Shlikevich was, apparently, also quite concerned with the danger of overexploiting natural resources. Rather, he believed that bringing fisheries (and other resources) more fully into Russian hands was a first step toward conservation: "In no case can we ignore the consideration that allowing wide access to capital creates predatory exploitation. The phase of predatory exploitation is, to a degree of course, unavoidable in a new territory . . . and without a doubt it is better to come to terms with the loss of a known percentage of natural capital because of predation than to lose that capital, along with territory that contains it."[46] Shlikevich's statement is striking in his cognizance of the impact of human activity on the environment ("natural capital") and also his belief that exploitation was a means to conservation.

In very similar terms, V. Romanov, who participated in a fact-finding expedition along the Amur in 1909–1910, advocated for a combination of modern fishing methods and conservation measures. He too argued that the state needed to subsidize refrigerated steamships, assist settlers with shipbuilding, strengthen enforcement organs, support individual initiative, order the military to buy up salmon, and ensure an adequate supply of Russian workers to nascent industries. Freeing the fishing industry from Japanese monopoly, in his view, demanded such urgent measures. He warned that Russia's "recent political enemies"—the Japanese—had taken a lively interest in the Far East and its "aquatic riches." The government's main goal, he argued, was to protect the territory, "economically and politically, for the Empire."[47] At the same time, Romanov also warned against the possibility of fisheries exhaustion, citing a decline in the salmon catch in the United States and Japan. He advocated fish-farming as the "most cultured . . . method to keep these riches [*bogatstv*] from exhaustion," and to "restore those [fish] reserves that have already gone into decline" and even populate rivers that they had not inhabited previously, citing the work of the ichthyologist V. K. Soldatov.[48] Without hatcheries and protective measures, Soldatov had warned, the situation would be dire: "the soil for colonization would disappear, life itself on the lower Amur would be swallowed up, and it would become a desert."[49]

A commission on resettlement, meeting throughout 1909 and 1910, also looked to a combination of capital-intensive production and state-led conservation. Commission members decided that, without "large, capitalist production," Russian fishermen sold their fish piecemeal to Japanese buyers and that further measures to populate the coastline should be carried out with a view to staffing large industries, as only this could combat the "seizure of the fishing industry by Japanese buyers," freeing the lower Amur from Japanese capital.[50] The commission also resolved, however, to send ichthyologists to the lower Amur and other vital areas to investigate fisheries so that development could occur within reasonable limits. Against the threat of fisheries exhaustion, the commission recommended regulation of capitalist industries, the development of fish-farming and, in areas "especially important for the protection of fish," a full ban on fishing. For his part Governor-General Unterberger stressed the necessity of settling the mouth of the Amur only with "Russian people experienced in fishing," so that "capitalist enterprises will have the necessary quantity of fishermen from the local, and not from the alien [*prishlago*], population." Unterberger also advocated limits on the new industries that would "guarantee the conservation of fish resources," and to forbid it in places where it might harmfully affect the migration of fish into the Amur.[51]

During last years of the tsarist era, growing concern with fish conservation dovetailed with the creation of a "modern" fishing fleet—one operating on a large scale and employing new technologies—that officials believed would be able to compete with Japanese fishing firms. Given their anxiety about the future of fish stocks and the outcome of the 1904–1905 war, such a view was understandable. Any step toward conservation, moreover, would require greater knowledge of fish populations and of the activities of fishermen and gatherers. The chaos that prevailed during the revolutionary period deepened ecological and political anxieties. Not surprisingly, then, Soviet officials adopted a similar emphasis on "modern" methods as a solution to both environmental and strategic challenges.

FISHING IN THE SOVIET ERA

The revolutions of 1917 and the upheavals that followed brought turmoil, bloodshed, and divided authority to the Russian Far East. During the civil war (1918–1921), intervention forces from Japan, the United States, Great Britain, and other nations supported White (anti-Bolshevik) governments in Siberia and the Far East. Partisan violence was endemic throughout the region. At the end of the war Primor'e came under the control of the Far Eastern Republic (Dal'nevostochnaia Respublika, or DVR), a buffer state that nominally ruled much of the Far East until 1922. The DVR operated under the thumb of the Japanese, though Moscow influenced its internal workings behind the scenes.[52] Throughout the war and the DVR period, Japanese fishermen expanded their operations in Russian waters, including in those zones from which they had been excluded by the Convention of 1907.

Exploitation of fisheries by the Japanese continued to prompt criticism from Russian authorities. In May 1921 the Chamber of Trade and Industry in Primor'e lodged a complaint with the Vladivostok-based Russo-Japanese Conciliation Commission, arguing that Japanese fishermen were illegally operating in the Amur, threatening the "exhaustion of fish reserves . . . which have noticeably diminished in recent years as a result of intensive fishing."[53] In 1922 a fishing cooperative wrote to the DVR's Economic Ministry to express their anxiety about a steady decline in chum salmon in the Amur, for which they faulted both "thoughtless destruction" of marine riches, partly through use of Japanese gill nets, which impeded migration to spawning grounds.[54] The DVR also published a scathing indictment of Japanese exploitation in the United States, arguing that they had begun to "loot" and "ravage" fisheries without any control or regulations.[55] After the Japanese withdrew, Soviet officials railed against the "predatory seizure by the Japanese of Russian national riches," though with a twist of class conflict: now it was "Japanese capitalists" perpetrating the "plunder of . . . natural fish reserves."[56]

When the Soviet government took full control of the Far East in 1922, fish conservation once again became a state priority. Soviet fisheries officials, like their tsarist-era predecessors, believed that with proper management and modern methods, fish populations could be conserved even in the context of rising output. Officials of the Far Eastern Fishing and Hunting Administration (Dal'nevostochnoe upravlenie rybolovstva i okhoty, known as Dal'rybokhota, or simply Dal'ryba) were aware, for instance, that "the exhaustion of natural [salmon] reserves" was apparent in 1923, threatening the Amur fisheries in particular, but they remained hopeful, that "with the correct economic approach," fish output would be colossal.[57]

Similarly, Dal'ryba inspector V. O. Kolobov published an article in 1923 in which he argued that overfishing and economic underperformance in the Far East could both be fixed with the right approach. Kolobov contended that irrational and "merciless" fishing practices were to blame for the rapid decline in salmon on the lower Amur but a more rational approach could yield sustainable output. He argued for a shift from salmon to other species, such as herring, which offered "riches of a colossal scale" if exploited by proper means. He wrote that Far Eastern fisheries suffered from a "provincialism and amateurishness," that was "specifically Asian." With an industrial fishing fleet and the support of the masses, however, Soviet fishing industry could "shed its centuries-old clothes of amateurishness and yellow provincialism and enter the wide road of global, concentrated production."[58] Although Kolobov's article was hyperbolic, his faith in new production methods and his association between "Asian" fishing and overexploitation were both well established before the revolution.

In the 1920s Soviet authorities invested resources in the study and regulation of Far Eastern fisheries. In light of declining yields and fish size among sturgeon populations, the Far Eastern Revolutionary Committee (Dal'revkom) imposed a seven-year ban on sturgeon fishing on the Amur in 1923. To gather more information on existing fish stocks, a Pacific Scientific-Industrial Station (Tikhookeanskaia nauchno-promyslovaia stantsiia [TONS]) was created in 1925, which later became known as the Pacific Scientific Research Institute for Fisheries and Oceanography (Tikhookeanskii institut rybnogo khoziaistva i okeanografii). TONS representatives warned that output from Far Eastern salmon fisheries could only be sustained through protection of forests around spawning grounds, regulation of logging along salmon rivers, artificial fish-raising, and a shift to unexploited regions and species. Dal'ryba and other state agencies supported most of the TONS suggestions, including the protection of forests around spawning areas. In 1926 the Far Eastern Executive Committee (Dal'kraiispolkom) mandated controls on certain types of nets in the Amur and Ussuri watersheds (including gill nets and Japanese purse seines), empowered Dal'ryba to set catch norms and institute bans on any fishing plots, barred peasants from fishing in known spawning areas, and outlawed trade in undersized sturgeon.[59]

Enforcement of conservation laws seems to have become somewhat more robust in the 1920s and 1930s than before the revolution but still suffered from a lack of personnel and resources. In 1928 there were guard stations covering the area between the Korean border and Soviet (formerly Imperial) Harbor, in the Tatar Strait. According to Dal'ryba's records, in 1925 guards made 282 arrests for fishing violations, and increased their "catch" in 1927 to 363. However, guards had only three motorized patrol vessels at their disposal, each responsible for hundreds of kilometers of the coast, and the boats were often too slow to catch poachers. Overseeing Japanese fishing and crabbing operations proved particularly challenging—and sometimes violent. In 1930 a Soviet patrol boat near Kamchatka fired on a Japanese vessel, killing one man, and inspectors killed three during an encounter off Primor'e's southern shore in 1933. In another case, a Dal'ryba patrol attempted to stop a crabber allegedly in nonconvention waters, only to be forced away by the appearance of a Japanese torpedo boat. Not surprisingly, the incident prompted calls for heavier armaments to be mounted on patrol vessels.[60]

RACE TO THE BOTTOM

Dal'ryba and other elements of the Soviet government were certainly aware of a decline in certain fish stocks and were concerned with conserving marine life but lacked the resources to do much about it. Moreover, economic and strategic imperatives trumped any inklings of restraint. As before the war, further development of the fishing industry had a strategic dimension. Tensions between the Soviet Union and Japan rose with the latter's occupation of Manchuria in 1931 and reached their dénouement in armed clashes along the Manchurian border in 1938 and 1939. And throughout the 1920s and 1930s, Japanese firms continued to operate in Soviet territorial waters, as the Convention of 1907 remained in force and the Soviet Union was in no position to contest it.

In response, Soviet officials looked to the same policies that had prevailed before the war: directed resettlement and capital investment. Once again, resettlement organs attempted to fill a socio-ecological "niche" with several thousand fishermen from central Russia. However, echoing Busse's first experiments in the 1890s, Soviet resettlement officers found that the new arrivals were "not sea fishermen, but river [fishermen]" and, moreover, "a significant part of them were simply dock workers," unfamiliar with ocean conditions and unwilling to use the ocean-going fishing boats common in Primor'e.[61] Another wave of maritime settlement followed in 1931, employing whole kolkhozy of fishermen and fishnet makers, numbering roughly eight hundred households, relocated from the Volga-Caspian and Azov-Black Sea regions.[62] Ultimately, these resettlement initiatives brought roughly fifteen hundred fishermen and their families eastward between 1928 and 1930, and an unknown number—perhaps as many as ten thousand—during the following decade.[63]

Renewed migration came alongside capital investment aimed at the creation of an industrial fishing fleet, much as the tsarist-era Resettlement Organization had wanted before the war. In the context of Stalin-era industrialization, increasing output was the order of the day, and Far Eastern fisheries were no exception. Dal'ryba gradually built up the state fishery sector, beginning with a modest herring operation in Peter the Great Bay, off the southern coast of Primor'e. In the 1920s, Dal'ryba also oversaw the creation of fishing arteli (work collectives), tolerance for Russian commercial fishing operations (run by the same men who had owned them before the war), and the formation of joint stock companies for which the state served as the majority stockholder. According to N. A. Bilim, output from Far Eastern fisheries of all sorts more than tripled between 1923 and 1928.[64]

As the Soviet fishing fleet grew, the presence of Soviet fishermen increased relative to their Japanese counterparts. In 1932 Soviet fisheries officials chartered one hundred thousand tons of Japanese fishing vessels to work in Soviet waters, but the following year they took less than one-third of this amount, and by 1934 supplanted Japanese boats entirely with vessels from Norway or those recently purchased from abroad.[65] State fisheries were organized into nine fishing trusts operating primarily along the region's northern shores and in the Amur estuary. The steady growth of state fishery trusts, collectivization, and capital investment during the first and second Five-Year Plans enabled rapid growth in Soviet fishing tonnage and output. By 1937, the Japanese rented only 14 of 578 plots in the Sea of Japan. Japanese vessels were excluded completely from the lower Amur, the Khabarovsk area, and northern Sakhalin.[66] Manipulation of the yen-ruble exchange also gave Soviet fishermen some advantages in bidding on fishing plots.[67] By the end of the decade, Soviet fishermen had gone some way toward displacing the Japanese from coastal waters.

Increasing output while conserving fish stocks, however, proved more of a challenge, simply because fishermen ran up against natural limits. The combination of Soviet and Japanese fishing put greater pressure on Far Eastern fisheries, particularly on salmon, the most important commercial species. Output from both Soviet and Japanese fisheries operating around Primorskaia oblast increased from 95,140 to 129,260 metric tons between 1929 and 1931, with salmon making up the majority of fish caught. The Sea of Japan region alone produced 184,490 metric tons of fish (of all types) by 1937.[68] In response to Soviet pressure, Japanese fishermen moved toward pelagic fishing beyond Soviet coastal waters, using giant "mother ships" that carried smaller vessels.[69]

However, by the end of the decade both Japanese and Soviet fishermen ran into diminishing returns. Japanese deep-sea fishing produced increasing yields through 1934, but in light of signs that stocks were becoming rapidly depleted (the average size of salmon caught, for instance, was the smallest ever recorded in 1934), the Japanese Ministry of Agriculture and Forestry mandated a 30 percent reduction in the catch and forced a reduction in the number of vessels deployed. Meanwhile, Soviet fishermen's catch of humpback and chum salmon declined in the late 1930s and early 1940s, leading to bans on salmon fishing throughout the region in 1945. Similarly, the catch of herring in Peter the Great Bay fell rapidly after 1926 from a high of 20,000 tons to barely 180 tons ten years later, while output from northern Primor'e also collapsed.[70] The decline in fish production, particularly of salmon, necessitated a reorganization of Far Eastern fisheries after the war and, ultimately, a shift toward pelagic fishing.[71] The effort to displace the Japanese had gained its own momentum; in the fight for a common resource in contested waters, conservation was clearly subordinate to exploitation and economic growth.

CONCLUSIONS

In considering the changes that occurred in human relationships with the fisheries and aquatic environments of Primor'e during the late imperial and early Soviet eras, one is struck by the coexistence of divergent tendencies. On the one hand, we see efforts on the part of state officials to protect fisheries from foreigners and, on the other, active promotion of fishing by Russian and European settlers and/or commercial fishing firms. In both cases, the competition posed by foreign (mainly Chinese and Japanese) fishermen was a key consideration, alternately motivating protection and expanded use.

In the mid-nineteenth century, Chinese, Korean, and Japanese fishermen and gatherers were active in Primor'e's waters, seeking fish, seaweed, and other marine products. For tsarist officials, the presence of thousands of seasonal migrants represented a threat to imperial control and an affront to national (and imperial) pride. Foreigners' overexploitation of the waters around Primor'e also inspired apprehension for future prosperity, providing another argument—ecological concerns—for the ousting of non-Russians from Primor'e's waters. Making Primor'e's coast "Russian" was not simply about maps and treaties; it also required a change in *who* was exploiting aquatic environments, and how they did so.

In order to supplant foreigners and assert imperial claims, administrators encouraged migration to coastal areas, seeking to attract supposedly "natural" fishermen from elsewhere in the empire (whether ethnic Russians or not). In the years after the Russo-Japanese War, Primor'e's administrators shifted their approach to one aimed at state-supported development of capitalist industries. A "capitalist" fishing industry, regulated by the enlightened state, seemed to offer a means to at once more effectively compete with foreign fishermen while exercising more oversight over fisheries, with a view to conservation.

The same thinking undergirded Soviet efforts to build up a fishing fleet while simultaneously seeking to conserve valuable species like salmon and sturgeon. Ultimately, the Soviets' efforts to assert control over Far Eastern fisheries were largely successful (partly because US forces destroyed much of the Japanese fishing fleet during the Second World War). However, in the process, the quest to utilize marine resources put significant pressure on fish populations in and around Primor'e.

The tension between protection and exploitation that runs through the period under study here suggests significant continuity between tsarist and Soviet eras. Soviet leaders, like their predecessors, were drawn to industrial, "modern" methods of fish extraction (ostensibly informed by science) in part because of regional challenges and historical experience—foreign competition and localized ecological decline—rather than an ideologically grounded antagonism toward nature. Their faith in (European) science and technology to reconcile human needs with natural limits, moreover, was not unusual in the early twentieth-century world. They shared this belief with their counterparts in China (as noted above) and in North America, where fisheries experts put their faith in technological fixes (including hatcheries) that failed to restore salmon populations.[72] Primor'e acquired a "Russian" fishing industry in the twentieth century, but it shared much of its experience with other parts of the globe.

14 The Tragedy of Captain Ligov

THE IMPERIAL AND SOVIET LITERATURE OF WHALING, 1860–1960

RYAN TUCKER JONES

SOMETIME IN the 1860s the fictional Captain Oleg Nikolaevich Ligov lay sick on his whaling ship sailing through the North Pacific. Ligov was the hero of the Russian Far Eastern writer A. A. Vakhov's wildly successful 1951 novel *Tragedia Kapitana Ligova* (The tragedy of Captain Ligov; see figure 14.1). The captain had doggedly pursued the Pacific's whales for most of his life—in one scene even poaching them from shore without any hopes of recovering their bodies—but he had pushed himself too hard. As his wife held him close, Ligov drew his last breath and died. With him, Vakhov's book made clear, died Imperial Russia's hopes of establishing a profitable and steady whaling industry of their own. Ligov's tale encompassed two tragedies: one personal, the other national.

Western readers may have a difficult time imagining a positive hero emerging out of the history of Russian whaling. The Soviet Union, after all, was the target of an American boycott and the dramatic Greenpeace antiwhaling demonstrations of the mid-1970s, which helped launch the "Save the Whales" movement.[1] Years later it emerged that the Soviet Union had, astonishingly, even exceeded the worst nightmares of environmentalists. Despite being signatories to the 1946 International Whaling Convention, the Soviets had illegally

FIGURE 14.1. Captain Ligov.

and covertly killed hundreds of thousands of extra whales beyond what they had reported in the Antarctic and Pacific.[2] Such excesses have helped cement the Soviet Union's legacy as one of the modern world's worst environmental despoilers.

However, as captured in Vakhov's frequently purple prose, Russians themselves had very different ways of imagining their country's relationship with whales. In the course of three massive tomes, Vakhov, a writer for the Vladivostok paper *Proletarskaia pravda* (Proletarian truth) and a budding author of fiction, attempted to encapsulate nearly the entire history of Russian whaling. In the two companion volumes to Ligov's tale, *Shtorm ne utikhaet* (The storm does not still [1957]) and *Fontany na gorizonte* (Spouts on the horizon [1951]) he moved from imperial whaling through the first decades of Soviet Far Eastern whaling in the 1930s.[3] Largely through this trilogy, Vakhov garnered a measure of fame in the Soviet Union, becoming an active member in the All-Russian Writer's Union.[4] He was the closest thing Russia has ever had to its own Herman Melville. Nor was Vakhov the only Russian author interested in whaling. From the 1860s through the 1960s dozens of Russians took up pens to describe their country's engagement with the sea. A trickle of reports and some scattered fictional treatments appeared in the late imperial era and early Soviet decades, followed by a flood of popular literature in the 1950s celebrating the Soviet Union's entrance into modern industrial whaling. These accounts traced Russia's sporadic and mostly unsuccessful attempts to join Europeans in their chase of leviathans before the Second World War and then described the beginning of the Soviet Union's massive harvesting of Antarctic and North Pacific whales.

Popular whaling literature can provide environmental historians with the rare chance to gauge common Russians' relationship to the natural world, especially the ocean. Understanding Russian environmental culture is important in part because Eurasian people have played an integral part in the history of the world ocean, a fact almost completely ignored by marine environmental history.[5] Many of the works discussed here—especially Vakhov's—are nearly forgotten now but were enormously popular and reflect an increasingly literate Russian populace entranced by the sea, its scientific study, and its exploitation. Given that much of this whaling literature came from the pens of state-sponsored scientists, that it was meant for popular consumption, and that proletarian writers such as Vakhov echoed scientists' themes, all suggest there was broad cultural concordance between the state, scientists, and reading Russians, one that grew through the twentieth century.[6] These common cultural concerns lubricated the machinery of the Soviet whaling industry—if not with the legendary clarity of sperm oil, then at least with the second-grade whale products the industry proved adept at producing in fabulous quantity.

What emerges in the printed sources analyzed here are two consistent themes that help illuminate some of the reasons behind the Soviet Union's reckless destruction of whales in the late twentieth century. First, Russian writers exhibited a continuous preoccupation with the presence and impact of foreign whalers in national waters, a concern that reflected state policy and also attempted to influence it. In part because of its delayed commercialization and industrialization, Russia failed to establish a successful whaling industry before the 1930s.[7] As a result, Russians long perceived whales in their waters as problems, valuable but evasive creatures that lured well-capitalized, free-wheeling foreigners to their shores, threatening imperial sovereignty and their indigenous subjects' well-being. Once Soviets themselves mastered the craft, especially after the Second World War, the tone shifted to include a second theme. Now writers attempted to tie Russian whaling to the earlier, heroic era of sail whaling. Even in the face of increasingly overwhelming technology and an officially collectivist culture, Soviet-era writers stressed the heroic nature of the individual harpooner matched against noble beast. As in other spheres of Soviet literature, whaling writers could not fully reconcile these tensions, but they could draw on classic Russian figures, such as the passionate hunter, to lend an increasingly reckless industry a romantic, if not entirely orthodox Soviet, sheen. Western whaling literature offers some parallel themes, but the particular currents of national resentment and collectivist culture help illuminate some of the idiosyncrasies of Russia's environmental history.

RUSSIAN/SOVIET WHALING HISTORY

Russians have been hunting whales off their shores for at least five centuries. The White and Barents Seas support a rich cetacean population and the experienced marine-mammal-hunting Nenets and Russians around some northern monasteries (and the Pomor as well) took belugas and dolphins from their small boats. The take was small and played a very limited role in the national economy.[8] The same can be said about early Russian hunting expeditions to the Arctic islands of Spitsbergen in the seventeenth and eighteenth centuries. In the early eighteenth century Peter the Great, determined to emulate European commercial whaling practices, hired Dutch whalers to train Russians in open ocean commercial whaling. The plan met with no success, nor did similar plans concocted during the reign of Catherine the Great.[9] The most successful nineteenth-century attempt at commercial whaling came from the Russian American Company, a colonial joint-stock company with offices in St. Petersburg, eastern Siberia, and Alaska. The company at first hired American harpooners to teach Aleuts modern hunting techniques but abandoned the idea fairly quickly. Instead, in the 1850s the company used connections among its Finnish employees to outfit several ships from Turku (Åbo) under the aegis of the Russian-Finnish Whaling Company (RFWC). For several years the Finnish whalers made a nice profit hunting whales in the North Pacific, but during the Crimean War (1853–1856), British ships captured and burned several of the fleet's vessels. Others were kept in harbor for the duration of the war, and mounting expenses ruined the company. The rest of the century saw several government-funded ventures in eastern Siberia, none of which made a profit.[10]

As a result of its failure to join the whaling rush of the sail age, in contrast to the history of the United States and most European nations, Russia's whaling heyday came in the twentieth century, not the nineteenth. In 1923 the Soviet Union granted Norwegian whalers a concession to hunt off Chukotka for fifteen years, but the contract was cancelled after two years (a controversial decision that would exercise the imaginations of novelists). Instead, the Soviet government resolved to begin industrial whaling on its own, and in 1930 it formed the Kamchatka Stock Organization for this purpose. The organization ordered a whaleship built in Norway, and in 1932 the first whale flotilla, the *Aleut*, sailed from Leningrad complete with a crew of Norwegian hunters and Russian apprentices.[11] After the Second World War, the Soviet Union seized several German whaleships, with which they were able to expand their whaling into the Antarctic, by then one of the last oceans where whales could still be found in abundance.[12] While European whalers dropped out of the Antarctic hunt because of declining productivity in the 1950s, the Japanese and the Soviets expanded their fleets, hunting the major whale species to near extinction, a process replicated in the North Pacific. Soviet whaling continued until 1987, when it was discontinued in the face of international pressure and declining stocks. While some whale species have subsequently recovered nearly to pre-exploitation levels, others—such as blue whales—remain perhaps at only 10 percent of former abundance, and the effects of the unprecedented removal of large predators from the world's oceans in the twentieth century still ripple through marine ecosystems.[13]

Most research on Russian whaling has focused on the postwar period, especially the massive fraud perpetrated by the Soviet Union regarding catch numbers. Soviet motivations for this fraud are not entirely straightforward to decipher. Certainly, the destructive logic of producing for an arbitrary Five-Year Plan, the subordination of individual whalers to bureaucrats charged with fulfilling these plans, the socialist system's disregard for the living object of the hunt, and simple corruption at all levels of the process, all played a role.[14] Russian scientist Alfred Berzin, in his post-Soviet memoir that exposed the falsification, claimed that Soviet whaling was not even profitable.[15] One wonders, in that case, why the Russians even bothered with such an expensive and ultimately embarrassing venture. Indeed, one writer has even claimed this was the "most senseless environmental crime of the 20th century."[16] Without denying that Soviet whaling possessed irrational aspects, the literature Russians produced about their relationship with the oceans shows that many found good reasons for whaling in defiance of Western desires. Furthermore, a look back into the prerevolutionary history of Russian whaling reveals that explanations focusing entirely on the Soviet bureaucratic and economic systems are inadequate.

FOREIGNERS AND FAILURE

The tsarist literature of Russian whaling was not abundant, but there was a great interest in the topic. Journals such as *Morskoi sbornik* (Naval collection) published foreign accounts of dramatic whaling voyages and shipwrecks and included several Russian accounts of the limited whaling then being done in the Pacific.[17] An anonymous Finnish sailor aboard the RFWC's first whaling voyage delighted the journal's readers with descriptions of exotic Honolulu and dramatic pursuits of whales in the North Pacific. One of his shipmates, the Swedish-Finn Otto Lindholm, persevered through the RFWC's inglorious end, and afterward almost singlehandedly established shore-whaling near Vladivostok in the Russian Far East in the 1860s. This same Lindholm would become the model for Vakhov's Captain Ligov.

It was also, ironically, the Scandinavian Lindholm who would establish the earliest and most persistent theme in Russian whaling literature—the menace of outsiders drawn to Russian shores by whales. As Mark Sokolsky notes (this volume), resentment toward foreigners—especially Americans, Chinese, and Japanese in the Far East—had spurred imperial investment in Russian settlement and fishing and provoked some of the region's first conservationist decrees. Lindholm was one of the first of these settlers. He not only established a primitive shore station along the Sea of Okhotsk coast in 1862 but also wrote about his experiences in tsarist maritime journals. There he explained his motives for pursuing whaling in nationalist terms: "if such wealth [of whales] had given people in other parts of the world the possibility of establishing cities, why couldn't Russia have its own Schmeerenburg [a Dutch whaling town in Spitsbergen]?"[18] No amateur, Lindholm studied his chosen field intensively, making the first scientific description of the Sea of Okhotsk's oceanography and attempting to describe whale movements' relationships to ocean currents and plankton distribution.[19] But he reported serious difficulties. Because of the large numbers of foreign whalers (mostly Americans) already in the Sea of Okhotsk, the whales had becomes skittish, harming the efforts of Russians upon whose very shores these whalers often landed and carelessly chopped down forests. Thus, Lindholm cast his venture in what would soon be the familiar terms of whaling fiction; to take the whales that were rightfully theirs, Russians would have to fight the incursions of foreigners as well as the environmental consequences of outsiders' intensive whaling. Several decades after Lindholm's venture also faltered, Russians began finding literary inspiration in his quixotic pursuit.

In the first known work of Russian whaling fiction, A. Ia. Maksimov's 1898 "Kitoboi [Whaler]," Captain Ligov's later mixture of the patriotic and pathetic found an imperial precedent. In Maksimov's short story, a Vladivostok whaler, Savelii Nikitich Zaitsev, takes a crew to sea. For the first time, and against his wife's wishes, Zaitsev's fourteen-year-old son comes along. Zaitsev had run into a spell of bad luck, faced with the empty seas caused by "American whalers who had managed to kill or disperse the whales that had gathered to mate or rear their young."[20] The reckless killing of breeding whales by these "uninvited guests" had made the simple Russians desperate, ready to take unadvisable risks. When Savelii finally sights a whale, he ignores the gathering dark clouds on the horizon and pursues the animal anyway. The crew manage to kill the creature, but such disregard for danger soon has consequences. Before long the small ship is tossed in a terrible North Pacific storm that continues through the night. As Maksimov wrote, "Nature itself laughed at its unhappy victims, who had spent the night under the fearful pressure of the heavy expectation of death."[21] Thus, nature and foreigners conspire to ruin the Russian whalers, and one-by-one the crew members die of exposure during the night and the following day. Finally Zaitsev's son, too, dies and slips beneath the waves. Savelii himself is the only survivor and returns home bearing the terrible news to the boy's mother. The story ends with an apostrophe to the cruelty of nature: "Only the blowing wind will sing the boy's endless funeral pyre, as the waves throw his body onto the sandy shore and birds of prey carry his white bones to the four corners."[22]

A far less melodramatic account of whaling in Russian waters ten years later suggests that Maksimov had no monopoly on his themes. K. N. Nosilov's "Na kitoboinam parakhode iz puteshestviya po severu [Aboard a whaleship on a voyage in the north]" describes a far different theater of whaling—the White Sea—but picks up Maksimov's resentment of foreign incursion. Nosilov shipped as a journalist aboard a British whaleship out of Murmansk in 1911 and observed the crew catch at least one whale. After bringing a whale on deck, the Russian journalist had a long conversation with the British captain, conducted "in French, half in English, with horribly mangled Russian words he had learned in Murmansk mixed in." After discussing various whaling grounds throughout the world (mostly through gestures and a few commonly understood scientific names for whale species), Nosilov asked about whaling in Russian waters. "Oh!" was the captain's excited response, followed by gesturing and pointing at a map of the Arctic Sea. As the Russian interpreted: "It indicated an uncountable richness used only by them [foreigners]—these dexterous, entrepreneurial people—and by no one else because we Russians fear the sea."[23] Like Maksimov, Nosilov concluded by evoking the power of the sea. This time it inspired awe rather than terror, though tinged with similar regret: "How wonderful it was to be in our Arctic Sea! In those moments I comprehended how much our ancestral northern sailors loved this sea, who even risked sailing up ice flows on their leather *shitiki* [boats sewn together with leather] when hunting whales. And it seemed to me, that Russians already had gained all of this experience, and if foreigners were using these seas now it doesn't have to be that way—it is only because we have grown too used to the land."[24] It would take a return to this historic spirit, Nosilov opined, in order for Russia to retake its rightful place as chief catcher of its own whales. Paradoxically, as Julia Lajus notes regarding Pomor fisheries (this volume), it would also require the adoption of foreign technology and scientific research.

The themes expressed by the late imperial literature on whaling go a long way toward helping the historian understand some of the contours of Soviet whaling. It is surprisingly rare in the imperial literature to encounter economic reasons for hunting whales. In contrast to American whaling and its literature, which celebrated the growth of towns around the whaling industry (an idea faintly echoed in Lindholm's account) and its whalers' contributions to national prosperity, imperial Russians saw the whales off their coasts principally as problems. Whales appeared far less as an opportunity for adventure and profit than as the unfortunate lure bringing money-grubbing foreigners to Russian shores. Furthermore, those foreigners behaved irresponsibly toward Russian nature, destroying treasures that rightfully belonged to the empire. When Russians themselves did venture out to sea they found a denuded environment full of perils, quite the opposite of the American whaling writers who, with some dark shadows, mostly celebrated the abundance of the ocean.[25] Of course, such impressions largely reflected a reality Russians faced and whalers of other nations (except Japan) mostly did not—whalers had come to Russian shores and Russian whaling ventures through the imperial era all resulted in failure.

Just as in other spheres of life, becoming Soviet meant that Russians could open an entirely new chapter in their history of relations with whales. Early Soviet writers produced little concerning the whaling industry, largely because Russians had ceased whaling altogether with the First World War. In the 1930s, however, a new and very popular genre of Russian whaling literature was born—the scientific voyage journal. These journals expressed a new Soviet optimism but did not discard imperial concerns about foreigners. In 1936 the zoologist B. A. Zenkovich accompanied the first Soviet whaling flotilla, the *Aleut*, around the world from Leningrad to Vladivostok, during which whales were killed from Panama to Chukhotka. In his description of the voyage published in 1936, *Vokrug sveta za kitami* (Around the world for whales), Zenkovich had little negative to say about foreign whaling, largely in part because nearly all the whalers aboard the *Aleut* were Norwegians. He did announce proudly that the honor of being the first Soviet man to kill a whale went to a certain Naum Serebrennikov.[26] Indeed, much of Zenkovich's account, and the flotilla itself, displayed a growing Soviet pride and confidence (see figure 14.2). Boats bore revolutionary names such as *Entusiast* (Enthusiast) and *Trudfront* (Labor front), and Zenkovich reported that while docked in Europe such was the enthusiasm for the "new Soviet man" that all his posters of Lenin were stolen. Native Hawaiians encountered at Honolulu also seemed attracted to the anti-imperialism reflected in the red flags of the Soviet ships. Honolulu was a particularly poignant stop for the Soviets; as the hub of the nineteenth-century Pacific sail whaling industry, the Soviet arrival there seemed to tie this modern venture to the heroic era Russia had bypassed. Zenkovich promised a great future for Soviet whaling, no matter its inexperience, for "Our Soviet life every day is presenting clearer examples of brotherhood, heroism, self-sacrifice, perseverance, and selflessness for the sake of the cherished goals of our nation: free, happy, and beautiful life."[27]

In the aftermath of the Second World War, when the Soviet Union inherited several former German whaling vessels capable of whaling in the Antarctic, popular enthusiasm for the industry exploded. A coterie of breathless journalists and scientists accompanied the first Antarctic voyages in the early 1950s, and Andrei Solyanik, the flamboyant first commander of the *Slava* (Glory) fleet, penned a widely distributed account of the event. Solyanik's account was mostly cribbed from those of the scientists (though with less imagination and verve), but it found the most international distribution, appearing in English translation in 1956.[28] In *Cruising in the Antarctic*, foreigners made a dastardly reappearance, this time with a stronger emphasis on their exploitative environmental practices. Foreigners, Solyanik informed his readers, had exterminated much of the Antarctic's marine mammal life in previous centuries, and the reckless whaling of the prewar decades had reduced whale populations to dangerously low levels. Without irony (and we now know with gross underestimation), he also boasted at the end of the season that the Soviets had captured far more whales than the foreign fleets.[29] The Soviets had finally turned the old whale problem on its head; instead of foreigners killing Russian whales, Russians were killing whales that foreigners thought of as their own. And while Solyanik did not publicly admit it (though his honesty reportedly increased when he was drunk), Russians were now leaving oceanic deserts in their wake.[30] Zenkovich's prediction that Soviet aptitudes would let Russians instead of Americans remove whales from the sea had come true.

Such an impression found rich confirmation in the account of the *Slava*'s chief scientists, V. A. Arsen'ev and V. A. Zemskii, authors of an account entitled

FIGURE 14.2. Soviet whaling fleet *Aleut* with the boats *Entuziast* and *Avangard*. From Zenkovich, *Vokrug Sveta za Kitami* (1936).

V strane kitov i pengvinov (In the land of whales and penguins) published in 1951.[31] Besides devoting several chapters to the biology of whales, the scientists dwelled at length on how the Soviets had quickly bettered the Norwegian harpooners who handled the harpooning on the first Antarctic voyages. Since the late nineteenth century, when the Norwegian Svend Foyn had invented the exploding harpoon, humans had finally been able to consistently kill the faster rorquals (whales with baleen) that had eluded previous whalers. However, the harpoon guns required new skills, both for operating the guns and aiming them effectively. For some time only Norwegians had mastered the technology, and even in the 1940s it was common for other European whalers to hire Norwegian gunners. As noted, the Soviets had done the same in the 1930s North Pacific. But in the context of a developing Cold War, this reliance on Western experts became more distasteful. Now, Arsen'ev and Zemskii were proud to report, the era of Soviet dependence had come to an end. The Soviet harpooner Vasilii Tupkov, for example, had killed 357 whales in one season, which surpassed the best foreign results. After 1949 the Norwegians were not invited on board Soviet ships again.[32]

At the same time, the Soviet scientists repeated the same cautions as Solyanik had about the reckless depredations of foreign whalers. Whales had previously been found in abundance in the Southern Ocean, they reported, but recently had faced the prospect of total extinction. Although in the 1930s Europeans had agreed on some conservation measures for the Antarctic, the Soviets viewed their motivations with skepticism. Such measures had not come about, claimed Arsen'ev and Zemskii, because foreigners had had any concern with protecting the animals but, instead, from thoughts "about protecting their capital."[33] The Second World War had then given whales an accidental and far too short reprieve. With the war's conclusion, despite the establishment in 1949 of the International Whaling Commission (IWC) as the industry's first permanent regulatory body, there were abundant signs that whales again faced an existential threat. Rare were the whales encountered by the *Slava* that had reached their maximum size, since so many of the adults had already been killed. Despite such potentially worrying signals, the *Slava* caught whales without restraint, and the ship that caught the most received the honor of being the first to enter Odessa port, to huge cheering crowds.[34] Neither the *Slava* scientists nor Solyanik bothered their readers with explanations for this contradiction. Nor did they inform anyone that the *Slava* had begun violating even the pitifully inadequate limits the IWC had set for Antarctic catches. Not coincidentally, the deception had begun the same year the Norwegians had been left on shore.[35]

If comprehensible reasons are to be found for this duplicity and hypocrisy, they are tied to a mixture of real need for edible fats, the perverse incentives of the centrally planned economy, the problems of the IWC's regulatory schemes, and a Cold War atmosphere that encouraged cheating on both sides of the Iron Curtain. In the 1940s Soviet planners such as Economics Minister Anastas Mikoyan thought Antarctic whaling especially crucial to meet the "urgent need for meat" in the war-torn regions of the Ukraine and Belorussia and a general need for fats in the country.[36] After the crisis passed, however, Soviet whaling catches continually increased despite the lack of substantial increases in demand, one result of an economic culture that demanded only increases in production.

Of course, the Soviets were hardly the only nation to take advantage of the IWC's lack of enforcement mechanisms and the relative ease of misreporting catches, but the scale of the cheating that developed was unique. As other scholars have noted: "every environmental issue between 1945 and 1991 took place in an evolving geopolitical context dominated by the Cold War," which in the Soviet Union was tied up in a longer history of resentment, suspicion, and envy toward Western powers, precisely those who had thwarted Russian ambitions to whale for so long.[37] While the West suspected that geopolitical considerations such as the desire to claim an Antarctic presence also drove Soviet whaling's expansion, the clearest connections between Soviet enthusiasm for whaling and difficult international relations can be seen in the production of whaling literature.

The excitement generated by the *Slava* and its accounts stormed into the rest of Soviet life in the 1950s. The literary magazine *Ogonek* ran a series of articles about the fleet, reprinting issues of the newspaper produced on board and announcing that the harpooner Georgii Ovsiannikov had set a new record by killing twelve fin whales in one day.[38] Meanwhile, several novelists incorporated the whaling mania into their work, including most thoroughly, the Far Easterner Vakhov. In fact, Vakhov got his start in the genre through writing popular factual accounts of Soviet whalers. In 1949 he wrote a profile of the "best harpooner" on the Far Eastern *Aleut* fleet, F. D. Prokopenko, which was part of a larger series of short pieces about or by Soviet whalers that mainly focused on the Antarctic.[39] He began work on the whaling trilogy in 1952 and finished *The Tragedy of Captain Ligov* three years later.

It is worth noting that these were some of the most repressive, and most aggressively nationalist, years in Soviet life, as an aging Stalin saw enemies all around, both abroad and at home. Additionally, it had become popular to resurrect triumphant national stories after the enforced internationalism of the early experimental days of Soviet literature.[40] The state's policy toward literary artists reflected these trends, and it encouraged artistic vigilance against subversives. This does not mean, however, that authors lacked all freedom to portray life as they saw it. The press and the Party also attacked nationalism in literature, and especially "Great-Russian chauvinism."[41] Against such cautions when interpreting Vakhov, it should be noted that the some parts of his whaling trilogy appeared after Stalin's death, when a limited thaw encouraged a more critical examination of the past, and that he wrote and published in Vladivostok, in relative isolation from Soviet central power.[42] Vakhov, in other words, was working in a minefield when he decided to turn to prose, but not a minefield without a few clearings.[43]

The question of foreigners vexed Vakhov's heroes. In fact, the opposition between heroic Russian whalers and capitalist/monopolist Europeans and Americans drives all three trilogies (the love stories never manage to feel as passionate as the national rivalries). The novels' central conflict is between the crafty and violent but ruthlessly successful American whalers and a succession of Russian captains whose idealistic refusal to exploit their men or the natives of distant shores puts them at a competitive disadvantage. These themes did not originate in the Cold War, but they found new resonance as the Soviet Union challenged Western maritime dominance.

Ironically, however, in *The Tragedy of Captain Ligov* Vakhov chose one of the least successful, least national, and least socialist eras of Russia's whaling history to explore his themes. Captain Ligov was modeled after the aforementioned Otto Lindholm, who had started his Pacific whaling company from scratch in the 1860s and 1870s. Lindholm was not ethnically Russian but, instead, a Swedish Finn who recruited a highly polyglot circle of companions (including two American citizens and two "negroes" from South Carolina!) to his whaling outpost on the shores of the Sea of Okhotsk.[44] After some initial success selling whale meat and blubber to the Japanese, Lindholm acquired a steamship and requested a monopoly on Russian whaling on the Pacific. St. Petersburg, however, rejected the petition and instead subsidized a Russian, Akim Dydymov, in a rival venture.[45] Lindholm wrote an angry memoir, and Dydymov's company failed disastrously.

In an authorial sleight of hand that also speaks to the importance Vakhov placed on national conflict in his stories, he evaded the complexities of the story by casting Dydymov as Captain Klementev—not a rival but an equally brave and noble successor to the late Ligov/Lindholm—in the trilogy's second volume.[46] Most crucially, the two patriotic whalers shared the same enemy, a grasping octopus of an American whaling company controlled by the devious Robert Dalton and his henchmen, William Gillard and Stardson. Dalton's company represented everything evil about the capitalist system in all its nineteenth-century vigor and recklessness. Infiltration of Russian institutions—a recurrent theme in Soviet history and again in the public spotlight in the 1950s—also contributed to the Russians' failure. Dalton's company's agents bought off Russian officials, who even sold Ligov's boat to the Americans—"Dollars will get anything."[47] Historical reality allowed Vakhov to heighten the American threat; the Yankees had indeed been whaling in Russian waters, and in Vakhov's rendering they practiced brutality toward their prey and Russia's indigenous subjects with equal lack of restraint. Landing at an Evenk coastal village, Stardson led a massacre of the natives, abducting their women and burning their houses (see figure 14.3). Having already secured an agreement to buy Alaska from the Russians (more tsarist betrayal), the Yankees transformed Sitka into a den of thieves and reduced the local population to wage-labor slaves. Striving to obtain a monopoly on Pacific whaling, Dalton's men attempted to intimidate and bankrupt their last Russian competitors and an international harpooners' union that offered stiff resistance.

It is easy to see the influences of official Soviet culture on Vakhov's work. Grasping and murdering American capitalists/monopolists, for example, were stock-in-trade figures of Socialist Realism.[48] Perhaps more importantly for whaling, though, Vakhov's preoccupation with foreign whalers was nothing new and drew on a long tradition in Russia of interpreting whaling as national competition. Only Vakhov's Americans' cartoonish brutality and obsession with monopoly were true Soviet-era inventions.

Intriguingly, and in contrast to the official Soviet emphasis on the con-

FIGURE 14.3. American whalers destroy an Evenk village and kill its inhabitants.

quest of nature, Vakhov also managed to impart to his heroes an unconventional environmental sentiment. What set the Russian whalers apart from the Americans—aside from a commitment to the working class, of course—was a deeper connection to the ocean and a respectful rather than exploitative attitude toward it. The trilogy's second installment, *The Storm Does Not Still*, dwelled on Russian attempts to conserve the fur-seal colonies of the Commander Islands, a motif in fact taken from decades of tsarist- and early Soviet-era conservation decrees protecting marine mammals. Just as often, the Soviet attitude toward the ocean was sentimental. As Ligov's scientist compatriot is made to say: "The sea is in my soul," a blending of human and oceanic profundity familiar from the Romantic era and a continued source of inspiration for European and Russian nature lovers.[49] Other writers went to great pains to establish the linkages between modern and premodern Russian whaling, which was thought to have existed in an era and in places that exhibited a more primitive harmony with the natural world. For example, Nosilov in the early twentieth century waxed eloquently about the Pomors' traditional love of the sea and the adventure of pursuing sea mammals, a sentiment common among urban Russians, who at the same time also sought to replace this harmony with scientific knowledge.[50]

Vakhov, as a Far Easterner, traced the Russians' environmental ethos to the native peoples of that region, especially the Chukchi. These were ancient and expert whalers whose (not entirely willing) participation in the empire meant that "there was no need whatsoever for foreign instruction."[51] That said, the Inuit people of the Arctic were indeed some of the world's oldest whalers, but the historical link between Siberian and Russian whaling was quite weak.[52] Nonetheless, Vakhov devoted a large portion of the novels' plots to detailing cooperation between the two people. The Russians learn a lot about whaling ethics from the natives. These "trusting children of nature" used the entirety of the whale instead of just selling off its parts for profit.[53] Then, in the final book, *Fountains on the Horizon*, the favor is returned. A Chukchi man who is married to a Russian woman from an earlier whaling voyage tells the Soviets of his people's lack of food—in response, the *Aleut* fleet puts its industrial technology to the benefit of its premodern subjects, lending the Chukchi outboard motors that enable them to catch enough whales to save them from starvation.[54]

Soviet whaling officials championed this supposed harmony of interests between Russia and its indigenous subjects on the international platform, repeatedly insisting at the International Whaling Commission's founding meeting in 1946 that exceptions to the gray whale hunting ban be made for the Chukchi.[55] The Soviets also presented the Chukchi's supposed total use of the whale as a key component of their own environmental sensibility. Soviet delegates to the IWC's founding convention were ordered to stress this point to other representatives, and it indeed became a part of the IWC's charter.[56] Such an insistence on a relatively minor issue as far as conservation is concerned says more about Soviet representations of their own environmental sensibility than their real concerns about the future of whale stocks.

Vakhov brought the contrast with the American whaling ethic home in a strange scene on the Far Eastern seaboard, based loosely on one of Lindholm's experiences. Led to a forgotten shore by their native allies, Ligov and the crew come across a "whale graveyard" filled with thousands of pounds of baleen. American whalers, "chasing after riches, had killed off the whales, taking only one piece of baleen and throwing out the rest."[57] In contrast to this waste, the Russians respectfully gather the discarded baleen and sail off to Sitka where they sell it and accumulate the necessary capital to purchase a larger whaling boat. This fictional episode was replete with historical ironies. There is no doubt that American whalers made use of only a few parts of their kills, but the evidence is more ambiguous both for Chukchi and Russians. The Chukchi likely killed mainly nursing mothers and their calves—much safer targets, but much worse for the overall bowhead whale population.[58] Less ambiguously, Soviet whaling was notoriously wasteful, as the refrigerators and tinning operations strained, and often failed, to keep up with the ever-increasing kills. In some years, more meat spoiled than was successfully delivered back to Odessa. In *Fountains on the Horizon*, Vakhov did acknowledge that Soviet whalers also wasted much of the meat they attempted to process, but their shortcomings came from the steep technological learning curve they faced rather than from greed. Such concern for conservation, in all its senses, was only natural among Russians, Vakhov emphasized. It had long characterized their behavior toward the well-loved, but much tormented (by foreigners) sea, and contact with nature-loving Siberians had only strengthened the Russian dedication to the ocean. This, at least, was the picture drawn by Russia's whaling literature.

In a hundred years of whaling literature, Russian authors had almost universally bound their stories up with the presence of foreigners in national waters. Imperial-era stories stressed the tragedy of disappearing Russian resources, showing the consequences of environmental havoc on coastal populations. At the same time, they depicted the development of Russian whaling not as part of a global capitalist story but, rather, as the continuation of indigenous practices in the North and Far East. The first Soviet whaling fleet's name—the *Aleut*—came from such imagined links. These concerns added a uniquely Russian bent to the national history of whaling, and even once the Soviets managed to establish the first sustained industry, old preoccupations remained. In various ways the explosion of literature that followed the Soviet whaling ventures of the 1950s looked at the success not as an economic breakthrough but, rather, as a much delayed victory in an older struggle with outsiders bent on suppressing Russian industry. All identified Russia's superior relationship to the natural world as an important lesson. The *Slava* writers stressed the Soviet Union's enlightened approach to conservation, while Vakhov delineated a longer tradition of Russian environmental sensibility derived from its colonial subjects.

Certainly, Russian whaling literature partook of some of the themes common to such works around the world—the journey of personal and social transformation encountered during the whaling voyage, the terror and romance of the open ocean, and the importance of whales for national identity can all be found from American Herman Melville's sail-whaling classic *Moby Dick* to New Zealander Witi Ihamaera's postcolonial *Whale Rider* (though here the social bonding occurs through protesting mistreatment of whales).[59] Whaling's threat to oceanic ecosystems and nearshore communities appears far less frequently. While Japanese officials frequently complained about this in the days of America's Pacific whaling, it is interesting to note that the only known Japanese novel to take up the theme is far more critical of homegrown capitalism than of foreigners.[60] In his 1936 novel, *The Pod Stayed Put*, Osaka Keikichi describes the disastrous personal consequences for the members of a rogue whaling ship that violates conservation standards.[61] Seen comparatively, then, Russians' association of whales with historical missed opportunities, indigenous imperial subjects, foreign threats, and nativist environmental sensibility seems a unique blend whose characteristics were only partly Soviet. This mixture was born from the historical reality of Russia's maritime weakness of the nineteenth century and the Soviet Union's oceanic ascendance during the opening years of the Cold War, with the whaling industry perhaps best representative of these developments. The contradictions the Soviets produced for sentiment around whaling do not entirely explain the Soviet Union's recklessness. But, as a creature that both damned foreigners' long history of environmental recklessness and whose mass killing proved Russia's ascendance, whales offered themselves as appealing targets of both the harpoon and the imagination.

MONSTERS AGAINST THE HUNTER

Another theme confronted whaling literature with contradictions—the whale itself and the struggle between human and whale. And, if much global literature around whaling was transfixed by the mysterious creature (*Moby Dick* above all) and sometimes agonized around its killing (much more so in the late twentieth century, but with important early exceptions), again Russians added distinct features to these concerns, though this time the most important variations on the theme came from Soviet ideology.[62] Little Russian literature focused directly on the whale. However, the animal was always present, either as the magnet drawing Russians through the world's oceans or as a sudden—sometimes impressive, sometimes disappointing—object of human lives. Examining the evolving descriptions of the whalers' quarry reveals the development of a simultaneously greater scientific understanding of the creature as well as a distancing from the living entity itself. Whether engaged in primeval combat with the beast or calmly eliminating these thoroughly comprehended zoological specimens, however, killing whales invariably ennobled the Russian hunter and this persistent alchemy provides another clue to Soviet whaling practices.

Early descriptions of whaling provide very indistinct pictures of the animals themselves. The Russo-Finnish sailor whose letters were published in *Morskoi sbornik* gives the whale vague appellations such as "black whale" or simply "fish." Additionally, the influence of his cosmopolitan shipmates is evident, as he frequently resorts to American terminology, especially "savage fish."[63] The Russians possessed an equivalent term, very frequently used in accounts from the time—*chudovishche* (monster). As applied to whales, the name appears not to have been derogatory (some accounts refer to a "beautiful monster"), but certainly expressed distance from the foreign beast.[64] In Nosilov's account of Arctic whaling, he describes examining a whale that has been brought on board: "It is strange how the whale—which very recently had drawn the curiosity of everyone, which not long ago had heated our blood while fleeing into the open, unbridled ocean, which not long ago had shocked us with its appearance swimming next to our boat—now evinced very little interest and appeared much smaller than it had just a few minutes before to these self-same eyes. . . . Soon the interest in the dead animal completely disappeared."[65] Nosilov's disinterest extended to almost all aspects of cetology; he gave no indication of what kind of whales were being taken, nor did he give any insight into their behavior or characteristics.

As Nosilov's disappointment attests, a dead whale slumped on board did not present the animal in its best light. A live whale, however, was something else, and it drew more enthusiastic descriptions. Before most of Maksimov's heroes died on the open sea, they had a chance to observe whales in the distance: "Frolicking whales—a charming and amazing picture! It is difficult to describe the powerful yet grandiose movements of these giants, with the water churning around them as if boiling in a pot, seemingly battling the whales. Magnificent high jets of water . . . from the whale's nostrils shatter in the air like millions of brilliant emeralds."[66] By the end of the century, even dead whales were attracting interest. Otto Lindholm suggested that the Russians try to gather some of the monstrous skeletons in the Sea of Okhotsk and send them back to St. Petersburg, as the animals were quickly disappearing and ought to be preserved for science. Lindholm also suggested that the live whales receive some sort of government protection, a proposal that long went unheeded.[67]

During Soviet times the whale's beauty faded, in both fiction and nonfiction. Instead, scientific descriptions took increasing prominence, exhibiting a growing concern with differentiating between species and detailing their behaviors. This evolution in whale imageries coincided with the development of a professional body of Soviet cetologists and the general ascendance of science from late imperial times.[68] Zenkovich, aboard the *Aleut*, provided scientific lectures to the crew about the classification and behavior of whales, with stress placed on the animals' interesting biology.[69] The sperm whale, for example, was presented as the "strangest animal in the world," with an oversized head full of strange substances. Soviet novels, Vakhov's in particular, attempted to educate the Russian public on cetology, part of a belief that the Soviet public must be scientifically literate in order to be good socialists. One of Vakhov's main protagonists in *The Tragedy of Captain Ligov* is a scientifically inclined Russian interested in the whales themselves. He obsessively draws pictures of the whales in his sketchbook. In contrast, however, a scientist in the later *Fountains on the Horizon* is depicted giving long, boring, technical speeches about whales to crew members who were passionately interested in the subject but put off by the arrogance of the so-called expert.[70] This was classic Bolshevik suspicion of bourgeois specialists, and perhaps a criticism of the mediocre cetological work produced on the *Aleut* fleet.

When Vakhov describes whales themselves, they are seen chasing prey or mating, all in accordance with scientific descriptions.[71] While there was nothing to compare with Melville's long chapters on nineteenth-century conceptions of whales and their behavior, nonetheless Vakhov narrated in detail whale biogeography, feeding habits, and migratory patterns. The popular *Slava* literature went much further, recounting the scientists' tagging operations and offering crash courses on cetology for the average Russian reader. Some literature delved into the individual personalities of the prey; especially prominent was the wizened whale who had survived previous attacks, such as the blue whale Solyanik described catching or the sperm whale bull initially too wise to fall victim to the whalers but finally captured when defending his harem.[72]

While Soviet writers mostly informed their readers about the unique behaviors and ecology of whales, they sometimes expressed pity for their plight. True, the sympathy remained vague and mostly anodyne, but there was a distinctive repetition of emotional engagement with the animals. "The poor whale" was a common phrase used to portray the death agonies of a harpooned giant.[73] Arsen'ev and Zemskii described the "remarkable" motherly devotion sperm whales show to their offspring when the latter are harpooned, which often led to the entire herd being captured in one blow.[74]

Modern Western whaling fiction often uses the grim financial necessity of the hunter as the justification for killing these gentle creatures, while Japanese sometimes begged forgiveness of whales they had killed and erected shrines to appease whale gods.[75] The Soviets—living in a secular workers' paradise—did not have such recourses available. Instead, Russians usually reconciled their enthusiasm and sympathy for whales with their brutal slaughter by stressing the resourcefulness and nobility of the hunter. This theme had been more common in preindustrial Western literature but largely disappeared with the invention of the diesel-powered catcher boat and the harpoon grenade.[76] The scientist authors from the *Slava* expedition, on the other hand, claimed that "the hunt for whales—that is a real hunt."[77] Both they and Solyanik devoted much of their accounts to the drama of the pursuit of the whale, inventing much doubt for the nearly inevitable outcome. In an unusual conflation of human and ship, Arsen'ev and Zemskii wrote: "the fight between the injured giant and the ship continued until the animal started to tire. . . . The fight ended, man had won."[78] One might protest that the ship had won, but that would rob the hunt of its heroic dimensions. In fact, the real heroic labor of the modern whaleship lay in keeping the complex floating factories functioning and locating dwindling numbers of whales in oceans that, at first, appeared featureless. This was the work of engineers and the professional oceanographers aboard, whose painstaking measurements of currents and biomass helped them predict where whales might be abundant; killing the creatures once found was not particularly difficult.

Such collective, scientifically informed, industrial labor might have formed the basis of a truly Soviet whaling literature. However, despite the presence of hundreds of other Russians onboard the gigantic floating factories that made up the *Slava* fleet, the harpooners received the lion's share of the attention, though these men did little until the moment of the kill. The scientist observers detailed the lives and exploits of several harpooners of the fleet, mentioning their unique contributions to Soviet whaling. One, Fedor Prokopenko, had invented night-hunting for sperm whales, illuminating the waters with a giant searchlight. Arsen'ev and Zemskii then described how another harpooner made his kill, and they imagined what the haughty Norwegians would have thought if "they had seen the broad, stocky figure of Regushevskii, standing firmly with legs spread wide on the whaleship's deck, which was pitching in the waves. . . . They hardly would have said that a human could kill a whale all by himself after only a year of training."[79] "All by himself" reveals a remarkable (and frankly anticollectivist) need to eliminate the monstrous technology that played the dominant role in the Antarctic hunt.

Another hunter onboard the *Slava*, P. A. Zarva, who became something of a celebrity during the Soviet era, contributed his own account of his participation on both the *Aleut* and *Slava* fleets. Zarva began his story by relating his father's difficult work as a fisherman from Kerch in the Crimea. The son began his own transformation into a hunter while assisting Norwegian harpooners hired by the Soviet Union to do the hunting on the *Aleut*. As Zarva saw the first whale killed, "on that day my heart first beat with the hunter's passion [*azart*—it is instructive that a specific word for "hunter's passion" exists in Russian, but in few other languages], and it still beats like that twenty years later whenever I throw a harpoon into a gigantic animal of the sea."[80] The struggle with the whale had made the Ukrainian boy firing iron harpoons from the skyscraper deck of the industrial whaling ships the equal of his fisherman father unfurling his sails and nets from his small boat.

Zarva's story reveals something essential about the Russian experience with whales—a continued desire to connect the animal's industrial killing with folk maritime practices that in Russia had never been extended to whaling, along with a peculiarly Russian conception of the hunter. The harpooner, for the Russian writers, was a noble figure, partaker of a long Russian tradition celebrating the hunt, the solidarity of the hunting group (*artel*), and the appreciation of the hunter for his prey. Zenkovich wrote glowingly about one shipmate's love for his native *krai*, the Far East: "'For a hunter there is nothing more beautiful,' said Ginter, a passionate hunter for all animals and birds." Zenkovich admitted that, "Although I am not a hunter, during my travels around the Far East I am able to make my way successfully through a hunting conversation."[81] The universally positive *Slava* literature described the harpooners as "cold-blooded," meant as a compliment.[82]

Fiction made the same points. Ligov's brave crew was composed of "honest hunters, sailors, and good compatriots."[83] One might say that the first and last of these labels were nearly synonyms in whaling literature. As William Husband has pointed out, Imperial and Soviet literature both broadly connected "hunting and concern for nature." This juxtaposition persisted—improbably—even through the profligate 1930s of rushed industrialization, as brutal scenes of hunting coexisted with scenes of tenderness for animals even in children's literature.[84] The beauty of nature as revealed by killing it was perhaps not unique to Russian literature, but its staying power through the late twentieth century attests to a unique strength of this theme.

An obvious problem with the idea of harpooner as heroic hunter was the extensive machinery that surrounded him, giving him every advantage against the whale and protecting him from any counterattack. These were not the romantic days of open-boat whaling, which the Russians had mostly missed. It is not difficult to detect discomfort in the Soviet writers. Zenkovich, for example, wrote that "There is no hunt to compare with that of the giants of the sea, even if the hunters have many modern advantages and are protected from accidents."[85] One way Soviet hunters could overcome the evils that paved their success was to attain outsized numbers of kills—thus the emphasis on individual world records for whales taken that appears in all of the literature from the *Slava*. Vakhov, too, inserted the idea of Russians as record whale killers in his second novel, an anachronistic addition to a nineteenth-century industry that did not keep such records.[86] Thus, the mental stage was set for the reckless Soviet exploitation of Pacific and Antarctic whales. An industry sustained by the idea of the worthiness of the traditional hunt, equipped with high-modernist technology that hopelessly tipped the scales between humans and whales, justified itself in part through superhuman orgies of killing.

One of the strangest scenes in Soviet whaling fiction is best explained through the persistent Russian approval of the hunter. Vakhov's Captain Ligov, frustrated and nearly defeated by his inability to purchase a whaleship, is gripped by anger. Standing on shore he sees a sperm whale break the surface of the ocean. "The captain's eyes flushed with a hunter's passion [*azart*]"; Ligov grabs his Winchester and shoots hopelessly at the distant whale. "Soon we will not have to hunt them like this" he growls. This scene, in all of Russian whaling literature, approaches nearest to Melville's descriptions of Captain Ahab. Ligov's burning, insensible passion to kill leviathan (it is even a sperm whale here) glows nearly as hot as Ahab's. And yet Ahab, for all his moments of pathos, is for Melville unambiguously an antihero, pursuing in his wrath his own destruction. Ligov also drives himself too hard and perishes at the end of the novel in pursuit of his dream. However, through it all Ligov remains a hunter rather than a demon, someone who kills in noble battle and for the good of others and his country. Ligov's death finally comes not under a whale's tail but from the hounding of foreigners.

Alfred Berzin, quoted earlier as the author of a revealing memoir about Soviet whaling, hammered the Soviet system for its contribution to the environmental catastrophes of the twentieth century. "I was not and did not want to be a member of the Communist Party," he wrote, and he generally absolves the individual Soviet whalers of guilt, instead faulting the state apparatus. Yet, he also prefaces his exposition by debunking what he considers the myth of the hunter in the twentieth century: "The last few decades of Soviet large-scale whaling had nothing in common with greatness or heroism. The powerful catchers with artillery-like harpoon guns, harpoons that pierced a whale, and sonar to follow the whale underwater, all left the whale with no chance. And there was nothing resembling single combat. Whalers acquired the ability to kill all the whales that they found, and they did not miss this chance, often leaving a desert in their wake."[87]

CONCLUSIONS

In this chapter, I have suggested that the attitudes Berzin locates as specifically Soviet and statist in fact permeated a much larger cross-section of society for a far longer time, from imperial whaling enthusiasts to expedition scientists, to Soviet novelists. To the persistent romanticization of the hunter can be added a long-standing and passionately held belief that Russians should replace foreign whalers and that whales had brought the scourge of foreigners to Russian waters. These attitudes were rooted in Russia's unique intersection with the history of whaling, missing sail whaling while becoming one of the world's dominant industrial whalers. While these attitudes certainly did not themselves kill the whales, they both provided an entrenched cultural background in which Soviet officials could form destructive policies and provided historically resonant justifications for such destruction. This, in retrospect, was the true tragedy of Captain Ligov.

PART V

Bodies and Disease, Health and Environment

15 Strengthening the Tsarist Empire's Immune System

Environmental Cures along Crimea's Coast of Health

George Lywood

"What can treat consumption?" asked a leading tuberculosis researcher, Dr. F. T Shtangeev, speaking to the Yalta division of the Society for the Protection of the People's Health in 1898.[1] "As neither a rational nor radical treatment exists yet, and no specific method for the extermination of the tuberculosis parasite has been found," he continued, "what is left is only to heighten the immune systems of patients, to promote the counteractive strength found in tissues and blood."[2] To do so, Shtangeev recommended that patients visit a place such as Crimea's coastline, where a series of environmental factors allowed, or could be marshaled, for region-specific treatments designed to improve the body's immune systems. Crimea, he and others suggested, was home to a "Coast of Health, [where] nature was somehow united in one focus: on the life-giving strength of the sun, air, and sea." Crimea, they claimed, actually gave "clear-cut miracles" to the Russian people.[3]

Shtangeev, in a fashion typical of Crimean physicians, adhered to a predominantly environmentalist vision of pathology and medicine, despite the onset of the bacteriological revolution.[4] Crimean physicians were practitioners and proponents of medical geography (or medical topography) in the sense that they based their practices around an understanding of how humans could live in and benefit from various environments for health purposes.[5] Although the Russian state, when developing policies to combat epidemics such as cholera leaned far toward contagionism and, therefore, most often sought to battle disease through cordons, disinfection, and quarantines, what we see in Crimea is a physician-driven emphasis on environmental cures backed by scientific research.[6]

Physicians in Crimea observed the environment and wrote vast reports and scholarly studies as well as the most popular tourist guidebooks. Rather than going on about micro-organisms, they discussed landscape and pointed to the ways in which the environment as a whole could be harnessed as a healing entity. Physician discourse about environmental cures then shaped planning and the built environment of the region's coasts. They guided individuals with both minor and serious afflictions to carry out treatments individually, as part of their daily routines. And, whenever funding was sufficient, physicians pushed to create spaces, such as sanatoria, where the nexus between modern human habitation or convenience and natural forces could best be used to treat "the people."

Physicians and the public discourse they helped to produce established the myriad connections between health and the environment that would ultimately give Crimea its reputation as the "Coast of Health." The close links between medical advice, medical practice, and the environment in Crimea served as key advertisements for the region and strove to distinguish it from other parts of the empire and resorts abroad, shaping images of Crimea in the popular imagination across tsarist Russia. More than simply a destination for leisure, Crimea drew in the seriously ill and offered possibilities for the generally healthy to take advantage of the temperature, humidity, mountains, seaside, and other environmental factors that served to strengthen immune systems and provide comfort and an escape from filthy industrializing cities.

THE ENVIRONMENTAL CURES AND PUBLIC IMAGES OF THE COAST OF HEALTH

The rise of germ theory at the end of the nineteenth century has for some historians represented a turning point away from environmentalist thinking and toward a more scientific understanding of "modern" medicine.[7] Crimean medicine, however, did not yet extensively draw upon the newfound evidence, developed by Louis Pasteur, Robert Koch, and others in the 1880s concerning the transmission of contagious illnesses. Other historians have compellingly argued that the supposedly "unscientific" environmental medicine of the nineteenth century persisted in many places, despite overbearing narratives of contagion and increased emphasis on laboratory work in the field of medicine.[8]

The influence of germ theory and new understandings of disease transmission did not supersede environmentalist thinking whether Crimean physicians were compiling masses of data into multivolume scientific reports or placing emphasis on rigid, institutionalized sanatorium routine. Indeed, the majority of visitors who sought health in Crimea, certainly before 1917, regulated and treated themselves rather than entering sanatoria. They pursued water and solar therapies with little or no direct supervision from trained professionals but routinely based their activities upon the recommendations of doctors in travel literature. Authored (or coauthored) by physicians based upon their work in the field and laboratory, guidebooks carried on the tradition established by romantic writers in the 1820s and 1830s of taking the exotic landscapes of the Black Sea coast and turning them into something accessible for Russians.[9] Authors gave the coastline special significance in the empire, drawing upon European aesthetic principles, geographical knowledge and a growing concern about national identity in the late nineteenth century.

But the medical benefits almost always remained central, and the "exotic" treatments available often only in Crimea drew readers to the region. Crimea's south coast boasted a confluence of key environmental factors that contributed to the peninsula's therapeutic potential, including fresh maritime air, moderate temperatures, and high annual rates of sunshine. Visitors bathed in the Black Sea, sought treatment at altitude, and purchased fresh produce on a daily basis. Mineral deposits in lakes on the west coast of Crimea made the region ideal for the construction of mud baths. Most Russians came to the south coast of Crimea to bathe in the sea, breathe its fresh air, eat therapeutic grapes, drink wine, kefir, koumiss, and enjoy the benefits of the warm sunshine. In other words, they practiced the local therapeutic methods: hydrotherapy, fangotherapy, physiotherapy, solar treatments, and various forms of balneotherapy.

Crimea offered relief and strength to victims of a wide range of illnesses and afflictions. The spectrum of visitors ranged from the terminally ill to those with minor and perhaps temporary maladies. Russians came to Crimea to treat a wide range of ailments and diseases such as tuberculosis, syphilis, hemorrhoids, chronic rheumatism, neuralgia, hysteria, emphysema, asthma; respiratory or pulmonary diseases; ailments of the liver, spleen, stomach, genitals, or skin or simply if they experienced a recurring cough, got headaches, felt depressed, or felt tired.[10] And this is far from an exhaustive list. Many Russians in fact came to Crimea for its therapeutic environment without ever having experienced any symptoms at all, seeking simply to revitalize and strengthen their bodies prophylactically.

As each sort of treatment could be carried out in the confines of a medical institution or entirely individually, the extent of physician supervision varied greatly as well. The degree to which individuals sought out care from doctors depended upon personal preference, the severity of symptoms, the sort of affliction requiring treatment, and of course cost. Just as crucial, however, was the state of medical knowledge among Crimean doctors and the ways in which they presented that information to the public.

Climatotherapy

On perhaps the most basic level, many Russians visited Crimea's south coast simply in order to undertake climatotherapy (*klimatoterapiia*), which was defined as the therapeutic "influence of fresh air."[11] The way that the mountains stretching along the south coast of Crimea protected south coast resorts from the cold winds of the north did make them ideal spots for such therapy and the region's promoters took advantage. N. Golovkinskii wrote in his 1898 guide that "you have to come to the conclusion that [Crimea] is the best place in Russia in terms of its therapeutic properties." He continued, "so incomparable are its merits, that for almost the entire year, almost every single day, the ill can be out, unharmed, in the fresh air."[12] Dr. K. A. Belilovskii claimed, as more than just part of a sales pitch in guidebooks, that Yalta could be imagined throughout Russia as "a miraculous city in the soft, warm south, flooded for eternity by the shining sun."[13]

The entire peninsula possessed essentially two climate zones that concerned the public. Most of Crimea was steppe, including all of its northern regions, Simferopol, and right down to the northern environs of Sevastopol in the west and Feodosia in the east. The Crimean mountains, which sloped right into the sea, protected coastal towns from the colder prevailing winds from the north, while opening them to the sun shining from the south, creating a second climate zone along the south coast. Often known as the *Iaily* (Crimean Tatar for mountain plateaus), the highest peaks all rose parallel to the sea in southern Crimea and rose over five thousand feet above sea level.[14]

Those towns located at the base of the mountains and right on the sea, therefore, had the deserved reputation of possessing the most favorable climates. V. I. Kosarev wrote in 1913 that, "Yalta, protected from northern winds by the high wall of the Crimean mountains, and cleansed by the warm waters of the Black Sea, was in possession of a climate so warm that it would be impossible to find anything similar throughout European Russia."[15] Similarly, Alupka, "situated in front of the height of the *Iaily*, the mountain 'Ai-Petri,' protected from the east by the cape of Ai-Todor and from the west by the 'Black Mountain,' is open only to the sea and the southern sun, making it the warmest point along the Russian Riviera."[16] A guide to the town claimed that the mountain seaside climate of Gurzuf acted "charitably" toward anyone suffering from breathing disorders.

The warmest parts of the Crimea extended from just east of Balaklava to the point at Ai-Todor and the town of Alupka. It was along this part of the coastline that the mountain cliffs were closest to the sea, completely sheltering the resorts on the coast from the cold northern air, while the cliffs, most notably the famous Ai-Petri, acted as "natural condensers" for vapors coming from the Black Sea, which increased precipitation and created a natural source of cycling fresh sea air (which, unfortunately, became rather cold in the winter). Yalta, however, was open to winds from the east, further increasing the amount of sea air received but decreasing slightly its average daily temperature. Guides immediately pointed out the differences between the temperature along the Crimean coastline and other parts of Russia. Comparing Yalta (where the average yearly temperature—the first thing guidebooks typically mentioned—was 13.4°C) to other parts of the country was commonplace (Moscow, 3.9°C, St. Petersburg, 3.7°C, Kiev, 6.8°C). Even in January, Yalta's average temperature was above freezing, while in the heart of the summer the average temperature was a reasonable 25°C and cooled only below room temperature by October.[17]

The moderate maritime climate was key to *klimatoterapiia* for, compared to mainland Russia or even most of Crimea, medical and meteorological studies were able to show less fluctuation in temperature, significant humidity in the air, high barometric pressure, strong air currents, and that the air itself contained a significant composition of ozone, a general lack of dust, and small parts of salt, bromine, and iodine. "Under the influence of maritime air," wrote the leading researcher and author of a comprehensive guide to all of Russia's therapeutic places Dr. N. P. Ivanov, "[one] breathes more easily, the pulse slows, while sleep and well-being increases." The maritime climate could further improve the blood flow and fortify the nervous system, he continued, and thus, "the overall effect of maritime climate is the saving of the breathing organs, blood-flow, digestion, and generally holds together the economic balance of the organism."[18] Simply spending time on Crimea's coast was naturally therapeutic and even enhanced the effects of mud bathing, sea bathing, and all other forms of treatment on the peninsula.

Generally, the climate was stable, moderate, and comfortable, but once in a while for one to three days at the height of the summer, dry winds from the northwest called *samiel* (from the Turkish word for a strong suffocating desert wind) could cause the temperature to rise up to anywhere from 34°C–39°C.[19] This was the exception, however. By and large the climate was humid and there were no such winds in Yalta. Temperatures gradually decreased as one moved east toward Alushta, Sudak, Koktebel, and Feodosia.[20]

Mild temperatures and opportunities for climatotherapy became the top selling point for Crimean resorts, and a key way in which doctors used guidebooks to help define the peninsula's image in comparison with other regions in the tsarist empire or Europe. Publishers looked to draw to the Crimean shores all those wealthy Russians who went abroad (and especially to Europe) for extensive and expensive climate therapies. Infused with nationalist rhetoric that in many ways set out to make Crimea a European place, guidebook authors would speak of how, "in all of Russia, Crimea's environment is the softest and most pleasing," or in the words of another prominent physician, how "Yalta, by its geographic position on the south coast of Crimea, in a warm climate, belongs on the list of the very best climate stations on the whole of the European mainland."[21] A 1910 guidebook began by claiming that "the Crimean peninsula is the richest region in our native land in terms of its beauty and its excellent climate conditions. It stands out not only from the rest of Russia, but from the whole range of world resorts."[22]

Balneotherapy

Beyond supplying fresh air, the Black Sea itself was responsible for much of Crimea's therapeutic capability. Surface waters along the coast hovered around 20°C during the summer months, and although the temperature dropped significantly in winter, the sea around Crimea very rarely froze (though parts of it did in 1911, creating anxiety among the locals regarding the upcoming season). Although comfortable during summer months, the Black Sea therefore was nearly twice as cold as the Mediterranean and had half the salinity. Nonetheless, its moderate temperature, spread over a long bathing season (which lasted from mid-May to October or November, even) was convenient and allowed for extended treatment regimes. Bathing in the saltwater, they claimed (which also contained a variety of other minerals not found in freshwater in other regions of the Russian Empire, for example), was shown to have a more therapeutic effect on the body than bathing in lakes or rivers.

Sea bathing (as well as the koumiss, kefir, and grape treatments discussed below) fell under the broad category of medical treatments known as balneotherapy, defined by Dr. Ivanov as the "science of the therapeutic properties of mineral waters, or, more specifically, the science of the particular properties and types of action external and internal use of mineral waters for the treatment of chronic illnesses."[23] Balneotherapy placed an emphasis on chronic illnesses, which typically, Ivanov continued, "resulted from dietary issues, bad habits and customs, excess, effeminacy [*iznezhennost'*], overwork, neurasthenia, and so on." Each the result of lifestyle choices and the realities of city life, he wrote, chronic illnesses were those most treated in Crimea in general, and the various forms of balneotherapy were likely the most used by visitors to the Black Sea coast. Essentially, he continued, urban lifestyles were causing a great deal of stress on the human body, "causing it to become less able to defend itself from becoming sick, [often] causing permanent damage to the organism, which is why we say 'chronic.'" Balneotherapy, however, according to medical knowledge in the nineteenth century, could be extraordinary successful in helping to treat chronic illnesses, and Ivanov emphatically stated, countering claims that bathing was but a leisure activity, that it was *not* simply "an act of fashion."[24]

The primary form of balnoetherapy was to actually drink waters. Because such a large portion of the human body is composed of water, ingesting waters with a variety of minerals or chemicals caused the "healing agents" (in the words of Ivanov) to spread throughout the body through the bloodstream and to nerve centers and organs. Few came to Crimea to drink waters, however. Patients requiring these treatments were far more inclined to go to the North Caucasus, where mineral waters were far more diverse and abundant and where effective treatment facilities had existed since the early nineteenth century.

Sea bathing, however, was naturally the most popular form of balneotherapy in Crimea, and Crimea was the most popular site for sea bathing in tsarist Russia. Although the sea's minerals were not directly ingested, doctors spoke publically about how bathing did have a number of key therapeutic effects on the human body. Depending upon the temperature of the water, its chemical elements, its movement, the climate along the coastline, and other surrounding conditions, they claimed, bathing could have profound effects.

The leading authority on the methods and effects of bathing in the Black Sea was Dr. V. N. Dmitriev (1839–1904). A practicing physician in Yalta for thirty-five years, he published the standard reports that visitors were advised to read and that guidebooks summarized when considering any of these topics from 1869 onwards. His seminal work on bathing, *Healing through Bathing on the Coast of the Black Sea*, was published posthumously by his son, also a doctor, in 1913, but multiple editions of his treatise on bathing, *Sea Bathing in Yalta and in General on the South Coast of Crimea* represented the core of research on the therapy by the 1890s.[25] Excerpts from his publications appeared in tourist guidebooks, pamphlets, and advertisements for resorts across the peninsula.

Bathing, according to Dr. Dmitriev, was therapeutic for almost every part of the body, and for a wide variety of conditions. It worked to improve everything from the internal organs to the brain, to the immune system, to the digestive system. It simply gave energy. Filling the needs of both the seriously afflicted and those with moderate or no symptoms at all, bathing, he argued, could have positive effects for both the ill and the healthy.[26]

Much of the water's therapeutic potential, Dmitriev wrote, derived from the fact that it gave the body a sort of shock. Although the shock was "the same for the cowardly or the fearless," as the pulse increased by several beats per minute, one breathed with greater energy, and the nervous system would become "excited" upon entering the water. "It is exactly these effects that give bathing its healing potential," he continued, "when the nervous system becomes more impressionable and excited, it prepares the body for a strong response." By repeatedly training the body to respond with strength, he argued, it would heighten the immune systems of individuals suffering from a range of possibly otherwise untreatable health issues.[27]

More particularly, he suggested that there were several effects that the water would have when an individual went into the sea: the skin became cooler, tightened, and whitened, and the nervous system became generally excited, which affected the other parts of the body, especially the muscles. The heart slowed, the lungs increased the frequency and depth of breathing, the pumping of blood to the extremities was reduced and directed to the internal organs. When the skin is first submerged and contained less blood, he argued, it was at this point that the nerve ends became as if they have been subject to a light anesthetic, which in a sense meant that they have become more able to withstand the shock of outside phenomena.[28]

Then, Dmitriev went on, as the skin warmed and the influx of blood returned to the nerve ends, as the body reacted to the shock of entering the water, nerves became even more responsive and strengthened against external phenomena. This strengthening effect did not affect just the skin but all parts of the body. Internal organs were able to "feel" better. The liver, digestive system, and the intestines became more "alert" (which was why hunger arose after swimming), as did the heart and lungs. Vision became more acute, as did hearing and muscles. Bathing even affected the mind. It allowed ideas to develop more quickly, for clearer thinking, and for "all psychological abilities to be enhanced." After bathing the "brain worked more easily," its "impressions became livelier," and bathers, "generally felt better." Combined, "these effects are the greatest influence that bathing can have," Dmitriev concluded.[29]

Because bathing could range from a social and leisure activity to a methodical and therapeutic one, Dmitriev set out cautions and guidelines that were publicized in order to ensure that visitors, no matter their state of health, were not harmed through bathing in the Black Sea. In fact, the main advice that he and virtually every guide gave was that any therapeutic regime be approved by a doctor beforehand. He warned in particular that the shock that the body felt when entering the water—the primary source of bathing's positive effects—could do more harm than good. "The general rule," he advised the public, "must be to always be especially cautious [with bathing]. You need to choose the proper time of day, plan how many minutes you will spend in the water, what you are going to do before and after bathing. It's necessary, for both the sick, who are planning to use the health benefits of the seawater, and the healthy, who are just bathing for pleasure, to be very attentive." However, he argued, if bathing was done properly, and daily, for four or six weeks in a row, it would have lasting positive effects on the body's ability to combat disease.[30]

Solar Therapy

Bathing was not the only therapeutic activity undertaken on the beach, of course. Exploiting the effects of Crimea's abundance of warm sunlight was also recommended by authorities. During the summer in Yalta there was an average of ten hours of direct sunlight (that is, no cloud cover, no shade from the mountains) per day. The sun in Evpatoria, guides advertised, would "revitalize the entire body," especially along its "natural velvety sand beaches," while its coastline was "untouched by any sort of filth."[31]

Sunbathing could be carried out on any of the coast's beaches but was also a key form of treatment at the Children's Beaches, specifically maintained for solar therapy by Dr. V. G. Lapidus in Yalta. Doctors noted that the sunlight could have very positive effects on the body and particularly the skin, regardless of age. They recommended "sunbathing" for skin conditions such as eczema and lupus, saying that the sun could act as a disinfectant since it kills bacteria on the skin's surface. The sun was also understood to be healthy for internal organs, as it was seen to liven the work of the nervous system, speed the digestive system, improve blood flow, improve metabolism, and generally improve the state of the entire body. Sunbathing came with its own rules, the most predominant being that a doctor's supervision or at least advice was necessary, and that it should always be done for only between twenty and thirty minutes, with the use of an umbrella, hat, or white sheet to block direct sunlight.[32] However, even by 1909 journalists lamented the underdevelopment and underuse of solar therapy, suggesting that the results of the efforts of Lapidus had been excellent and warranted the construction of new facilities along the coastline.[33]

Grape Therapy, Milk, and Koumiss

One of the most unique sorts of treatment, and one that drew in visitors particularly to Crimea, was the consumption of grapes. Grape treatments were a distinctive and defining feature of resort life in Crimea, and viticulture was one of Crimea's most lucrative industries. The Massandra brand, produced on a vineyard on the outskirts of Yalta, was made into one of the best-known sorts of wine from the Russian Empire. Not all of the grapes grown were used for wine, however, and as was the case with dozens of other vineyards in Crimea, many were sold at markets in resorts, to be consumed for their healing potential. Most of the grapes in Crimea—and there were over two hundred unique varieties—were in some way unique to the region.

Doctors and guidebooks suggested that eating grapes according to their regulations could reduce phlegm and improve breathing ability, reduce the effects of liver diseases, relieve the engorgement of the abdominal cavity, eradicate kidney stones, hemorrhoids (it acted as a laxative), and was considered useful for hysterics, hypochondriacs, heart problems, and generally for those whose diet had caused weakness or fatigue. Remarkably, doctors recommended that patients carry out grape treatments continuously over the duration of from four to six weeks. Patients were advised to begin by eating one pound per day, then increase each day by a half pound until they were eating eight pounds of grapes per day, and continue at that level for some time, after which they were to decrease their consumption at the same rate down to nothing.[34] Other guides suggested that some treatments even recommended that up to ten to twelve pounds per day could be consumed, part on an empty stomach and the remainder over the course of the day.[35]

Perhaps even more so than with bathing, it was important to follow the correct procedure if visitors wanted to benefit from eating grapes. Not doing so was almost certain to harm the body. A doctor's consultation was always recommended before embarking on such a treatment, especially since it was difficult for a nonspecialist to determine exactly which sort of grapes to eat. The grape variety mattered a great deal since Crimean health professionals agreed that certain types could have a decidedly negative effect on certain types of illnesses.

In selecting the right sort of grape, patients were advised to pay particularly close attention to the amount of sugar contained within them. Apparently 12 percent was typically ideal for the generally healthy individual. Doctors could also provide advice on where to buy grapes, as certain salesman might claim to be selling grapes with healing properties (and therefore at a higher price), when in fact they might have no benefits whatsoever. Grapes in the center of Yalta could be bought on any number of street corners and cost up to twelve kopecks per pound—but about half of that in the smaller resorts. The most commonly consumed grapes were from either the *shasla* family (a common variety in western Europe, known as chasselas in English), or the *chaush* family. It was, as a general rule, considered inadvisable to consume dark or red grapes for treatment. Often, visitors preferred to drink pure grape juice in place of the regular fruit, and many sellers owned machines that produced fresh grape juice for consumers from grapes they selected themselves.[36] In addition to grapes, all sorts of other fruits were grown locally and also recommended for health reasons, including apricots, peaches, pears, plums, and apples.[37]

Adding to the balneotherapeutic solutions available in Crimea, doctors also encouraged the consumption of locally produced cow's milk, which, like grapes, they felt was part of a healthy diet. Because milk was an emulsifier, and because of its composition (fat together in an aqueous solution with sugar, protein, casein, and salt), guidebooks praised it as a critical component of an individual's diet. Koumiss, however, had a far greater curative effect than simple milk.

A beverage traditional to the Mongols and other peoples of the steppe, local varieties of koumiss had been produced by Crimean Tatars for generations. Koumiss was made from mare's milk and fermented, giving it a "pleasant, slightly sour taste." When under fermentation, the sugars in the milk produced alcohol, carbonic acid, and lactic acid, which were the ingredients that gave koumiss its ability to heal. Koumiss fluctuated according to the quality of the milk, the energy and length of fermentation, and the skill of the master and the cleanliness of his dishes. The amount of spirit could range from 9 to 33 percent, sugar from nothing to 39 percent, lactic acids from 2 to 12 percent, fat from 12 to 26 percent, proteins from 16 to 37 percent, and salt from 3 to 5 percent. Generally, the longer the koumiss was fermented, the more alcohol, carbonates, and lactic acids it would possess, while having less sugar and protein. For the varieties with the highest alcohol and carbonate content, koumiss would be fermented for seven days. In other cases, fermentation may have lasted only for an hour, retaining high sugar and protein content.

Because of its properties koumiss did not only have a dietary function, doctors believed, but also "therapeutic properties in that it increased blood flow, strengthened perspiration and the rate of urination, and stimulated the organism."[38] In comparison with regular milk, it offered many of the same curative properties, but guides argued that it worked much faster, improving the immune system on a daily basis and could have between 50 and 100 percent more of an effect. Koumiss was known in Crimea to be especially good for young adults (whose bodies could handle the intensity better) who had been weakened by a variety of afflictions, especially with ailments in the digestive system. However, guidebooks never recommended it for tuberculosis (and certain other) patients as it had a decidedly negative effect on them by actually accelerating the progress of the disease.[39]

Because koumiss brought about decidedly positive or negative outcomes, guidebooks advised patients to stick especially closely to doctor's recommendations and to proceed with caution at all times. Koumiss treatments required some acclimatization, and for those not used to drinking it, stomach pain and diarrhea were often the result. Therefore, doctors typically recommended that patients take just one glass per day to begin with. As with the grape treatments, patients would increase the dose daily and eventually take on stronger varieties of the beverage. Some patients would go as far as to drink six bottles per day as their maximum during the regime. More typically, koumiss treatments in Crimea were far less intense, and the treatments were seen largely as supplemental (*dobavochnye*) to other medical advice.[40]

Similar regimes were devised for the consumption of kefir, another beverage local to the North Caucasus and Crimea, but made from cow's milk. Doctors also recommended it for those with afflicted internal organs, anemia, or scrofula (tuberculosis of the neck).[41] Kefir differed from koumiss in that it had significantly less alcohol but significantly more carbonic acid. Prepared by placing kefir grains in warm water for up to three hours and then pouring it into hot fresh cow's milk, which fermented for several days (guides suggested it was best to leave for eight days), kefir was determined to be a very good dietary supplement. Unlike koumiss, doctors did not claim any substantial therapeutic effects for kefir.[42]

Mud Baths and Sanatoria along the Coast of Health

It was town and city officials at Crimea's resorts, drawing on the recommendations of physicians when budgets allowed, that were most responsible for constructing and maintaining the boardwalks, parks, beaches, pathways, and other physical infrastructure necessary to carry out the Coast of Health's treatments. Physicians were more influential over the activities carried out within more discrete structures such as the mud baths and sanatoria. Private sanatoria were constructed, but most were philanthropic in nature, requiring partnership with wealthy Russian nobility (often women). In either case, physicians set out to create spaces where patients would go voluntarily (often at great cost, though wings of beds were set aside strictly for the poor in philanthropic institutions) as part of a way to discipline themselves, separate from the general public, and get increased supervision over the same sort of treatments discussed already in this article. There were two phases of development: that of the mud baths, which began much earlier, and that of many resort sanatoria along the coast (see below).

Despite there long having been healing centers for Crimean Tatars, in 1826 Russians built structures to visit and maintain the mud baths at Saky Lake, located just fifteen miles to the southeast of Evpatoria, under the supervision of the facility's future head doctor, S. N. Ozhe. The baths (known as the *sakskaia griazelechebnitsa*) constituted the first formal fangotherapy facility in the Russian Empire and were the first physical structures designed for health purposes in Crimea. The baths at Saky were the larger of two major mud bath facilities that could be found in Crimea in the early twentieth century. The second was built much later on the western outskirts of Evpatoria, in 1886, on the small Mainakskii Lake, where the properties of the mud itself were advertised as being nearly identical to Saky as were the ways in which it was used with patients.[43]

Based upon empirical research and analysis of the mud's chemical composition, Ozhe concluded that "radioactive elements" in the mud would be most effective for treating tuberculosis, infected bones, rheumatism, different sorts of trauma, gynecological diseases, infertility, and urogenital disorders.[44] Although such claims universally derived from experimentation, and although there were several subsequent studies on the mud's physical properties, the actual reason behind the mud's healing properties was not fully understood during the tsarist years.[45] Despite physicians' uncertainty, medical professionals at Saky and Evpatoria intensively studied the contents of the mud baths, made clear what sorts of afflictions could be healed, and published the results in public reports that found their way across tsarist Russia.

The primary visitor's guide to the mud baths listed bone infections and pain, scrofula, syphilis, rheumatism, arthritis, peripheral nervous system disorders, and feminine diseases, especially diseases of the uterus as having been consistently successfully treated with the mud baths. Success rates were noted regularly in publications, and the range depended on the affliction, with "women's diseases" reportedly being cured entirely 60 percent of the time, compared with rheumatism and bone disorders having been reportedly cured at least 90 percent of the time.[46]

Formal sanatoria, ranging widely in purpose and methods, began cropping up across the Crimean peninsula beginning in the late 1890s. True, the real growth in sanatorium construction and attendance took place following the revolution, but the genesis of Crimean sanatoria as facilities designed to draw the most therapeutic elements of the Crimean environment to patients took place before 1917. Tsarist era sanatoria not only worked to treat individuals but also served as crucial sites of research that legitimized environmental cures, which resulted in their widespread popularization in the Russian imagination.

Perhaps the best-known sanatorium in or around Yalta and one that specifically dedicated itself to treating breathing disorders and tuberculosis was initially opened in 1897, based upon the donations and direction of Princess M. V. Bariatinskaia, a wealthy landowner, philanthropist, and member of the Red Cross and the Sisters of Mercy. Beds there were immediately in high demand, but it had an official policy of giving beds first to those who could not afford private facilities, naming specifically students, doctors' assistants, and low-level servicemen, making it one of the few places where those outside of the nobility and bourgeoisie could receive treatment in Crimea. The construction of the facility—and its operation subsequently—was made possible by public funds, the philanthropy of Princess Bariatinskaia, the cooperation of state and local governments, and after 1907, the Red Cross, which built a new wing into the facility.[47]

Yalta also boasted the philanthropic Alexander III Sanatorium, which, like Bariatinskaia's, was one of the few in Crimea that admitted patients whose tuberculosis had reached advanced stages, as the risk of contagion was too high for entrepreneurial owners.[48] The Red Cross constructed its own sanatorium, named after Empress Maria Fedorovna, which opened in 1910. It was unique in admitting only residents of Yalta whose afflictions had not yet reached advanced stages. The final predominantly charitable facility in Yalta was the Ai-Danil Climatic Colony for Weak and Sick Children. It opened in 1912 and operated year-round; its treatment regime involved bathing, exercises, special diets, treatment for some mild forms of TB and nerve disorders. The goal of the facility was explicitly stated: to harness Crimea's therapeutic natural characteristics in order to strengthen children's immune systems and physical strength.

Other, more upscale sanatoria possessed the requisite facilities to provide hydrotherapy and air and sun treatments, with exercise hall, foot baths, massage tables, and other amenities including costly menus and latest medical technologies. Advertisements stressed treating exhaustion and patients would be able, again, to strengthen their bodies and immune systems, especially those who had recently overcome a disease that left them in a weakened state.[49] Guests at sanatoria at altitude were able to breathe the sea air scientifically shown to have the highest oxygen content. Specifically dedicated to researching and treating lung disorders, especially those not at advanced stages, sanatoria were becoming, in the tsarist era, a significant feature of the Coast of Health because they provided spaces in which visitors and physicians could more acutely focus their treatment regimes—regimes that were nonetheless based almost exclusively upon environmental cures.[50]

RUSSIAN ENVIRONMENTAL HEALTH IN COMPARATIVE PERSPECTIVE

The most analogous situation to that in Crimea can be found with the American West, where (as in Crimea) migrants with breathing disorders and immune system deficiencies forever altered the region's history, its place in popular imagination, and its physical environment. Mark Twain published his thoughts on the curative powers of the region around Lake Tahoe in the 1870s, and the result of his publications was that consumptives and physicians alike poured to the Rocky Mountains in search of cures.[51] By 1900, one in four migrants to California came in search of health.[52] In Colorado, where an estimated one-third of the population in 1880 consisted of health-seekers and their families, hospitals sprang up to meet the needs of the sick in such great number that the state was dubbed "The World's Sanatorium." In New Mexico, tuberculosis treatments paralleled those in Colorado, while Nevada and Utah also saw an—albeit smaller—share of immigrants seeking medical treatment.[53]

Sheila Rothman has shown how widespread and deeply held fantasies about the American West, originating well before Twain, fostered an image of a region that was pure, wholesome, and reinvigorating. Travel accounts advertised the West as an El Dorado, an earthly paradise, as a site of both wealth and health. Drawing upon a romantic image of the Native American, physicians and literature alike encouraged outdoor activity in the mountains and a lifestyle separated from urban communities. Legends grew where dead bodies were brought to the West and returned to life in dry climates at altitude. Doctors published scientific treatises in local pamphlets and articles in the *Journal of the American Medical Association* espousing the West's benefits to consumptive easterners. They also initially pointed to the ways in which it was beneficial to live in what were known as open health resorts, places where a region and a routine, as opposed to a physician or more formal institution, could offer optimal results.[54]

Rothman's excellent work focuses solely on tuberculosis patients, while I am concerned with a broader spectrum (in smaller numbers) of afflicted travelers, but the story of the American West parallels that in Crimea by demonstrating the connections between the environment of a particular place, public perception, and medical literature. As in the American West, guidebook authors and doctors alike depicted Crimea as a sort of paradise. The medical literature pointed to the ways in which visitors to Crimea could exploit the natural environment for cures or relief for both serious and marginal ailments. Treatments in Crimea, in or outside the sanatorium, were studied rigorously and scientifically, gaining legitimacy with the public both through proven results and mass publications. Fangotherapy, climatotherapy, solar or light therapy, sea bathing, physiotherapy, and grape, kefir, and koumiss therapies were all based upon this growing body of knowledge that doctors publicized in reports and subsequently in guidebooks. Therapeutic institutions where patients could receive highly regulated and supervised care were built. The result was a region in many ways characterized by its therapeutic methods generating a public image of a coast of health.

16 Reshaping the Land, Chasing the Mosquito

SOVIET POWER AND MALARIA IN TAJIKISTAN, 1924–1938

LISA K. WALKER

THIS IS the story of the Soviet encounter with malaria, and how it both precipitated and reflected the Soviet effort to understand and control its far-flung territory. The story begins in the eastern marches of the Bukharan Emirate, in the area of present-day Tajikistan, a land of narrow valleys and churning rivers that cut through the rocky western foothills of the Pamir Mountains before spreading over the dusty plains that descend to the Amu Darya and the Afghan border.[1] Although the region is generally arid, Russian reports beginning in the 1860s indicate that malaria was endemic, particularly in lowland areas and valleys: the behavior of the water and the practices of the people, especially in times of upheaval, created the necessary conditions for malarial mosquitoes to breed and transmit their deadly disease.

Malaria figured prominently in the discussions of Soviet authorities as they tried to occupy and administer the Bukharan Emirate between 1920 and 1924. People in Bukhara suffered from both the more severe and often fatal form of the disease, caused by the *Plasmodium falciparum* parasite (referred to as "tropical" malaria in Russian), and the less deadly form, caused by *P. vivax*, often called "tertian [*trekhdnevniaia*]" for its two-day cyclical fevers. Amid early efforts to grasp the magnitude of the problem after the Red Army incursion into Bukhara in 1920, Soviet experts claimed prevalence rates as high as 96 percent, with falciparum malaria resulting in case fatality rate as high as 30 percent.[2] Indeed it was against this stark backdrop of the magnitude of the problem in the early years of the Soviet Union that the fight to achieve the eradication (*likvidatsiia*) of malaria came to function later as a central heroic story in Soviet public health following the 1960 declaration of the disease's demise.

The case of malaria control in Tajikistan illustrates how one kind of transformative agenda—embodied in the localized health improvement efforts of physicians and engineers in the 1920s and framed in terms of the ecology of malaria—lost out to quite a different agenda, shaped by the crude economic priorities and extractive goals and embodied in the massive scale projects of the 1930s. Examining Soviet malaria control with a focus on engagement with the physical environment sheds new light on historiographical debates about the Stalinist period: we find that the continuities in philosophy and rhetoric across 1929 are stronger than the contrasts.[3] The transformative projects of the Stalin era were not so much a break with or a repudiation of revolutionary ideals but an exaggeration, a stretching and expanding of the original Bolshevik ideals to new, epic, and—in most cases—grotesque forms.

By drawing our attention to the relationship of government rule to the environment, this story also helps us consider how aspects of Soviet history were very much in line with trends in other parts of the Western world in the twentieth century. Soviet efforts at malaria control—and more broadly Russian notions about malaria and ideas about how to control it—fit squarely within the global context, and attention to the health and environmental pressures on the new Soviet state can help us to understand the links and common context that unite Soviet and Western history. The Soviet approach corresponded very closely to malaria control strategies pursued by Western experts at the time. Environmental management for vector control was the central pillar of the strategy of the League of Nations Health Organization, for instance, with an underlying goal of improving health conditions more broadly in rural less developed areas. Soviet practice was in line with international historical trends in malaria control, but a shift away from this approach in the Soviet Union began even earlier than in the West. Environmental management as a means of vector control was largely abandoned in many places throughout the world with the introduction of DDT and synthetic malaria drugs after the Second World War. The Soviet malaria campaign also changed drastically after 1945 when DDT and the drug *akrikhin* (atabrine) became widely available.

Likewise, the urge to radically transform the land and the population's relationship to it—by forcibly moving people en masse from a mountain home to a plain being harnessed for cotton production, for example—in the name of development also did not render the Soviet project exceptional. Although the economic priorities of collectivization and the specifics of bureaucratic competition under Stalin gave particular shape to the role of the environment in malaria control in the Soviet Union, the general trajectory elsewhere on the planet was the same. In both East and West, by the 1930s and certainly by the postwar years, environmental management and small-scale modification of the landscape to limit malaria's insect vector ultimately gave way to the application of new tools and technologies within the sphere of public health, synthetic antimalarial drugs with DDT most prominent among them worldwide.

The history of malaria in Central Asia has been written and narrated primarily by medical and scientific researchers. Very often this history serves as a backdrop for descriptions of present efforts to eliminate malaria for the second time.[4] In a scientist's history of malaria, humans—physicians, scientists, and public health workers as well as, more anonymously, patients—are naturally foregrounded; some attention may also be paid to the plasmodium parasite or the mosquito that transmits it. But few existing approaches place the malaria story within the wider context of the landscape and ecological setting in which malaria thrived.[5] Applying an environmental perspective helps us to consider the history of malaria in this region in more of a comparative context with other areas of the world. While the focus of this article is still very much on policy and politics, the environmental perspective also allows us to ask new questions about old issues, to consider certain orthodoxies in a new light. As Christian Teichmann argues in this volume, analyzing more closely the relationship of Soviet rule to the environment helps us to see the improvisational nature of the Soviet project. Paradoxically, perhaps, it also allows us to understand Soviet decisions and strategies in parallel with some of the approaches used in the West.[6]

THE LAND, WATER, AND DISEASE IN EASTERN BUKHARA

Nineteenth-century European observers, including those from Russia, drew a distinct tie between fevers and territory, and the territory of Eastern Bukhara was generally known as a place of disease and ill health. Malaria in the Bukharan Emirate had been endemic for generations, a normal fact of life. It was the newcomers in the 1920s and the instability that resulted from their incursion that framed malaria as an essential problem to be solved, with science. When the Soviets entered Eastern Bukhara—the land that would become Tajikistan—in 1920, they brought with them a distinctly new point of view that not only placed greater emphasis on environmental factors but took this as a starting point for trying to solve or control malaria by actively engaging with and altering the environment.

The Soviets who arrived in Eastern Bukhara in the 1920s were informed by the picture they inherited from late nineteenth-century European travelers to Bukhara, including its eastern province. In the published accounts of their travels, these observers offered vivid descriptions of endemic malaria. The Hisor Valley, stretching from Regar (present-day Tursunzoda, close to Tajikistan's western border with Uzbekistan) to the east of present-day Dushanbe, was actually known as the "the valley of death [*dolina smerti*]". Prominent Russian Turkestan expert Nikolai Maev wrote in 1897 that "Gissar [Hisor] at this time is a city that has earned its description as sickly [*boleznennyi*]. It has a reputation for its malignant [*zlovrednyi*] climate and for an abundance of swamps and fevers."[7] A Red Army physician overseeing malaria control during the civil war reported that, "even before the war, Gissar was nearly uninhabitable in the summer months for the billions of mosquitoes and severe tropical form of malaria."[8]

Other European visitors to Eastern Bukhara had been virtually unanimous on the severity of malaria, not only in Hisor, but throughout the region, including the lowland Vakhsh River valley and its principal town, Qurghonteppa; the middle Amu Darya (Panj) River valley and those of its tributaries, near the town of Kulob; and the towns of Shahrtuz and Qabodiyon on the lower reaches of the Kofarnihon River, in the southeastern corner of present-day Tajikistan. Newly arrived Soviet officials in the 1920s, who learned about Tajikistan from these earlier observations, got the impression that in this harsh country disease was distinctly tied to the rough, unimproved environment.

The Soviets began their encounter with malaria in Eastern Bukhara with the notion that, to control the disease, you had to control the environment. The Soviet medical establishment of 1920 of course had considerable overlap with the Russian medical establishment of the late nineteenth century, whose view that intermittent malarial fevers were tied to the land where they were prevalent was not unique. Europeans had long viewed the cause of disease in terms of miasmas and "bad air" (note the etymology of malaria: "mal aria"), which in turn were often linked to climate and location. Long before the discoveries at the turn of the twentieth century that confirmed the role of the mosquito in transmitting the disease, drainage and engineering projects represented one common method of controlling malaria in many parts of the world.[9] By 1920 the germ theory of disease had basically displaced miasmatic theories in Russia and the West, but it was still not a far-fetched concept to emphasize the connection between place and health.[10]

European observers (primarily Soviet Russians) held fast to the idea that Tajikistan was a sickly place well into the twentieth century. Soviet observers' perception of malaria in Tajikistan was framed by a complex and evolving sense of its causes. It was only two decades prior to the Soviet incursion into

Bukhara, after all, that malaria's route of transmission involving the mosquito and at least two different malaria parasites had been fully elucidated by medical researchers. By the 1920s these ideas had still not yet been fully accepted or understood by many laypeople or even among all medical practitioners.

Yet malaria was arguably more than simply tied to place or terrain in Tajikistan: it was connected to the climate, to the waterscape, and to the practices of the people. Malaria represented an ecological problem, and correspondingly the Soviets' approach to malaria was part of a larger ecological problematic that they faced in this new territory. Everything posed a problem, from the land and its environmental features to the people who lived there and knew the land well and lived—and fought—effectively on it. The Soviet struggle between 1920 and 1924 to take firm hold of Eastern Bukhara involved defeating a violent insurgency waged by the local population. The struggle also entailed overcoming very fundamental physical and environmental hardships, including the mountainous terrain and the lack of roads and other infrastructure needed to traverse and administer the highlands and valleys.[11] Amid this array of obstacles, debilitating malaria infection also kept officials in sickbeds, put laborers out of commission, felled soldiers, and incapacitated local populations. Soviet officials who came from outside seemed to wish that all aspects of the local environment—from the land and water to the people—were more passive and malleable, or at least that each aspect would present itself more readily for counting and cataloging, if not outright manipulation.

According to European visitors, the Muslim Central Asian population dealt with the threat of malaria not by attempting to change anything fundamentally or to combat the disease or any factors that may have facilitated its spread; everyone simply tried to avoid infection by leaving. Maev noted that the mountain retreat closest to Hisor named Karatag (Qaratog) was one place where locals went to escape the summertime fevers. Not only the beys and wealthy people went there, but anyone "who can find even a mangy little donkey [*plokhon'kii ishachok*] to transport their domestic belongings."[12] The German explorer Willi Rickmer Rickmers noted the same summertime "exodus" to higher ground. "At the approach of the warm season, when malaria and mosquitoes render the place almost uninhabitable, [the bey of Eastern Bukhara] migrates with his goods and chattels and a great part of the population to Karatagh."[13]

If the Bukharan response to malaria involved seasonal migration and a perception of the disease that was framed by religion rather than science, a point of view that had reigned for generations, if not centuries, the Red Army and the Soviet tropical public health professionals entered and approached things from a fundamentally different point of view. For them, malaria control was part of a program to improve basic health conditions in the territory of former Eastern Bukhara, underpinned by two important elements: a desire to consolidate control over the territory and a genuine adherence to the Bolshevik revolutionary aim of improving health conditions for all Soviet subjects. The Soviet project was distinctly modernist, and as such it fit squarely in a broader international context, with state officials and technical experts seeking to reshape the raw geological and environmental materials they had been given in an effort to achieve development and progress.[14] Specifically in terms of malaria, Soviet efforts closely paralleled the efforts by other modern states to control disease for the dual purposes of improving lives and clearing the way for economic development or administrative rule.[15] Often this meant manipulating society or entire populations, and at times it entailed overhauling both territory and landscapes. When we think of Soviet officials' manipulation of the landscape in Central Asia, we quickly hit upon the history of large-scale irrigation projects and the cultivation of cotton. This was an important part of the picture, to be sure, but in the earliest years of their activity Soviet specialists aimed to reshape the landscape—and the waterscape—in the service of much more fundamental security and political aims than purely and explicitly for the goal of expanding cotton cultivation. It was only in the late 1920s that cotton entered the picture in Tajikistan.

L. F. Paradoksov, an ophthalmologist who played a prominent role in organizing public health in Soviet Tajikistan before the Second World War, summed up the Soviet view. A good Bolshevik health reformer, Paradoksov was perturbed by the essentially fatalistic view of disease that had held sway among the ruling class under the emir and the lack of any sense of responsibility to protect or improve the health of the general population under his rule. The fact that Allah's will was invoked as the ultimate factor affecting an individual's health—and human agency and rational organized action were absent—was particularly irritating to Paradoksov. "The past three centuries of the emirs' rule are characterized by the view that state property is equivalent to [their] personal property," he commented. "Under such a ruling order, naturally there was not even a hint of discussion about raising the cultural level of the population, and likewise there existed no safeguarding of public health. The state's attitude toward public health is summarized well in the following words of one of the emir's administrators: 'Those who are predetermined to die should perish from disease, and those who according to Allah's judgment should continue to live should recover.'"[16]

The Soviets, when they made serious attempts to administer their rule and normalize life in Bukhara in 1924, found the malaria situation particularly dire. It is likely that conditions in Eastern Bukhara were exacerbated by political unrest and economic problems in the emirate in the late nineteenth century and the chaos surrounding the Red Army incursion and the insurgency in the early twentieth. The physical destruction of Eastern Bukhara brought about by over half a decade of fighting—whether Red Army or insurgent forces were more to blame—was dramatic. Observers noted the sheer lawlessness that reigned in the territory, especially in the early 1920s. Some estimated that 70 percent of the livestock herds in Bukhara were destroyed in the conflict.[17] The breakdown of order included interruptions in water management practice, leading to the virulent mosquito breeding grounds and stagnant overgrown pools that Soviets deplored. The complex system of canals that irrigated and moved the water in the country were left untended and in some cases actively destroyed. "The irrigation system that supplies the plains of Bukhara and gives them life . . . fell into complete disrepair during the civil war, both as a result of the lack of necessary upkeep, as well as from the destruction of the walls of the canals [*aryk*] caused by time and water."[18] As one Soviet malariologist described the aftermath of the civil war, writing in the 1960s, "many populated sites and a huge portion of the irrigation system were completely destroyed, and in any case were already primitive. In the valleys, over a great area, swamps and lakes formed, which made the malaria situation even worse."[19]

THE EARLY SOVIET ENCOUNTER WITH MALARIA

Red Army units arrived in Old Bukhara city, the seat of power in the emirate, as early as 1920 and straight away began their efforts to fight the insurgency, secure a civil administration, and combat malaria. Four years later their comrades marched into remote Eastern Bukhara province and endeavored to do much the same. The priority during the 1920s was to improve the general health of the population, with the objective of stabilizing the situation and enabling the young Soviet state to rule Tajikistan from its newly established regional capitals (even as the exact location of these administrative posts were continually shifting from Bukhara to Samarkand and then to Dushanbe). Another priority was to combat the insurgency and to help convince the mass of the local population to support and participate in Soviet rule. One reason the Red Army established an outpost for malaria control in Kulob as early as 1924 and did not, for instance, focus its efforts in another town that suffered from malaria, such as Hisor, was the importance of securing the border with Afghanistan.

The physician and parasitologist Leonid Isaev, a representative of Moscow's newly established Institut tropicheskoi meditsiny (Institute for Tropical Medicine; today the Institute of Medical Parasitology and Tropical Medicine named after E. Martsinovskii), was the first medical expert sent to organize a response to the malaria problem in Bukhara.[20] When he arrived in 1922 he found the city in disarray, with malaria flourishing at every turn. Soviet officials who had not been evacuated for fear of succumbing to the disease were forced to leave work for the sickbed for days at a time. Water carriers, on whom whole neighborhoods continued to depend for their water supply, frequently fell ill and left off hauling their sheepskins of water into the *mahallas* (neighborhoods). The function of the city at all levels had broken down.[21]

When Red Army medics from the TurkFront division, some of them trained at Petrograd's Voenno-meditsinskaia akademiia (Military Medical Academy), set up a medical station in the Kulob area in 1924, their attention also went straight to the water conditions and the climate. Under wartime conditions, Soviet efforts were limited to little more than distributing quinine among the troops. When they had the opportunity, the Soviet authorities aspired to do much more in reshaping the land and the natural and human-made waterways they found in the newly occupied territory. They aspired to control the vector, the mosquito, and in doing so to make a longer-lasting impact, but much of the time they were only able to address the parasite in the human body.

Two malaria brigades were assigned to Eastern Bukhara as the troops moved in. All along the rough route from the end of the railroad line in Qarshi, east over the sandy plains of Surkhandarya and up into Regar in the Kofarnihon Valley, the troops were given a preventive dose of quinine at each bivouac. Soldiers' food rations were withheld unless the record showed each man had taken his dose. The military physician A. A. Makarin recalled with disappointment how "the entire fight against malaria in early 1924 boiled down to prophylactic quinine distribution. The mosquito, as the vector of malaria, remained outside of our focus."[22]

In Makarin's estimation, naturally existing climate conditions, paired with the practices of the local population, formed the complex of root causes of malaria. The natural waterscape of Eastern Bukhara facilitated the spread of malaria, and the state of the irrigation systems aggravated these conditions, creating swampy areas with poor drainage. "The irrigation system and the circumstances of its use and [other] water use, in direct association with the system of rivers, led to the formation of large areas of swamp and standing water, which in turn created fertile ground [*blagopriiatnaia pochva*] for the breeding of mosquitoes."[23] Tajikistan has a naturally dry climate, Marakin noted, but local residents tended to dig basin pools (Taj. *havz* or Russ. *khauz*) in which to store water, which were left for long periods without being cleaned, facilitating the generation of mosquito larvae. Pits (*chigiry*) for storing water and irrigating fields and the moats surrounding the old feudal city walls in places such as Regar, Sary-Asyo, and Hisor also collected standing water and served as breeding grounds for mosquitoes.[24]

The physician and civilian public health administrator Paradoksov also explained the high prevalence of malaria—the chief infectious disease problem in Tajikistan, in his estimation—by pointing first of all to natural climate conditions that facilitated the proliferation of mosquitoes (dry climate paired with the tendency to allow or cultivate pooled water). But human failure, including neglect of existing water systems, was close behind as a second major factor. Paradoksov complained of "the inferiority of the irrigation and water drainage systems, and their complete absence in some localities; a lack of attention to the network of irrigation canals; and the existence of natural swampy places, which occupy a significant area."[25]

Makarin and his Red Army colleagues represented Isaev's counterparts in Eastern Bukhara, laying the foundations for the fight against malaria at their medical station near Kulob. Although a large proportion of military medics were themselves laid low by the disease, they partnered with the Rossiiskoe obshchestvo krasnogo kresta (the Russian Red Cross) to begin organized efforts in 1924.[26] The aim of the station was to treat malaria among both the military and the local civilian population and to conduct some basic scientific study of the disease, as much as the security situation would allow.[27] These two sets of efforts—in Bukhara city and Kulob—together formed the model upon which Tajikistan's republican People's Commissariat of Health (*Narodnyi komissariat zdravookhraneniia*, or Narkomzdrav) would establish its program and institutions for fighting malaria.

When their situation was stable enough, Soviet specialists took aim squarely at the mosquito vector and its habitat, focusing particularly on limiting the development of larvae in standing water. Isaev's approach was to target breeding grounds that had formed in the stopped-up canals and the tangled weeds of the reservoirs inside and outside the ancient city walls.[28] Today this approach would be termed environmental management; public health and tropical medicine specialists at the time called this vector control. Soviet specialists themselves termed their measures "hydrotechnical works." These included repairing and patching existing canals and building new ones, building proper drainage for canals, cleaning stagnant *havz* reservoirs, filling in pits and drying up bodies of standing water, draining marshy areas, straightening riverbeds, laying embankments and levees to contain the flow of rivers, and capping springs whose waters simply pooled and had no active use or drainage. They also made use of larvicides such as kerosene oil and a bright poisonous powder called Paris green to treat bodies of standing water, but shortages hindered attempts to fully integrate these measures.[29]

After a few initial years of stopgap work, this approach evolved to include more research into the nature of the territory and the distribution of malaria infection and how local habitats supported the various species of anopheline mosquito found in the region.[30] Trained investigators (*razvedchiki*) worked under a senior physician to survey the land, recording the horizon and features of the landscape and where standing water existed; systematically observing these bodies of water over time; making a thorough search for places where mosquitoes bred; collecting entomological material, which was studied in Tajikistan and shared with colleagues at the Military Medical Academy in Leningrad; and organizing, and even participating in, small-scale water management and health improvement works undertaken on the canals and riverbeds.[31]

The logic of this environmental approach stemmed from relatively recent medical discoveries about the role of mosquitoes in malaria transmission and the accumulation of a greater understanding of the life cycle of plasmodium parasites. This was still relatively new science in the 1920s, and Western-educated physicians based in Bukhara and Turkestan, not to mention non-medical Bolshevik officials in Bukhara, were skeptical of the ideas Isaev espoused: their understanding of malaria and its control began and ended with quinine, and its distribution as both a preventive and a treatment.[32] For this reason, a great part of Isaev's work in setting up a branch of the Tropical Institute (TropIn) and drawing up plans for malaria-control measures included educating the officials and administrators of what was then the Bukharan People's Socialist Republic about the scientific research that underpinned an approach focused on the mosquito and its environment.

To the lay audience of Soviet officials in Central Asia, TropIn staff initially appeared to emphasize the environmental element almost exclusively in order to hammer home the importance of the mosquito. Yet if we observe their program over time we see that TropIn representatives and the Red Army malaria teams, as well as the civilian Narkomzdrav specialists whom they later advised, advocated a complex multipronged approach. Environmental management techniques focused on the mosquito vector and formed the central pillar, but these were combined with patient treatment and prophylaxis, broad-based health education efforts, and measures such as the use of screens and netting to limit contact between adult mosquitoes and humans.

Education efforts and patient focus at times were explicitly combined with environmental management and one could fortify the other. Gaining trust by administering quinine, and introducing its seemingly miraculous effects, could help turn the local population and made it more feasible to enlist their support and help in engaging in the laborious work of environmental management. By 1926 active work among the rural population had become more organized. Red Army staff now used both print and speech, in Tajik, to teach the *dehqan* (peasant farmer) how malaria was spread, and in particular the role of the mosquito. TropIn staff were posted to the remote area near Kulob, perched on the border with Afghanistan, among thickets of reeds in the riverbed of the Panj River. Malaria was particularly severe in this stark landscape of rough wattle-and-daub huts. Malaria workers later recalled how quinine and its quickly visible results convinced the local population to trust the foreign medical personnel. "We had to be sure that each grain of medicine reached its goal," one staff member recalled. "We administered quinine literally into people's mouths."[33]

For the Bolshevik medics in Eastern Bukhara, health improvement—including malaria control—was an important element of replacing what they considered backward cultural and political institutions with modern ones, replacing "the old lifestyle [*staryi byt*]" with new socialist forms of living.[34] Paradoksov described malaria within the broader context of general public health in Tajikistan. In his view the Red Army had brought culture to Tajikistan, which for him included Western medicine. As with other efforts, this was an uphill climb, in tough conditions: Soviet public health workers were trying to introduce concepts of Western medicine while being countered by those who had benefited under the former regime, the wealthy and the *tabibs* (local doctors) and other folk medicine practitioners. For Paradoksov, the revolution in public health, including the program to combat malaria, would enhance basic health conditions in the country not only by improving care but also by raising the cultural level of the population.

At work at the grassroots level, Makarin also saw the relationship between these two spheres. He and his Red Army colleagues recognized that in their work among the *dehqans* in and around Kulob they were introducing European medicine into Eastern Bukhara. Their work fighting malaria continuously benefited from and in turn fed into the gradual process of earning the trust of the local population toward European-style medical professionals. "As the population's trust of physicians grew, so did their trust in what they taught. The population got the connection between the swamp, the mosquito and malaria."[35] Whatever larger goals of statecraft in Central Asia were served by these doctors' malaria control efforts, they also had a genuine desire to improve lives. As Makarin implies, it was empowering to teach someone—especially someone with so little previous knowledge of the cause of his suffering—the link between his surroundings and disease, and to enable the local population to take more responsibility for their health. In all spheres, and particularly in public health, education and propaganda represented an important element of the revolutionary program to reshape society.

THE TURN AWAY FROM ENVIRONMENTAL VECTOR CONTROL IN THE BUREAUCRATIC SOVIET REPUBLIC

Malaria control efforts evolved during the 1920s from emergency wartime measures to a burgeoning program of vector control, a central element of which was a comprehensive approach toward modifying the natural environment. For a brief period in the mid-1920s, malaria experts enjoyed a great deal of autonomy. Civilian specialists like Isaev were given a free hand to play out their scientific hunches, because of the severity of the problem. And Red Army medics like Makarin continued to combat a dangerous infectious disease on the remote southern border, far away from their superiors.

After 1926 attention turned away from environmental management measures for malaria control because of three related factors. The first was the institutionalization of a Soviet republican government in Tajikistan. With the establishment of a bureaucracy and public health administration came the goal of expanding very small-scale measures into a broad network of malaria control stations throughout Tajikistan. This proved virtually impossible to carry out effectively, however, with limited resources and qualified staff. Authorities also sought to legally compel a broad swath of society to participate in the program. This approach backfired, creating infighting and lower quality work and making it more difficult to carry out effective programs. Finally, the general shift in Soviet priorities with the start of intensive industrialization and collectivization in the late 1920s meant that public health occupied a lower priority amid the range of goals to which Soviet officials could devote their energies. The environmental approaches to malaria control that prominent public health specialists of the early 1920s had advocated fell out of favor amid this shift.

In Tajikistan, in the face of massive efforts to divert river water and resettle entire segments of the population to cultivate cotton in the Vakhsh River valley, Soviet officials lost enthusiasm for intensive focused environmental management for the sake of health improvement. It was more attractive to use chemical substances such as quinine (or malarial drugs, as they were developed) for treatment and prophylaxis or to apply larvicides to bodies of water in order to treat the problem, whether targeting the parasite or the vector. The vision of malaria control that had been formulated in the 1920s—centered on vector control through close study and modification of the environment in each locality—faded. In the late 1920s the reality of local malaria control was that few staff had the qualifications to implement anything more complex or technical than spraying or administering pills. They relied more heavily on a program of simple technologies, rote planning, and formulaic inspections.

The bureaucratization of public health in Tajikistan began in 1925, with the establishment of Narkomzdrav. A small group of physician administrators arrived in Dushanbe to establish a civilian public health administration for the newly created Tajik Autonomous Soviet Socialist Republic (Tadzhikskaia Avtonomnaia Sovetskaia Sotsialisticheskaia Respublika [TASSR]), then part of the Uzbek SSR. The conditions of their arrival give us a sense of what a remote outpost this remained. The officials flew in on a two-leg journey from Uzbekistan, having spent the night in Termez, and then set out, stocked with a handful of rifles and 250 cartridges to defend themselves in case of insurgent attack. Their eighty-minute flight put them down in the muddy Dushanbe airfield covered with slushy snow from the night before—nearly overturning the plane. The physician officials paired up and rode two to a horse to the Republican Narkomzdrav headquarters, a mud hut whose iron heating stove made the frozen walls sweat and sprout seedlings.[36]

Initially, while Narkomzdrav officials attended to the most basic health problems and found their footing, the Red Army continued to shoulder the weight of the malaria burden, with work at the Kulob station and with mobile detachments (*otriady*). By 1928, however, as the military decommissioned its units, Narkomzdrav TASSR officials began to take on malaria control. The director of the Central Asian Military Zone's health division warned in December 1928 that he and his colleagues did not wish to see "all [of their] achievements in malaria control reduced to nothing" by a lack of attention to the problem or by misguided policies. He complained that precisely such a setback had occurred in the vicinity of Old Bukhara city with the transition to civilian responsibility.[37]

TASSR health officials realized in late 1928 that, despite the foundation laid by the military with its Kulob work, they themselves had a very poor grasp of the republic's malaria problem. They had failed to clearly identify what approach would be deployed under the expansion under civilian administration. The deputy health commissioner noted with alarm that "the lack of a map of the focal points of malaria, and corresponding [lack of] a scientifically based plan for malaria control, hamstrings" prevention work.[38] Assistance from the Tropical Institute in Moscow and personally from its founder, Evegnii Martsinovskii, beginning in 1930 was instrumental in laying the basic building blocks of a republican program to combat malaria. The establishment of two Dushanbe research institutes—a local tropical medicine institute and an institute of epidemiology—did give some assurance that malaria control efforts were guided by evidence-based investigation.[39]

Despite this support and the temporary deployment of Moscow and Dushanbe TropIn staff to help with local malaria control efforts, a great distance still separated the research institutes in the capital and the remote malaria stations. According to the bureaucratic plan, a district-level malaria station was supposed to be directed by a qualified malaria specialist with a small support staff who would develop a plan for responding to the disease and a reporting system based on scientific expertise and careful study of the local conditions. The station would record patient cases of malaria and treat them with ample supplies of quinine. The relationship of the station to other local authorities would be smooth and cooperative. The district Communist Party executive committee (*raiispolkom*) and the malaria station head would work in collaboration to define how they shared responsibility for funding and implementing the malaria control program. There would be sufficient supplies and equipment from the center, including quinine and Paris green larvicide.

There would be enough resources to allow the station staff to travel to remote locations and to safely and effectively store their chemicals. The station head would be responsible only for malaria control; other staff would handle other public health issues.

These were the ideals, but judging from the evidence the reality, especially in southern Tajikistan, was very different. To begin with, it was hard to find qualified, responsible, and effective workers to staff remote malaria stations. Narkomzdrav official correspondence contains many discussions of inexperienced or ineffective malaria station directors. Even if directors were qualified, they were often overworked, acting not only as head of malaria control for the district but simultaneously covering the full breadth of local public health issues as chief of the district health department (*raizdrav*). Cooperative relationships with other local institutions were difficult to foster. Very often malaria station staff were hindered when local kolkhoz (collective farm) leadership refused to support their work or the party committee kept malaria control off its meeting agenda for weeks or months at a time. Many station heads were unable to stay above the fray of petty politics within the health department or with other local authorities. Resources were limited and public health did not constitute a first-order priority. The common refrain from inspectors sent out to evaluate the work of the stations was that many of the basic resources necessary to do their work were lacking, such as storage containers for treatment substances, uniforms, and transport not to mention shortages in quinine or Paris green. Inspectors in the mid-1930s sent repeated requests back to Dushanbe for the central Narkomzdrav's Malaria Group to disburse funds to buy horses, bicycles, oil drums, and safety dress.[40]

This combination of bureaucratization and resource-poor reality in Tajikistan took shape in the context of the other fundamental shift that served as a factor in the decline of the environmental focus of malaria control: the second half of the 1920s ushered in important changes in priority within the Soviet Union at large. Beginning in 1926 the central Soviet government revised its vision of Tajikistan's future. A "Special Commission for Tajikistan Affairs" was formed and made a study of the republic, its territory and population, to formulate a new strategy for administering and developing it. Two important recommendations came out of the report: the Vakhsh River valley was ideal for cotton cultivation, and segments of the population from mountainous regions should be resettled to provide the labor for cotton expansion in these more sparsely populated areas. This was of course just an opening salvo in the move toward collectivization and industrialization, and further shifts in priority at the center and at the republic level followed in 1929. It was difficult for expensive and labor-intensive malaria-control programs pinned on environmental management practices to compete successfully for attention and resources under the new focus on raw economic development, but malaria specialists made a valiant attempt.

Requests for malaria control support and funding had never been devoid of economic logic: the disease burden robbed the economy of worker capacity. With the start of collectivization in 1929 a new language emerged for talking about malaria and its remedies. Makarin gave a detailed quantification of malaria-related losses in terms of insured workers, number of labor days lost, and the expense incurred treating patients. He argued it was better to invest in preventive measures such as environmental management. "All branches of the economy suffer from malaria, especially agriculture," he argued. His vision of the agricultural cost of malaria interestingly did not include cotton. "Spring flare-ups of the disease take workers away from the sowing, and autumn ones—from the grain harvest [*uborki khleba*]."[41]

Malaria was no longer portrayed primarily as a basic public health problem representative of the generally low health conditions in the land, which could be improved by changing the surroundings and conditions of life so that all might enjoy better health. Malaria now represented a very specific problem: it was a drain on labor, primarily an economic hindrance. In 1929 Tajikistan's highest republican-level executive body, the Council of People's Commissars, couched the justification of credits to fund malaria control in the logic of economic costs: "The organs of public health of TASSR, having in their budget limited resources for malaria control work, are not in the condition to expand [*razvernut'*] [such work] in correspondence with today's real needs. . . . The spread of malaria is reflected particularly harmfully among the insured [*zastrakhovannye*], the great disease incidence among whom hinders the healthy [*normal'nyi*] course of work of enterprises and institutions and places a heavy burden on the budget of the insurance organs, who pay out each year huge [*gromadnye*] sums for temporary inability to work [*netrudosposobnost'*] because of malaria infection." The director of Dushanbe's young Tropical Institute put it more succinctly in the opening lines of a 1933 article on malaria control: "The economic damage caused by malaria in Tajikistan is extraordinarily great."[42]

In their complicated maneuvering with other commissariats and economic concerns, Narkomzdrav officials were starting to get the hang of the new justification. In a 1929 letter to Khlopkom (the cotton ministry), requesting funding for malaria control near cotton-processing plants and collection points, health officials drew their cotton comrades' attention to the economic costs of the disease: "Malaria infection robs the human organism of the capacity to work for an extended period of time, [and] in the summer months those taken out of the ranks of functioning workers in the majority of cases [have fallen victim to] malaria," the director of the division for public health and prevention wrote. He made a valiant argument for the cotton concern to release the funds necessary to "strengthen vector control work" in raions with cotton plants and to complete a combination of environmental modifications and oil treatments. Khlopkom requested further clarification and more information instead of approving the spending.[43]

By 1937, in a survey of malaria work in Tajikistan, U. M. Muminov introduced his subject with a comment about how "the rapid rise in socialist construction and the economic and organizational strengthening of the kolkhozes had significantly raised the well-being of the laborers and kolkhozniki to a substantial level." It was thanks to the construction of socialism and the "attention paid by the Party and the Government to the well-being of the laboring [population]" that malaria control had achieved such a rise in funding and real gains in the reduction of incidence rates. As always in Soviet public health, emphasis was on quantity rather than quality: funding went from 800,000 rubles in 1932 up to 3,954,000 rubles in 1936; over the same period the malaria network had mushroomed from fifteen tropical stations to thirty-two tropical stations, twenty-two malaria posts (*punkty*), and two roving malaria brigades. Not discussed were all of the problems rampant in the expanded network—and the fact that, in rolling out so many stations, the effectiveness of the work of limiting malaria and mosquito breeding had actually suffered.[44]

CONCLUSIONS

Official Soviet statistics show that gains were made in the fight against malaria through 1937, but beginning in 1938 the number of malaria cases in Tajikistan actually began to rise again. Observers have characterized the early 1930s, roughly coinciding with the start of collectivization, as a systemwide hiccup in Soviet malaria control. In the decades following the end of the Second World War the Soviet Union in fact found great success in eliminating malaria.[45] How did malaria control get back on track? In part, the experience of infectious disease during the war, including malaria, changed officials' minds about prudent investments in public health. There was also a general institutional restructuring of the Soviet public health system after the war, and in the 1950s there was a new determination to fundamentally eliminate malaria from all corners of the country. Coordination of this campaign from the center, headed by Petr G. Sergiev, helped solidify gains against the disease up to the declaration of its eradication in 1960. Sergiev, a parasitologist and epidemiologist who cut his teeth as a military physician in the First World War and honed his organizational skills heading the Soviet forces' malaria control efforts in the Second World War, proudly harnessed an array of methods in the postwar effort and paid ample attention to the environmental context.[46]

Another factor in the postwar successes was the introduction of two important new technologies in the armament of the malaria control worker:

the Soviet-manufactured antimalarial drug atabrine and the insecticide DDT, both of which entered into wide usage after the conclusion of the war. At the same time, research institutions in Dushanbe, in other malaria-prone areas such as the Caucasus republics, and in Moscow and Leningrad continued to make fundamental gains toward understanding the different species of mosquito prevalent in particular areas and the relationship of these vectors to their habitat. While the environmental modification model of the 1920s was never realized in full, research on the nuances of individual malaria vectors completed in the 1940s and 1950s helped make environmental strategies for eradicating various anopheline habitats more effective. The accumulation of more specific scientific knowledge about anopheline mosquitoes led to much more targeted methods and approaches of limiting mosquito breeding, and the application of very narrow spectrum tools and technologies to the waterscape.

Malaria control took a hit after the start of collectivization and the expansion of cotton cultivation in Tajikistan. The vision of the early Bolshevik health reformers and tropical medicine experts had been grounded in a perspective that tied disease and health to the physical environment. These public health workers also viewed health improvement as an essential part of their mission in Central Asia. The popularity of vector control and the emphasis on environmental management grew out of these two factors. One irony of the shift was that Soviet insistence on the human ability to bend the physical environment to human interests did not suffer in the least. Rather, public health improvement could no longer compete for priority as the aim meant to be served by manipulation of the environment. There was also a shift in approach, as engineering methods and careful localized and fundamental alterations were abandoned in favor of the application of chemical treatments. There were practical reasons for this: the Soviet state in Tajikistan did not possess the resources or the level of organization to support the kind of rigor required in environmental management for malaria control. When the Soviet state at the central and republican levels came to assign the highest priority to economic development and was willing to apply resources and systems to significantly alter the physical environment for the sake of this priority, health and this particular approach to malaria control lost out.

Examining public health policy and malaria control approaches from the point of view of the environment allows us to consider new interpretations of Soviet history and to reconsider the orthodoxies of Soviet historiography that we have inherited. Looking at Soviet rule through the lens of the environment sheds light on the messiness and the improvisational nature of state structures and programs that have been portrayed as uniform, monolithic, and uniformly successful. It allows us to see nuance and contestation in processes where this has not been visible in earlier studies. It also allows us to pinpoint and tease apart the specific ways in which Soviet policies of the 1930s differed from, and the ways in which they held continuities with, the policies of the 1920s.

This research helps push environmental history in a different direction as well. In this telling of the history of malaria in Tajikistan, we consider humans' relationship to the physical landscape, and we are able to parse out some distinctions among the different programs implemented to significantly alter the environment and people's lives within it. Programs aimed at changing how people in Tajikistan lived in their natural setting took many different forms and had various thrusts and justifications. We do not see a history in which humans were uniformly aggressive or destructive toward the environment, although, interestingly, socialist philosophy served as just as powerful a rationale for alteration of the physical environment as capitalism has done in other areas of the world. The Soviet case, examining approaches to transforming the environment for different competing aims, can help us to prod environmental historians' argument that human practices in the modern period have been inherently destructive toward the environment.

17 Conclusions

NATURE, EMPIRE, INTELLIGENTSIA

DOUGLAS WEINER AND JOHN BROOKE

THE CHAPTERS in this volume encompass a rich variety of scholarship on the environmental history of Russia and the Soviet Union since 1800, discussing conditions ranging from drought, irrigation, and disease on the steppe to fishing and water manipulations across the region's seas and rivers, to transformations of landscape, soil, and forests, to the efforts of scientific societies and imperial officials to shape the direction of national policy. There are two obvious questions that flow from a book such as this: How does an environmental framework change the way we think about the tsarist empire and the Soviet Union? And how should we understand Russian/Soviet environmental history in the broader context of global history?

A clear starting point for comparison must be the environmental history of landed and oceanic empires. Reviewing this history across the wider early modern and modern epochs, Kenneth Pomeranz suggests that we set aside a traditional focus on capitalism as the force driving environmental change over the past five or six hundred years and adopt a more capacious framework of the "developmentalist project."[1] Starting with the great global empires of the late Middle Ages and early modern period, the state has had a powerful role in driving "development"—the increased extraction of natural resources from increasingly large domains. Such a process, of course, does not frame the entire scope of anthropogenic earth-system change in the past half millennium, but it is a fundamental dimension, and it helps us organize our thinking about the environmental history of the vast region treated in this volume—namely, the northern, steppe, Central Asian, and Siberian peripheries of tsarist Russia and the USSR. Thus, while empire building in Eurasia had important roots in the specific problem of securing and expanding Russian boundaries against the threat of mounted warrior armies on the steppe, it shared with other imperial ventures the simple goals of appropriation.[2]

In the Russian case, a straitened agricultural base in the face of external military challenges from chiefly the west and the south led the early modern Russian monarchy to seek sources of revenue for defense. These they found in the enserfment of a large portion of the peasantry to support a cavalry, and in the relentless drive across Siberia to the Pacific in pursuit of high-end furs, a crucial component of export income for Muscovy.[3] Tsarist and Soviet rulers were later drawn into an increasingly capitalist-driven world system, first as suppliers of raw materials—timber, and then grain—but especially by the need to defend themselves against economically and militarily more dynamic, industrial capitalist empires. As Stalin famously said, "Russia was beaten for her backwardness."[4] Both before and after the advent of global capitalism, rulers of Russia continually confronted the injunction: develop or be conquered.

The spread of Russian/Soviet empires across Eurasia and their history of destructive extraction is thus one piece of a wider global pattern beginning in the early modern period, inflected by specific context. Three aspects of that context emerge particularly strongly in this volume. First, these chapters are centrally concerned with the tensions between the Russian/Soviet imperial venture and an intractable Eurasian nature. This is a bimodal relationship. Empire and nature were active agents in fundamentally incompatible and unsustainable tension: a rigid imperial project changed and degraded steppe, mountain, water, and desert ecologies, while the natural realities of those environments blunted and refracted the reach and success of that imperial venture. Second, these essays explore the ways in which the Russian imperial venture was constructed in an unequal dialogue between the state and a cultural and scientific intelligentsia, which sought to build influence in what can be called the public sphere of civil society and with the state itself.[5] Both were allies in the Russian developmentalist project, but each had different priorities and ultimate goals.

Third, as Alexander Etkind and others have argued, this entire process of expansion, radical transformation of newly acquired lands for imperial purposes, and rapacious resource extraction was accompanied by a triple colonization. After Peter the Great, the Russian elite were colonized by European models of culture. At the same time, the Russian and Slavic peasantries were transformed into a colonial underclass of taxpayers and serfs. And finally, there was a transformation of conquered peoples into a variety of types of colonial subjects, some having political and economic rights (Finns), some having only economic rights but no political or even linguistic rights (Poles), and some with no effective political or economic rights (the peoples of Siberia and Central Asia), drawing their condition close to that of colonial subjects of Britain and other overseas empires.[6]

The central question in any modern environmental history is the problem of sustainability and resilience. How elastic are natural conditions faced with the exploitive impact of empire building, whether driven by the macro policy of the metropole or the myriad micro impacts of settler populations? And conversely, how elastic are these imperial ventures, when faced with the realities of natural conditions? These are a more complex set of questions than would first appear, because all parties to the interaction are continuously changing with the encounter. The question is further complicated if there are human groups that already live in areas designated for settlement by conquering regimes. Consequently, it might be helpful to break down this question into observable stages.

At the moment when imperial states or settler groups began to penetrate new regions and to project new visions for its development, whether in the New World from 1492 or in Siberia from 1582, there was already a vast legacy of previous human occupation and environmental change in those places. Each wave of human occupation, in fact, had started with visions of how to use the area, had imposed transformations in the course of the area's use and occupation, and had been forced to make adaptations, both to the conditions the group found initially and to whatever changed conditions evolved as a result of the group's presence. In the process, the groups themselves experienced change in their own cultural and economic systems and practices. New incursions, displacements, or mixing groups only accelerated this never-ending process of mutual dialectical change between human groups and environments. At the same time, social groups in the metropole articulated and imposed their visions of what the imperial lands should look like. As Alfred Crosby argued in *Ecological Imperialism*, nineteenth-century global empires sought to create Neo-Europes. As described in this volume, the Russian intelligentsia and the Russian state worked together to create Neo-Russias—subsets of Neo-Europes—but alongside these, they created territories that strongly resembled the colonial dependencies of those other empires.[7]

THE CLASH OF EMPIRE AND NATURE IN THE EURASIAN INTERIOR

Much of the environmental impact of the Russian Empire was shaped by settlers, especially in the Caucasus–Black Sea region[8] and in Central Asia.[9] In recent scholarly literature, perhaps the best comparative cases illustrating the imperial impact of settler societies may be found in two works: William Cronon's *Changes in the Land* and Eric R. Wolf's *Europe and the People without History*.[10] While they concentrate on showing the profound changes wrought within indigenous societies and in the landscape itself as a consequence of European settlement in the Americas, these works chart how the settlers themselves were transformed in all sorts of ways as well. One such transformation was the settlers' adoption of corn, which became a central pillar of American agriculture, just to mention a salient example. This framework has only recently begun to be applied to Eurasian history. If the Mongol Empire may be viewed as the antecedent for the subsequent Muscovite, tsarist, and Soviet empires, it was, in its own way, a "settler" empire, even if its settlers were restricted to the steppe or prairielands. As with later Russian colonization in the reverse direction (west to east), both sides were transformed by the encounter. Works by Charles Halperin, Donald Ostrowski, George Vernadsky, Lev Gumilev, and Janet Martin, among others, illuminate how Russian culture, administration and politics, military organization, and other domains were transformed by Mongol-Turkic rule and exposure to steppe culture—and how politically and commercially entwined both Muscovy and the steppe were as early as the tenth century, despite ongoing episodic conflict.[11] In the seventeenth century, the Zaporozhian and Don Cossacks, autonomous communities on the "Wild Fields," epitomized a hybrid culture that married Slavic agriculture with Mongol-Turkic pastoralism and mounted raiding.[12] Later, when expansion proceeded to the east, Russian and Slavic settlers to varying extents adopted local foods, dress, and practices and intermarried with locals, even as they represented an imperial colonial presence.[13]

Additionally, we must consider the landscapes into which the Russian state and its minions expanded to be a mosaic of anthropogenic landscapes—from Ukraine, Crimea, and Caucasus all the way to Siberia and the Far East. Slavic (later, Great Russian and Ukrainian) and other imperial settlers—Germans, Serbs, Greeks, Jews, ending with the ethnicities that Stalin deported (Poles, Koreans, Germans, Chechens, Kabarda, Balkars, and others)—did not encounter pristine nature. They found communities that had participated in making those landscapes what they were through conscious or inadvertent introduction of new flora and fauna, hunting, extinction of older fauna, the impact of persistent grazing and animal traffic, controlled burning, accidental fires, extensive military use of fire (on the steppe), irrigation, river diversion (Chinggis Khan captured Samarkand that way), and urbanization along trade routes.[14]

These indigenous peoples had reformed Eurasian landscapes to suit their needs, in what William Cronon, John McNeill, and Simon Schama have described as the "second nature" of a humanized landscape.[15] They had also occupied the land in scales and modes that were at least minimally shaped by that landscape's extreme variability, in ways that the imperial project swept aside in its centrally authorized and occasionally directed developmentalist project. From a present-day perspective, the pressures of indigenous peoples on the landscape seem generally more sustainable than those of empires (although, of course, absolute sustainability does not exist). Many of the essays in this volume attest to the devastatingly destructive, unsustainable impact of imperial projects, driven by demands of state and population, on the often fragile and variable ecologies of the imperial periphery.

By the time the curtain rises on the episodes described in the chapters of this anthology, a number of important developments had already occurred in the geopolitics, political economy, and environmental history of the steppe and of Central Asia, the regions on which most of the chapters focus. Russian penetration of the Kazakh Steppe from southern Siberia advanced in the seventeenth and eighteenth centuries as a consequence of the need to defend and extend a grain belt. Grain fed the Cossack garrisons that protected and enforced the procurement of furs in Siberia. Two other developments set the stage for a recognizably "colonial" expansion of the empire, both coming to fruition under Peter the Great. First, the Russian Empire's new self-definition as a *European* polity transformed its elites' views of the peoples of the steppe, Siberia, and the Caucasus. Those were no longer rough equals—whether as potential ally or enemy—but more primitive groups in need of civilizing.[16] Second, by the nineteenth century the adoption of Western firearms and artillery eventually gave the Russian imperial troops the military upper hand in the steppe, Crimea, and Central Asia.[17] Russian settlement proceeded both at the initiative of the imperial regime but also through the agencies of individual settlers, impelled by their various hopes, dreams, exigencies, dispossessions, and other factors that brought them to the steppe.[18] The Russian Empire was pursuing a "frontier settlement" strategy that in many ways mirrored those of the United States in the American West, of Britain in Canada, Oceania, and southern Africa, and of France in the Algerian Maghreb. Here, as in these parallel empires, it is interesting to distinguish among the agendas of the state, of communities, and of individual settlers.[19] And it is important to remember that not all conquered peoples engaged in "resistance." There was a whole range of behaviors, including accommodation and collaboration.[20]

As John LeDonne and Ihor Stebelsky tell us, it was important for Russia to turn steppe grasslands into farmland, not just for income but, perhaps equally importantly, to deprive steppe nomads of any material base for threatening the empire ever again. In this, the Russian Empire and later the Soviet state (in Kazakhstan) were decidedly not unique. Almost exactly the same logic was followed in the multi-pronged war against the Indians of the Great Plains and American West—through the extermination of the bison, the parceling of territory by means of strategically placed forts, and the promotion of homesteading protected by the US Army.[21]

For voluntary settlers—whether gentry landlords or Mennonite communities—there was a keen desire, as David Moon shows us, to turn this arid unfamiliar place into a Neo-Russia by planting trees, which would also serve as a source of building materials and fuel. Certainly, Moon suggests that the cost-benefit ratio of tree planting, despite isolated successes (the biggest of these was no doubt the Great Stalin Plan for the Transformation of Nature) was weighted heavily in favor of costs. Indeed, some efforts, such as those of the Don Cossacks, were complete failures.[22]

If the area north of the Black Sea described by Moon had been cleared of Crimean Tatars after the conquest of the "Wild Field," then the steppes to the east were still heavily populated and used. Sarah Cameron describes how the imposition of European control over the Kazakh Steppe had a variety of consequences both for the land and for the Kazakh people. Russian imperial planners and politicians viewed the Kazakh way of life as primitive, located on a lower rung of the human scale of progress than sedentarism, let alone the rarified cultural atmosphere of the imperial court—a view similar to the perspectives of later Soviet rulers and educated professionals.[23] Whereas before Peter the Great a surprising number of Mongol and Tatar dignitaries were inducted into the Russian nobility (after, of course, converting to Orthodox Christianity), with post-Petrine Russia set on a course toward Europe there was little interest in bringing people into the ruling elite who clearly did not fit in.[24]

Instead, the backwardness of steppe people was blamed for their misfortunes (see Campbell, this volume). The severe damage to herds wrought by the freezing of pastures (*zhŭt*), which usually struck once every ten to twelve years, was seen as evidence that nomadic herding was an unstable and unsatisfactory way of life. As Campbell says, officials believed there were only two choices open to the Kazakhs: farming or flight. Many officials secretly hoped for the latter, so as to replace the nomads with Slavic settlers. Instead, they got revolt. What followed the failed revolt was a cascade of increasing vulnerability. A terrible combination of factors beyond the truncation of seasonal pastures conspired to undermine the long-term viability of Kazakh transhumance, not least being the arrival of large numbers of Slavic peasants—Russian or Ukrainian—from the mid-Volga or secondarily from left-bank and steppe portions of Ukraine.

From the perspectives of the Kazakhs, these settlers were the unbidden—"those who came without invitation"—ironically the same thing that was said by the peasants and town dwellers of Rus' about the Mongol invasion six centuries earlier. The history is also complicated by the fact that these Slavic peasant migrants were themselves long-standing victims of earlier chains of oppression. They had sold their property including seed and cattle to jump at the chance offered by the autocracy to start a new life outside of the crowded southern agricultural provinces of Russia, where the old landowning gentry still held sway even after the abolition of serfdom in 1861. The regime's cynical and ultimately bootless solution to land hunger and continued peasant poverty was to transport some of the surplus peasants to other people's lands: a new dispossession as solution to a previous one. Colonial control and the promotion of the in-migration of settlers was perhaps partly a response to another geopolitical issue—the challenge of an increasingly hostile British Empire in the second half of the nineteenth century, which had waged two (unsuccessful) wars to acquire Afghanistan, on the southern borders of the Russian Empire's newly acquired Central Asian territories.[25]

Thrown together into this crucible, both settlers and Kazakhs had to make essential adaptations. So, Ukrainian peasants needed to shift from oxen to horses as draft animals, and Kazakh clan leaders found new sources of income in supplying and caring for the horses of the new settlers. With increasing grain grown and milled in the larger region, the diet of Kazakhs shifted to grain-based foods, paid for by the increasing ties of Kazakhs with the market. As the noose tightened on the use of the steppe as pasture and with roads to Chinese pastures now closed by treaty, Kazakhs were forced to reduce their flocks and herds and to use pastures year-round, becoming semi-nomadic, storing hay for the winter, and shifting to cattle.

Settlers and Kazakhs made adaptations, but sometimes adaptations do not rise to the level of assuring long-term viability—or peaceful coexistence. Following the massive Kazakh revolt of 1916, the inheritors of the Russian Empire in 1917 had a second chance to achieve some sort of more sensitive policy on the steppe. However, this period of benign neglect came to an end with the collectivization drive of 1929–1933.

Despite their advertised differences, the Soviet state and the tsarist empire shared a basic identity of views about the steppe nomads: that they were both backward and a security risk. Stalin had to know where all the pieces were on the chessboard. Cut off from their grain merchants (who had been suppressed in 1929) and confined to collective farms without instructions or resources, in 1931 the nomads and their herds perished in unimaginable numbers. Percentage-wise exceeding famine mortality in Ukraine, the Kazakh ethnicity lost fully 40 percent of its precollectivization population—over one million people, with an additional six hundred thousand fleeing into emigration—and became a minority in their own autochthonous republic.[26] To fill the gap, the Soviet state moved in hundreds of thousands of deportees from among Polish, Korean, German, Crimean Tatar, and other ethnicities.

Kazakh dependence on marketed grain adds another piece of the puzzle of this terrible cataclysm. But of course, it does not absolve the Soviet state for making decisions of the level of magnitude of collectivization or for levying colossal tribute on the population without making contingency plans to feed the population in case of an emergency. It constitutes another example—a criminally egregious one—of humans making decisions in the absence of due diligence.

One of the most famous hallmarks of the Khrushchev period was the development of the "Virgin Lands," the subject of Marc Elie's contribution to this volume. Although a good chunk of this territory was in southern Siberia, the majority of the land was in Kazakhstan, especially in the northern third of the republic. In light of the multiple calamities that had befallen the Kazakhs, with losses to collectivization that almost rose to the level of ethnocide, Khrushchev's label "Virgin Lands" was beyond insensitive; it was a massive falsification. The steppe, formerly inhabited and used, was now orphaned, not virgin.

Its availability was a lucky break for Khrushchev who, trying to consolidate power after Stalin's death, needed to avert famine and secure grain supplies while preserving the collective farm system, which, despite conditions that approximated serfdom, could not feed the country. The political benefits that accrued to Khrushchev in the 1950s from the early success of the Virgin Lands, Elie suggests, were based on the same capriciousness of nature that brought drought and dust storms in the 1960s and 1970s. Certainly, as Elie notes, the methods originally employed on the fields, transported from the European steppe areas, exacerbated the loss of topsoil. But at base, the scheme was a game of chance with nature. Bringing Elie's data together, we can say that the overall costs to the state and Soviet society for Khrushchev's (failed) attempt to secure Soviet grain self-sufficiency were considerable subsidies. Were these costs worth it to the state? Soviet leaders evidently did not see any alternative.

On the other hand, Elie helps us to balance out a one-sided picture that focuses only on the inefficiencies of the collective farm/state farm system of Soviet agriculture, pointing to rulers' adaptive responses within the limits of the system. Most notable of these was the regime's decision during the period 1968–1974 to heed Viktor Kovda's soil science warnings and shift to non-inversing plows, terracing, and the planting of shelter belts. However, the benefits of these recommendations came with costs as well, most notably the increase of weeds, the reduction of acreage (and hence, of farm income), the cost of the appropriate plowing machinery to be bought from Canada, and the retraining of personnel. Yet both the farm directors and the rulers could not absorb short-term losses in the interests of long-term soil conservation, so pesticides and herbicides were aerially applied, crop rotation was cancelled in large provinces of the steppe, and the emphasis was placed on greater water delivery through irrigation. All of this reflected the predominance of short-term goals over the long-term perspectives and multifactor analysis of soil scientists and other agronomists. The time frame and its perceived exigencies determined the limits to adaptation of the Soviet regime. Of course, the same can be said about farming in the United States, where the viability of the wheat belt depends on two effectively nonrenewable resources: oil and the Ogallala Aquifer.[27]

None of the papers in this volume, it should be noted, deal directly with the patterns and causes of the droughts that were so fundamental in southern European Russia and Central Asia. Central Asian precipitation is primarily shaped by winter westerlies emanating from the Atlantic, piling snow on mountain glaciers. The distribution of this rain is determined on the grand scale of centuries and decades by the shift of the North Atlantic Oscillation (NAO): a positive mode of the NAO (strong low pressure at Iceland) during the Medieval "Warm Period" (circa 950 to 1250 CE) sent winter precipitation north into Scandinavia and shaped medieval drought conditions in Central Asia. A negative NAO mode (weak low pressure at Iceland) apparently brought ample rain to Central Asia during the Little Ice Age (approximately 1300–1800). Since the end of the Little Ice Age, a positive NAO mode has contributed to increasing aridity in Central Asia. Within this drier modern epoch, Central Asian droughts have been stronger between 1900 and the 1950s. These decades of more severe drought shaped conditions in late imperial/early Soviet Central Asia and put the Soviet experiment with different forms of agriculture, irrigation, land use, and interethnic interactions into its larger, climate context. In the more recent sequence of drought conditions, winter temperatures have been rising sharply. And right now there is controversy as to whether the positive mode of the Arctic Oscillation—the close cousin of the NAO—is intensified by industrial greenhouse gases, perhaps explaining the patterns of drought that Soviet scientists were working to explain in the 1960s. On the other hand, the precipitation and drought curves presented in these studies suggest that in fact, in the recent past since 1960, it was indeed the failings of Soviet agriculture systems rather than any necessarily overwhelming trend toward drought that has been afflicting Central Asia.[28]

It is difficult to separate out economic and political rationales when we examine Soviet resource policies, pursued in the context of these ongoing dry conditions. Soviet economic structures were designed to support and promote the political power of the Communist Party and its leadership. Still, political leaders like Lenin were motivated by visions of economic and social development as well as by the desire to retain power, and Lenin and others saw irrigation as an agent of "rebirth [that] buries the past and strengthens the transition to socialism."[29] In Christian Teichmann's chapter in this volume, we see how the Soviets tried to expand and administer a particular irrigation system along the lower course of the Amu Darya. Teichmann is quick to observe that states are much less modernist and efficient than usually credited, as James Scott has observed.[30] Nevertheless, elites and rulers command great power. Even when resisted or even thwarted, they can cause great disruption.

Owing to the dynamic nature of the topography of the lower course of the Amu Darya, a result of variations of water, ice, and silt flowing through the river from year to year, a flexible, locally based system of irrigation canal management had evolved in the region under the general authority of the khan of Bukhara. The khan was able to mobilize his subjects for the annual work of dredging and maintaining the canals. Soviet ideas about the political need to give each ethnicity its own administrative-political territory meant that ethnographers sent from the metropole helped the regime set the new boundaries, largely on the basis of linguistic criteria. As a result, Teichmann notes, canal heads were all too often located in different political units from the areas they administered. There were also much longer distances to administrative centers than when the delta was under the rule of the khan.

Although Soviet decisions had created these dysfunctions, authorities instead saw them as part of the general inefficiencies and backwardness of the locality. In their eyes, huge amounts of water and labor were wasted, and locals accepted endemic malaria with fatalism. Filled with feelings of cultural superiority, engineers sent in from the metropole tried to rationalize the system, using intimidation to compel people to work. This situation only worsened following collectivization, when the Soviet rulers had to depend on local administrative despots to achieve even minimal goals. The regime could supply neither needed materials and infrastructure nor a better working model. Soviet authorities found it easy to overpower old elites and to disrupt the "backwardness" they found. They could only replace those, it turned out, "with a permanent state of emergency."[31]

If there was a face to this "progressive," transformationist engineering in Central Asia, it was that of Viktor Dukhovnyi, one of the authors of the attempts to irrigate the Hungry Steppe of Uzbekistan beginning in the 1960s. As revealed in his memoirs, Dukhovnyi never doubted the goals he served. In this, he was one of an army of servitors of the Soviet state, beginning with its leaders, who never sought to apply due diligence and even ask what the likely, let alone possible, consequences of their policies might be.[32]

If, for the regime, the goals of irrigation—in the service of Lenin's desire for cotton self-sufficiency[33]—went unquestioned, the nature of expectations for plan fulfillment led to systematic misreporting by local political authorities, as Julia Obertreis discovered. What was covered up, for a time, were the causes of the (unreported) shortfalls in cotton production: salinization, soil exhaustion, and water deficits. The Soviet system could generate mandates, it could frighten politicians, but it could not manufacture cotton out of "whole cloth," as it were, when natural conditions were uncooperative, and it forced its minions to lie. This is where despotic regimes hit the wall.

Lisa Walker's chapter on Soviet malaria control in Tajikistan also shows how the regime's overwhelming drive for rapid economic growth and social modernization was unable to tolerate the slow, careful, locally sensitive efforts of public health workers to eliminate breeding grounds of Anophelid mosquitos that characterized efforts of the 1920s. Instead, like the rest of the world, the Soviets plumped for mass application of DDT, and the brief experimental moment was extinguished.[34] If the utopian impulse of the revolution led some to try to improve life for others using careful and responsible approaches emancipated from pressures of profit, it led others to sweep that first approach away in the name of economic growth, which was seen as the only guarantor of "progress." That brief moment of experimentation was also reflected in Aleksandr Fersman's precocious proposals for a Soviet extractive economy based on recycling of by-products, the subject of Andy Bruno's chapter in this volume. It was not an absolutely new idea; coal gas, a product of the coking of coal, had been used to light British houses, factories, and then streets since the late years of the eighteenth century.[35] Had Fersman's ideas proved successful, a largely closed-process model of the economy would have been pioneered, vastly reducing the polluting effects of by-products. Resonating with Bolshevik biases against capitalist wastefulness and inefficiency, Fersman's vision also implicitly embodied a concern for present and future communities, which would be burdened by legacies of pollution.

However, as Bruno demonstrates, these lofty ideas crashed against the chemical structure of rock. The pilot project was the recovery of aluminum from nepheline, a rock that represented most of the left-over tailings from the mining of apatite, a phosphate-bearing rock, in the remote Kola Peninsula. As it turned out, however, it was cheaper to import bauxite than to refine the nepheline. The nepheline tailings were allowed to leach into the lakes and rivers of the Kola Peninsula, polluting them and destroying much of the zoofauna.

Nonetheless, apatite production was not curtailed once the nepheline pollution problem came into view.

By emphasizing nature's lack of cooperation with the good intentions of Fersman and his associates, Bruno has tried to respond to the one-sided condemnation of Soviet economic practices as egregiously polluting. Yet, even had Fersman's plans succeeded, at least in the Kola Peninsula, we would still have to raise the question of whether due diligence was applied. There would still have been the disruption of reindeer-herding Sámi and Nenets communities by the mining itself, and questions would have remained about the purposes that both the phosphates and the aluminum would have served. Human planning has yet to catch up with the manifold, multidimensional implications of human actions. The environmental impact procedure (incidentally, which first appeared in incipient form in the USSR in the 1920s but which was terminated in 1931) is a first step in that direction.[36]

Pey-Yi Chu's chapter also highlights some nonnegotiable elements of the environment, noting that those who chose to live in regions with permafrost soils must give up the delusion that they can change the environment or the soil. Instead, they must adapt to conditions resistant to change. That the Soviets even in the Stalin period eventually developed adaptations—in the siting and construction of buildings especially—shows that the regime sometimes bowed to the reality principle. Had Soviet leaders been aware of it, Engels's dictum "Nature, in order to be commanded, has to be obeyed" showed the way to an effective Soviet presence in Iakutia, to say nothing of Eurasia in general.

Indeed, it was not as if there were no alternative visions of development available to political radicals and the technical intelligentsia. Mieka Erley's essay highlights the resonance of the ideas of Justus Liebig, first in the thinking and works of Karl Marx and Friedrich Engels and later among educated Russians. Essentially, Liebig postulated that soil nutrients constituted some kind of closed cycle. If the soil were mined, so to speak, phosphates and nitrates had to be returned to it in some way, by the application of fertilizer. This led Marx and Engels to their idea that urbanization constituted a "metabolic rift" that continually impoverished soils and that could be combated only by an ultimate reintegration of cities and farmlands—in Russian, a *smychka*. This was a precocious post–Industrial Revolution attempt to think about greater sustainability, and its spirit was reflected in a number of futuristic plans for *agrogoroda* or farm-cities both in the 1920s and in the Khrushchev period. However, like many other experimental ideas and programs, it was inundated by the overwhelming emphasis on economic growth, driven by imperial logics.

THE INTELLIGENTSIA, THE EMPIRE, AND EURASIAN NATURES

All of the chapters discussed above broadly stress the difficulties that the various environments of Central and Eastern Eurasia posed for the Russian/Soviet imperial projects. Here we see an intelligentsia that was working—as was the wider stratum of Russian and Soviet officials and professionals—to forge a semblance of the cultural authority and influence that their fellow European counterparts had created since the eighteenth century in the public sphere and civil society. They met in reform and scientific societies, founded journals, and attempted to assimilate the imperial project into an improving vision of reform. They would later try to play a similar role in the Soviet period. Since improvers, scientists, and intellectuals have left the lion's share of the documentation of historical nature, their projects make up a central part of the chapters in this volume.[37]

Thus, the forest-planting colonization schemes in the Pontic Steppe described by David Moon, were the result of botanizing exchanges among the Russian landowning classes, distantly shaped by the mid-eighteenth-century reports of the expeditions sent out by the Imperial Academy of Sciences but also by the tsarist regime's desire to populate the "Wild Field" and make it productive. At the other end of our chronology, Marc Elie's chapter is fundamentally concerned with the competing factions shaping Soviet agricultural policy enmeshed in a tense dialogue between the Soviet government and global civil society organizations. Policy is the central feature of the three papers on water and irrigation, in Teichmann's account of early Soviet administrators trying to reshape the hydrology of the Amu Darya, Obertreis's discussion of the swelling of a regime critique around the issue of Uzbek and Turkmen irrigation policy, or Walker's examination of the early Soviet efforts at malaria control in Tajikistan. The role of scientific knowledge, now embedded in the Soviet system, is obviously foundational to Andy Bruno's essay. Obertreis in particular draws attention to a small number of intellectuals in Central Asia who were able to compose a critique of both policies and the larger Soviet vision of development. In this they took advantage of the spaces opened by the decaying regime and by perestroika. Provoked by a perception of crisis, environmental criticism raises the questions of how we should live on the earth and what, in the end, gives life meaning? With the imposition of repressive dictatorships combined with the accompanying ideological confusion of the post-Soviet period, ecocriticism, notes Obertreis, has gone underground. However, the material contradictions remain in place.

But the role of the intelligentsia in this cultural construction of nature and empire becomes more central in the essays making up broadly the second half of this volume. Mieka Erley's chapter examines a world of Russian soil scientists with profound and perhaps unconscious commitments to the definition of a core Russian identity while they engaged with the wider intelligentsia and found their place in a wider European arena. Mark Sokolsky's chapter on the Russian Far East highlights two divergent faces of the Enlightenment: progress as rational study and use and progress as growth. Russian elites saw themselves as the vanguard of Europe and, hence, civilization on the Pacific, in contrast with the Chinese, Koreans, and indigenous groups of the area. They promoted settlement, which was seen as a step up the evolutionary ladder from nomadism and foraging. However, a few came to see the Russians as equally unenlightened, especially from the standpoint of stewardship of nonhuman nature. While it was convenient to see others' ways of gathering resources, eating, and otherwise organizing their lives as barbaric, it was more difficult to appreciate the cultural situatedness of those prejudices. It was even more difficult to come up with a workable definition of "rational" to begin with. And more profoundly, at the farthest reaches of empire, Russian educated people came face-to-face with the question: How should we live on the earth?

One way was to reap the environmental benefits of imperial reach. George Lywood examines the significance of the Crimea, annexed by conquest and treaty by Catherine the Great in 1783, to the perceived health and welfare of the Russian people. Once the railroad reached the peninsula during the Crimean War, Russians could reap a great reward from their defense against the Turks, British, and French by frolicking in the warm waters of the Black Sea, recovering in mud baths and sanitaria, and consuming the exotic products of a subdued steppe culture. Here an improving intelligentsia appears in the form of the Russian medical establishment among whom environmentalist medicine was orthodox doctrine: the benefits of a warm distant seaside on the Black Sea for those who had so long suffered the cold dark winters of the Russia homeland offered proof of their doctrine. The benefits derived from the specific geochemistry and maritime climate regime but also from a medically regulated appropriation of the core diet of an indigenous people, the fermented koumiss and kefir of the Crimean Tatars. Lywood develops an important comparative perspective with the similarly environmentalist doctrines of health in the American empire, specifically from the high, dry climates of the greater Mountain West. Indeed, since the eighteenth century, American empire builders had been on the lookout for therapeutic hot springs, locations that marked the opening of health tourism in the early republic, if not before. The history of the development of Florida is another particularly good comparison for Russia's Crimea.[38] And we might also note the enormous literature on exotic diet and empire building, one that stretches back for millennia but that had particularly important globalizing force in the early modern age of empire.[39] In Old World and New, appropriated and transformed First and Second Natures were deployed to improve the health of the empire builders.

Of all the discontent produced by tsarist rule, that of the intelligentsia is among the most poignant. Barred from effective political participation at the central level and hemmed in impossibly on the local level, treated as civic juveniles and subject to undignified supervision at work, educated Russians witnessed or imagined other kinds of relationships between regime and people and sought means to change or even remove the power of the tsar and his bureaucrats. Too small to act as the force of change themselves, civically engaged educated Russians tried to enlist more powerful and numerous social groups, particularly the peasantry and later urban workers. They did this, however, by framing their various schemes for change as universal ideologies. When those were not readily accepted, intelligentsia groups struggled to explain why their potential allies were not acting in their own "true" best interests.

This case has been told with respect to the political ideologies of Populism and Socialism (and equally for the ideology of Capitalism).[40] But the chapters by Stephen Brain and Julia Lajus in this volume show how this tutorial stance of the intelligentsia toward nonelite groups was reflected in attitudes toward the Pomors, the fishing villagers of the remote White Sea and Barents Sea littoral. In this telling, the Pomors had no need of a more developed ethical and political system; they practiced one that was very serviceable. Notably, it relied on a religiously based ethic of self-restraint regarding fish catch, which had the effect of allowing fish and marine mammal populations to remain viable. The loss of twenty-five of their small vessels in a storm served as the justification for St. Petersburg elites to create committees for modernizing, revamping, and "improving" the lives of the Pomors. Anticipating the Soviets, a Northern Commission for the Assistance to Russian Trade Navigation was formed to place fishing on an industrial basis, which, it was simply assumed, would improve the lives of the people of the northern coasts. Interestingly, the soul of the commission was a marine zoologist, Nikolai Knipovich, a radical and acquaintance of Lenin's—providing a direct link between the tsarist era and the Soviet one.

For the Soviets, who also partly came out of the intelligentsia tradition, the Pomor lifestyle was multiply noxious. It was materially "backward," with a shore-based orientation, and its productivity was low, leaving the open waters of the Barents Sea vulnerable to the more efficient penetration of foreign fishing fleets. Moreover, its culture was based on Christianity. While the Soviets were able to overpower the Pomor culture and substitute mechanized fishing, they ultimately created a crisis of overfishing in the Barents Sea and left an impoverished biota behind. It was another Pyrrhic victory for their vision of development.

Many of the above themes are repeated again in the chapter by Ryan T. Jones, who explores why Russia and the USSR, not a traditional whaling nation, emerged as one of the most aggressive, even "reckless" destroyers of whale stocks in the mid to late twentieth century. Here, we see narrow imperial calculations: that preemptive whaling was necessary to prevent the presence of foreign ships near Russian territorial waters. Added to this was the fact that whale procurements by the 1930s were subject to the five-year plans, and so there were inducements for captains to harvest more than the quota—to be able to sell whale meat on the side to the Japanese, for example. Finally, the overall "unprofitability" of whaling can be explained by the tsarist and Soviet attitude that inputs are free goods, and therefore capitalist notions of profitability do not apply in a situation where the opportunity costs to the state are as much political as purely economic.

THE TSARIST AND SOVIET EMPIRES IN GLOBAL ENVIRONMENTAL HISTORY

In sum, then, the chapters comprising this volume present a wide cross-section of the work being done on nature and empire, broadly conceived, across the Russian/Soviet domain in Eurasia. Most importantly, they contribute to three critical problematics in modern environmental history: (1) the specific material clash of imperial visions and environmental realities; (2) the discourse of empire, culture, and science bound up in that clash; and (3) the inseparable link between economic and political orders, on the one hand, and specific forms of land use, on the other. These are fundamental dimensions of Kenneth Pomeranz's "developmentalist paradigm" and Alfred Crosby's and William Cronon's broad descriptions of the construction of North American Neo-Europes on imperial, formerly indigenous lands.

If there is something distinct about the circumstances of Russian Eurasia, it lies in the following. Russian empire building took place under harsher and more challenging environmental conditions than in much of the temperate Americas. With the exception of the indigenous peoples of the North and Siberia, who were epidemiologically isolated and who therefore sustained large population losses to exposure to diseases brought by Russian conquerors, Russians did not encounter similarly vulnerable groups elsewhere. People in the Caucasus and Central Asia had the same set of immunities as the migrating Russians. So, in comparison with European empires in the Americas, it was difficult to replace people there with Russian and Slavic settlers. And, importantly, the agricultural potential of the newly acquired Russian imperial lands was lower than that in the temperate Americas owing to lack of precipitation and extensive desert and mountainous terrain. In addition, Russia faced the dual challenge of harsh arid environments and epidemiologically similar indigenous peoples. Unlike eastern North America, where temperate conditions and devastating disease impacts on American Indians combined to smooth the way to empire building, the challenges of ecology and indigenes made Russian empire building far more difficult. In light of the grandiose imperial geopolitical ambitions of both the tsarist empire and the Soviet state, these conditions forced those states into short-term tribute-taking behaviors both toward their subjects and toward resources, engendering further stresses in the system. This was one, but by no means the least important, paradox of Russia's contiguous colonial expansion across the vastness of Eurasia.[41]

Yet, it is in just this tributary attitude that profound similarities may be seen between Russia's empires and those of not just the Ottomans and Habsburgs, with which Russia is most frequently compared, but with the British and French overseas empires. As has been shown by Diana Davis, James Fairhead and Melissa Leach, and Ravi Rajan, among many others, scientists and other educated professionals (intelligentsia) worked hand in glove with the Western overseas empires to impose forestry and land-use schemes on colonial subjects. Although these were advertised as "rational" and "improving" measures, they were vehicles of dispossession, depriving the ruled of access to critical traditional resources in the service of economic and geopolitical interests of the colonizers. Such measures led to equally disastrous and dysfunctional outcomes as in the tsarist and Soviet cases.

Similar too was the degree of tributary exploitation: even as tsarist and Soviet exploitation led to famines (in 1891 and 1931–1933) that collectively took at least five million lives, British, French, Belgian, German, Dutch, Spanish, Portuguese, and American colonial regimes engineered the deaths of tens of millions.[42] Apart from their Asian colonial territories, the tsarist and Soviet regimes did not rule distinctly different colonial "others" to exploit in the headlong pursuit of "development" and so had to exploit their own internal "others"—mainly, the Russian and Ukrainian peasantries.

The tsarist empire and the Soviet Union also shared another attribute with the more vigorous capitalist Western empires that they feared, emulated, and with which they sought to compete—a short time horizon. Essentially existing in a reactive mode relative to their external geopolitical environment, tsarist and Soviet rulers felt they could hardly afford to wait to consider the costs and risks of development schemes. It is ironic that the Soviet regime disregarded the potentialities within Marxist thinking for an integrative view of the relationship between humans and our environments. Liberated as it was from the dictates of the capitalist profit motive, had it not been in thrall to the fetish of development the USSR conceivably could have approached land use with a more long-term perspective and placed an emphasis on due diligence. However, this was not to be. Its environmental practices showed the Soviet polity not to have differed substantially either from its tsarist predecessor or from its Western capitalist adversaries. The studies in this volume make it harder to make a case for Russian or Soviet exceptionalism, freeing us from the historiographical fetters that have burdened our thinking from the Crimean War through the Cold War and beyond.

GLOSSARY

aksakal (plural: *aksakaly*): canal communities along the Amu Darya River

arshin (plural: *arshiny*): measurement of length equal to 0.71 meters (2.33 feet)

artel (plural: *arteli*): a cooperative group of peasants, craftsmen, or workers; a work cooperative

aul (plural: *auly*): tent used by pastoral peoples of the Caucasus and Central Asia

bey: Central Asian nobleman

chernozem: black earth, soil characterized by its dark color, fertility, and high proportion of humus

desiatina (plural: *desiatiny*): a measurement of area equal to 1.9 hectares (2.7 acres)

funt (plural: *funty*): unit of weight equal to 409.5 grams (14.44 ounces), or 1/40 of a *pud*

kolkhoz (plural: *kolkhozy*): collective farm (Soviet Union)

kolkhoznik (plural: *kolkhozniki*): collective farm worker (Soviet Union)

krai: administrative division of territory; can also refer to an "edge" or a borderland region

kulak (plural: *kulaki*): lit. "fist," refers to a rich peasant; in the Soviet Union, often applied to any rural dwellers designated as anti-Soviet

Kulturträger: "culture bearer," one who brings or transmits culture and enlightenment, esp. to a colony (German)

nachal'nik: head, chief

narodnyi: people or people's (adj.)

nomenklatura: Communist Party functionaries

novosel (plural: *novosely*): a newly arrived settler, esp. in late-nineteenth-century Siberia, Central Asia, and the Far East *obkom*: oblast-level Communist Party committee

oblast (plural: *oblasti*): province

Pereselencheskoe upravlenie: Resettlement Administration

pochva: soil

Pomors: Russians residing along the coast of the White and Barents Seas

pomor (plural: *pomory*): seafarers; those who live near, travel to, or regularly work on the sea

pud (plural: *pudy*): unit of weight measurement equivalent to 16.38 kilograms, 36.11 pounds, or 40 *funty*

raiispolkom: *raion*-level Communist Party executive committee

raion: district (esp. in the Soviet Union and post-Soviet Russia)

raizdrav: *raion*-level health committee

rod (plural: *roda*): clan

sovkhoz (plural: *sovkhozy*): state farm (Soviet Union)

starozhil (plural: *starozhily*): "oldtimer": refers to a long-established settler, esp. in late-nineteenth-century Siberia, Central Asia, and the Far East

uezd (plural *uezdy*): district (esp. in prerevolutionary Russia)

ulu: elder (Kazakh)

versta (plural: *versty*): unit of length equal to 1.0668 kilometers (3,500 feet).

volost (plural: *volosti*): canton (or hamlet), the smallest territorial division in prerevolutionary Russia

zapovednik (plural: *zapovedniki*): state nature reserves earmarked primarily for scientific study and conservation (rather than tourism), especially in the Soviet Union and post-Soviet Russia

zemstvo (plural: *zemstva*): local unit of self-government, established in 1863 in the provinces of European Russia

zhŭt: failure of grass fodder harvest as a result of frost, esp. among herding communities of the Eurasian steppe (Kazakh)

NOTES

Chapter 1: Toward an Environmental History of Tsarist Russia and the Soviet Union

1. S. T. Aksakov, *The Family Chronicle*, trans. M. C. Beverley (E. P. Dutton, 1961), 14. On the Slavophiles, a nineteenth-century Russian intellectual movement, see also Nicholas V. Riasanovsky, *Russia and the West in the Teaching of the Slavophiles: A Study of Romantic Ideology* (P. Smith, 1965).

2. Aksakov, *Family Chronicle*, 12–13.

3. In her introduction to the novel, Kathleen Parthé calls the Village Prose movement "the largest and most coherent body of aesthetically interesting and ideologically significant literature to be published in the Soviet Union during the three decades following Stalin's death. . . . There was in Village Prose a search for—and a celebration of—the values and rituals of the disappearing Russian village." Kathleen Parthé, "Foreword: Master of the Island," in Rasputin, *Farewell*, vii–viii. See also David Gillespie, *Valentin Rasputin and Soviet Russian Village Prose* (Modern Humanities Research Association, 1986); Stephan Hirzel, *Ökologie und Öffentlichkeit: Untersuchungen zur Rolle der sowjetrussischen Schriftsteller in der ökologischen Bewusstseinsbildung der fünfziger bis achtziger Jahre* (Lang, 1996). My thanks to Julia Obertreis for alerting me to the latter book. In 1983 a film was made of the novel: *Proshchane*, directed by Elem Klimov (Mosfilm, 1983).

4. Nicholas Breyfogle, "At the Watershed: 1958 and the Beginnings of Lake Baikal Environmentalism," *Slavonic and East European Review* 93, no. 1 (2015): 147–80; Alan Roe, "From the Heroic Conquest of the River of Electricity to Despoiled Wonder: Narratives of the Angara River, 1955–1991," paper presented at the annual conference of the Association for Slavic, Eastern European, and Eurasian Studies (Boston, 2013), 2–3; and the epic poetry of Yevgeny Evtushenko, *Bratsk Station, and Other New Poems* (Praeger, 1967).

5. Mark Carey, *In the Shadow of Melting Glaciers: Climate Change and Andean Society* (Oxford University Press, 2010), 12. See also Gosudarstvennyi Arkhiv Respubliki Buriatii (State Archive of the Republic of Buriatiia) f. R-195, op. 13, d. 821.

6. Rasputin, *Farewell*, 119–20.

7. We hope that this book will serve as an agenda-setting introduction to the possibilities and insights of environmental studies for Eurasian history. There have been several recent excellent review essays that outline the tendencies and directions of tsarist and Soviet environmental history. See, for example, Brian Bonhomme, "Writing the Environmental History of the World's Largest State: Four Decades of Scholarship on Russia and the USSR," *Global Environment* 6, no. 12 (2013): 12–37; Stephen Brain, "The Environmental History of the Soviet Union," in *A Companion to Global Environmental History*, ed. J. R. McNeill and Erin Stewart Mauldin (Wiley-Blackwell, 2012), 222–43; Randall Dills, "Forest and Grassland: Recent Trends in Russian Environmental History," *Global Environment* 6, no. 12 (2013): 38–36; David Moon, "The Environmental History of the Steppe in a Global Perspective," in *Ekologicheskaia istoriia v Rossii: etapy stanovleniia i perspektivnye napravleniia issledovanii*, ed. E. E. Merzon (Izd-vo Elabuzhskogo instituta KFU, 2014), 8–31; David Moon, "The Curious Case of the Marginalization or Distortion of Russian and Soviet Environmental History in Global Environmental Histories," *International Review of*

Environmental History 3, no. 2 (2017): 31–50; Jonathan Oldfield, Julia Lajus, and Denis J. B. Shaw, "Conceptualizing and Utilizing the Natural Environment: Critical Reflections from Imperial and Soviet Russia," *Slavonic and East European Review* 93, no. 1 (January 2015): 1–15; Andy Bruno, "Russian Environmental History: Directions and Potentials," *Kritika: Explorations in Russian and Eurasian History* 8, no. 3 (2007): 635–50; Julia Lajus, Dannil Aleksandrov, and Franz-Josef Briuggemaier, eds., *Chelovek i priroda: ekologicheskaia istoriia* (Aleteiia, 2008); Paul Josephson, Nicolai Dronin, Ruben Mnatsakanian, Aleh Cherp, Dmitry Efremenko, and Vladislav Larin, *An Environmental History of Russia* (Cambridge University Press, 2013).

8. Authors in this book use the much debated and contested term "Eurasia" or "Eurasian" as a shorthand or synonym for the vast lands and peoples integrated into the tsarist and Soviet empires. In this, the authors follow the practice of the Association for Slavic, East European, and Eurasian Studies, which changed its name in recent years to include "Eurasian" as a descriptor for the organization's area of study. When the authors employ "Eurasia" or "Eurasian" here, they are not using the terms in their Russian nationalist and imperialist ideological meaning; nor are they referring to the long unpleasant history in the Eurasianist movement. On the debate over using the term "Eurasian" for the former tsarist/Soviet territories, see for example, Alexandra Hrycak, "Redefining 'Our Field': Is it Slavic, East European or Eurasian studies?" *NewsNet* 48, no. 3 (2008): 7–10; Katherine Verdery, "What's in a Name, and Should We Change Ours?" *NewsNet* 46, no. 2 (2006): 1–6; Michael David-Fox, Peter Holquist, and Marshall Poe, "Eurasian Studies?" *Kritika: Explorations in Russian and Eurasian History* 2, no. 2 (2000): 233–35. On Eurasianism, see for example, Sergey Glebov, Mark Bassin, and Marlene Laruelle, eds., *Between Europe and Asia: The Origins, Theories, and Legacies of Russian Eurasianism* (University of Pittsburgh Press, 2015).

9. Richard Pipes, "The Environment and Its Consequences," in *Russia under the Old Regime* (Charles Scribner's Sons, 1974), chapter 1; and W. H. Parker, *An Historical Geography of Russia* (Aldine, 1968). In part, the relatively slow adoption of environmental history has been a consequence of historians' preoccupation with other questions, such as explaining the roots of the Bolshevik revolution, Stalinism, war and violence, and empire-building. For comparison with other parts of the world, see Alan Mikhail, ed., *Water on Sand: Environmental Histories of the Middle East and North Africa* (Oxford University Press, 2012); Ian Jared Miller, Julia Adeney Thomas, and Brett L. Walker, eds., *Japan at Nature's Edge: The Environmental Context of a Global Power* (University of Hawaii Press, 2013); Bruce L. Batten and Philip C. Brown, eds., *Environment and Society in the Japanese Islands: From Prehistory to the Present* (Oregon State University Press, 2015); Christopher R. Boyer, *A Land between Waters: Environmental Histories of Modern Mexico* (University of Arizona Press, 2014).

10. Andrew C. Isenberg, ed., *The Oxford Handbook of Environmental History* (Oxford University Press, 2014), esp. Isenberg, "Introduction: A New Environmental History," 1–20 (quotation at p. 4); J. R. McNeill and Alan Roe, *Global Environmental History: An Introductory Reader* (Routledge, 2012); McNeill and Mauldin, *Companion*; Frank Uekoetter, ed., *The Turning Points of Environmental History* (University of Pittsburgh Press, 2010); Joachim Radkau, *The Age of Ecology* (Polity, 2014); Joachim Radkau, *Nature and Power: A Global History of the Environment* (Cambridge University Press, 2008); J. R. McNeill, "Observations on the Nature and Culture of Environmental History," *History and Theory* 42 (December 2003): 5–43; Richard White, "Environmental History: The Development of a New Historical Field," *Pacific Historical Review* 54 (1985): 297–335; Paul S. Sutter, "The World with Us: The State of American Environmental History," *Journal of American History* 100, no. 1 (June 2013): 94–119; and essays by William Cronon, Alfred Crosby, Carolyn Merchant, Stephen Pyne, Richard White, and Donald Worster in the *Journal of American History* 76 (March 1990): 1087–147.

11. In John McNeill's helpful formulation, environmental history is "the history of the mutual relations between humankind and the rest of nature. Humankind has long been part of nature, but a distinctive part, especially in recent millennia, when it acquired the power and the numbers to become a rogue mammal, exerting ever-increasing influence upon earthly

ecosystems. Human history has always and will always unfold within a larger biological and physical context, and that context evolves in its own right." McNeill, "Observations," 6.

12. In this way, the essays here reflect the rich and diverse works of environmental history that have appeared of late. A recent, but by no means inclusive or exhaustive, selection includes Frances Lee Bernstein, Christopher Burton, Dan Healey, eds., *Soviet Medicine: Culture, Practice, and Science* (Northern Illinois University Press, 2010); Brian Bonhomme, *Forests, Peasants, and Revolutionaries: Forest Conservation and Organization in Soviet Russia, 1917–1929* (East European Monographs, 2005); Stephen Brain, *Song of the Forest: Russian Forestry and Stalinist Environmentalism, 1905–1953* (University of Pittsburgh Press, 2011); Kate Brown, *Plutopia: Nuclear Families, Atomic Cities, and the Great Soviet and American Plutonium Disasters* (Oxford University Press, 2013); Andy Bruno, *The Nature of Soviet Power: An Arctic Environmental History* (Cambridge University Press, 2016); Jane T. Costlow, *Heart-Pine Russia: Walking and Writing the Nineteenth-Century Forest* (Cornell University Press, 2013); Jane Costlow and Amy Nelson, eds., *Other Animals: Beyond the Human in Russian Culture and History* (University of Pittsburgh Press, 2010); Laurent Coumel, "A Failed Environmental Turn? Khrushchev's Thaw and Nature Protection in Soviet Russia," *Soviet and Post-Soviet Review* 40, no. 2 (2013): 167–89; Marc Elie, "'Au centre d'un double malheur': Le séisme du 7 décembre 1988 en Arménie et l'expulsion des sinistrés Azéris de Spitak," *Revue d'études comparatives est-ouest* 44, no. 1 (2013): 47–80; Marc Elie, "Coping with the 'Black Dragon': Mudflow Hazards and the Controversy over the Medeo Dam in Kazakhstan, 1958–66," *Kritika: Explorations in Russian and Eurasian History* 14, no. 2 (Spring 2013): 337–67; Klaus Gestwa, *Die Stalinschen Großbauten des Kommunismus: Sowjetische Technik- und Umweltgeschichte, 1948–1967* (Oldenbourg, 2010); Zsuzsa Gille, ed., "Nature, Culture and Power," special issue of *Slavic Review* 68, no. 1 (2009): 1–94; Ryan Tucker Jones, *Empire of Extinction: Russians and the North Pacific's Strange Beasts of the Sea, 1741–1867* (Oxford University Press, 2014); Paul Josephson, *The Conquest of the Russian Arctic* (Harvard University Press, 2014); David Moon, *The Plough that Broke the Steppes: Agriculture and Environment on Russia's Grasslands, 1700–1914* (Oxford University Press, 2012); Julia Obertreis, *Imperial Desert Dreams: Cotton Growing and Irrigation in Central Asia, 1860–1991* (V&R Unipress, 2017); Nigel A. Raab, *All Shook Up: The Shifting Soviet Response to Catastrophes, 1917–1991* (McGill-Queen's University Press, 2017); Douglas Weiner, *A Little Corner of Freedom: Russian Nature Protection from Stalin to Gorbachev* (University of California Press, 2002); Douglas Weiner, *Models of Nature: Ecology, Conservation, and Cultural Revolution in Soviet Russia* (University of Pittsburgh Press, 2000).

13. Karl Wittfogel, *Oriental Despotism: A Comparative Study of Total Power* (Vintage, 1981); James Scott, *Seeing like a State: How Certain Schemes to Improve the Human Condition Have Failed* (Yale University Press, 1999); Donald Worster, *Rivers of Empire: Water, Aridity, and the Growth of the American West* (Oxford University Press, 1985).

14. On local knowledge, see for example, Piers Vitebsky, *The Reindeer People: Living with Animals and Spirits in Siberia* (Mariner Books, 2006); and V. K. Arseniev, *Dersu the Trapper* (McPherson, 1996).

15. For a recent selection of theoretical and methodological approaches, see the *American Historical Review* (hereafter cited as *AHR*) Roundtable, "History Meets Biology," *AHR* 119, no. 5 (December, 2014): 1492–629; B. Brown, "Thing Theory," in *Things*, ed. B. Brown (University of Chicago Press, 2004); Eduardo Kohn, *How Forests Think: Toward an Anthropology beyond the Human* (University of California Press, 2013); Bruno Latour, *Reassembling the Social: An Introduction to Actor-Network-Theory* (Oxford University Press, 2007); Timothy J. LeCain, *The Matter of History: How Things Create the Past* (Cambridge University Press, 2017); Timothy Mitchell, "Can the Mosquito Speak?" in Mitchell, *Rule of Experts: Egypt, Techno-Politics, Modernity* (University of California Press, 2002), 19–53; Linda Nash, "The Agency of Nature or the Nature of Agency?" *Environmental History* 10, no. 1 (January 2005): 67–69; F. John Odling-Smee, Kevin N. Laland, and Marcus W. Feldman, *Niche Construction: The Neglected Process in Evolution* (Princeton University Press, 2003); Edmund Russell, *Evolutionary History: Uniting*

History and Biology to Understand Life on Earth (Cambridge University Press, 2011); Peter-Paul Verbeek, *What Things Do: Philosophical Reflections on Technology, Agency, and Design*, trans. Robert P. Crease (Pennsylvania State University Press, 2005).

16. See also the insightful article on 1917 land use by Andy Bruno, "What Does It Mean to Liberate a Land? Toward an Environmental History of the Russian Revolution," in *Russia's Home Front in War and Revolution 1914–22, Book 3: National Disintegration and Reintegration*, ed. Christopher Read, Peter Waldron, and Adele Lindenmeyr (Slavica Publishers, 2017), 1–28.

17. Pey-Yi Chu uses the term "de-exceptionalize" in this volume. Paul Josephson, *Industrialized Nature: Brute Force Technology and the Transformation of the Natural World* (Island Press, 2002); J. R. McNeill, *Something New under the Sun: An Environmental History of the Twentieth-Century World* (W. W. Norton, 2000); J. R. McNeill and Peter Engelke, *The Great Acceleration: An Environmental History of the Anthropocene since 1945* (Belknap Press, 2014); and Radkau, *Nature and Power*.

18. Brain, "Environmental History," 225. See also Brown, *Plutopia*; Murray Feshbach and Alfred Friendly, *Ecocide in the USSR: Health and Nature under Siege* (Basic Books, 1992); Donald Filtzer, *The Hazards of Urban Life in Late Stalinist Russia: Health, Hygiene, and Living Standards, 1943–1953* (Cambridge University Press, 2010); Boris Komarov, *The Destruction of Nature in the Soviet Union* (M. E. Sharpe, 1980); Moon, "Curious Case"; D. J. Peterson, *Troubled Lands: the Legacy of Soviet Environmental Destruction* (Westview Press, 1993); Philip R. Pryde and Don J. Bradley, "The Geography of Radioactive Contamination in the Former USSR," *Post-Soviet Geography* 35 no. 10 (December, 1994): 557.

19. Maria Shahgedanova, ed., *The Physical Geography of Northern Eurasia* (Oxford University Press, 2002). For an easily accessible source, see "Russian Nature," online at http://www.rusnature.info/.

20. The same is true of the excellent environmental histories of Canada. See, for example, Liza Piper, *The Industrial Transformation of Subarctic Canada* (University of British Columbia Press, 2009). Ultimately, a disproportionate focus of global environmental history has been on warmer or tropical lands, with the exception of climate history and polar studies. On the climate history of the Little Ice Age, see Sam White, *A Cold Welcome: The Little Ice Age and Europe's Encounter with North America* (Harvard University Press, 2017); Dagomar Degroot, *The Frigid Golden Age: Climate Change, the Little Ice Age, and the Dutch Republic, 1560–1720* (Cambridge University Press, 2018). On the tropical focus, see as one of many examples, Richard Grove, *Green Imperialism: Colonial Expansion, Tropical Island Edens and the Origins of Environmentalism, 1600–1860* (Cambridge University Press, 1995).

21. Jones, *Empire*; Bruno, *Nature of Soviet Power*; Josephson, *Conquest of the Russian Arctic*; and John F. Richards, *The Unending Frontier: An Environmental History of the Early Modern World* (University of California Press, 2006).

22. John Davis, *Russia in the Time of Cholera* (I. B. Tauris, 2018); Charlotte E. Henze, *Disease, Health Care and Government in Late Imperial Russia: Life and Death on the Volga, 1823–1914* (Routledge, 2011); William H. McNeill, *Plagues and People* (Penguin Books, 1976); William C. Summers, *The Great Manchurian Plague of 1910–1911: The Geopolitics of an Epidemic Disease* (Yale University Press, 2012).

23. David N. Livingstone, *Putting Science in Its Place: Geographies of Scientific Knowledge* (University of Chicago Press, 2003), 5.

24. A spate of recent books in Russian/Soviet environmental history have made clear the importance of the local Eurasian environmental context in the production of scientific ideas (notably, scientific theories that diverged from or influenced Western models). Josephson, *New Atlantis*; Jones, *Empire*; Moon, *Plough*; Davis, *Cholera*. See also George Lywood, "Our Riviera, Coast of Health: Environment, Medicine, and Resort Life in Fin-de-Siècle Crimea" (PhD diss, The Ohio State University, 2012).

25. On Russian/Soviet science, see also Kendall E. Bailes, *Science and Russian Culture in an Age of Revolutions: V. I. Vernadsky and His Scientific School, 1863–1945* (Indiana University

Press, 1990); Paul Josephson, *New Atlantis Revisited: Akademgorodok, the Siberian City of Science* (Princeton University Press, 1997); Loren R. Graham, *Science in Russia and the Soviet Union: A Short History* (Cambridge University Press, 1993); Daniel P. Todes, *Darwin without Malthus: The Struggle for Existence in Russian Evolutionary Thought* (Oxford University Press, 1989); Daniel R. Todes, *Ivan Pavlov: A Russian Life in Science* (Oxford University Press, 2014); J. D. Oldfield and D. J. B. Shaw, *The Development of Russian Environmental Thought: Russian Geographical Perspectives on the Natural Environment, 1880s–1960s* (Routledge, 2016), J. D. Oldfield and D. J. B. Shaw, "V. I. Vernadskii and the Development of Biogeochemical Understandings of the Biosphere, c.1880s–1968," *British Journal for the History of Science* 46, no. 2 (2013): 287–310; J. D. Oldfield and D. J. B. Shaw, "Russian Understandings of Society-Nature Interaction: Historical Underpinnings of the Sustainable Development Concept," in *Understanding Russian Nature: Representations, Values and Concepts*, ed. A. Rosenholm and S. Autio-Sarasmo (Aleksanteri Institute, 2005), 237–52; J. D. Oldfield, "Russia, Systemic Transformation and the Concept of Sustainable Development," *Environmental Politics* 10, no. 3 (2001): 94–110; Ethan Pollock, *Stalin and the Soviet Science Wars* (Princeton University Press, 2008).

26. Doug Weiner and John Brooke elaborate further on this point in the concluding chapter. See also D. R. Weiner, "The Predatory Tribute-Taking State: A Framework for Understanding Russian Environmental History," in *The Environment and World History*, ed. Edmund Burke III and Kenneth Pomeranz (University of California Press, 2009), 276–315; Marshall Goldman, *The Spoils of Progress: Environmental Pollution in the Soviet Union* (MIT Press, 1972); Bathsheba Demuth, *Beringian Dreams: People, Nature, and the Quest for Arctic Energy* (W. W. Norton, 2018); Natalia Mirovitskaya and Marvin S. Soroos, "Socialism and the Tragedy of the Commons: Reflections on Environmental Practice in the Soviet Union," *Journal of Environmental Development* 4, no. 1 (Winter 1995): 77–110; Josephson, *Industrialized Nature*, 15–68; Kate Brown, "Gridded Lives: Why Kazakhstan and Montana Are Nearly the Same Place," *AHR* 106, no. 1 (February 2001): 17–48; Zsuzsa Gille, *From the Cult of Waste to the Trash Heap of History: The Politics of Waste in Socialist and Postsocialist Hungary* (Indiana University Press, 2007); Karen Dawisha, *Putin's Kleptocracy: Who Owns Russia?* (Simon and Schuster, 2014). Compare to Alan Mikhail and Sam White on state-political-environmental relations in the Ottoman case: Alan Mikhail, *Nature and Empire in Ottoman Egypt: An Environmental History* (Cambridge University Press, 2013); Sam White, *The Climate of Rebellion in the Early Modern Ottoman Empire* (Cambridge University Press, 2013).

27. Donald Worster, "Transformations of the Earth: Toward an Agroecological Perspective in History," *Journal of American History* 76, no. 4 (1990): 1087–106; James C. Scott, *Against the Grain: A Deep History of the Earliest States* (Yale University Press, 2017); John Robert McNeill and Verena Winiwarter, eds., *Soils and Societies: Perspectives from Environmental History* (White Horse Press, 2006).

28. See for example Marc Reisner, *Cadillac Desert: The American West and Its Disappearing Water* (Penguin Books, 1993); Ruth A. Morgan, *Running Out? Water in Western Australia* (UWA Publishing, 2015); Michael Cathcart, *The Water Dreamers: The Remarkable History of Our Dry Continent* (Melbourne, 2009).

29. For comparison, see the classic work by Mike Davis, *Late Victorian Holocausts: El Niño Famines and the Making of the Third World* (Verso, 2017). On tsarist and Soviet agroecological history, see N. M. Dronin and Edward G. Bellinger, *Climate Dependence and Food Problems in Russia, 1900–1990: The Interaction of Climate and Agricultural Policy and Their Effect on Food Problems* (Central European University Press, 2005); R. Davies and S. Wheatcroft, *The Years of Hunger: Soviet Agriculture, 1931–1933*, vol. 5 of *The Industrialization of Soviet Russia* (Palgrave Macmillan, 2003); Moon, *Plough*; Richard Robbins, *Famine in Russia, 1891–1892: The Imperial Government Responds to a Crisis* (Columbia University Press, 1975); Jenny Leigh Smith, *Works in Progress: Plans and Realities on Soviet Farms, 1930–1963* (Yale University Press, 2014); Robert Conquest, *The Harvest of Sorrow: Soviet Collectivization and the Terror-Famine* (Oxford University Press, 1986); Mark Tauger, "The 1932 Harvest and the Famine of 1933," *Slavic Review*

50, no. 1 (1991): 70–89; Nicholas Ganson, *The Soviet Famine of 1946–47 in Global and Historical Perspective* (Palgrave Macmillan, 2009); Aaron Hale-Dorrell, "Khrushchev's Corn Crusade: The Industrial Ideal and Agricultural Practice in the Era of Post Stalin Reform, 1953–1964" (PhD diss., University of North Carolina, Chapel Hill, 2014); Chris Miller, "Gorbachev's Agriculture Agenda: Decollectivization and the Politics of Perestroika," *Kritika: Explorations in Russian and Eurasian History* 17, no. 1 (Winter 2016): 95–118. My thanks go to Sarah Cameron for her thoughtful comments and suggestions on the importance of the environmental history of tsarist/Soviet agriculture.

30. Sigrid Schmalzer, *Red Revolution, Green Revolution: Scientific Farming in Socialist China* (University of Chicago Press, 2016).

31. Kenneth Pomeranz, *The Great Divergence: China, Europe, and the Making of the Modern World Economy* (Princeton University Press, 2001); Tom Kemp, *Industrialization in Nineteenth Century Europe* (Routledge, 1985); Karl Polanyi, *The Great Transformation: The Political and Economic Origins of Our Time* (Beacon Press, 2001); Marie-Janine Calic, Dietmar Neutatz, and Julia Obertreis, eds., *The Crisis of Socialist Modernity: The Soviet Union and Yugoslavia in the 1970s* (Vandenhoeck and Ruprecht, 2011).

32. Chris Otter, "Planet of Meat: A Biological History," in *Challenging (the) Humanities*, ed. Tony Bennett (Australian Academy of the Humanities, 2013); Sam White, "From Globalized Pig Breeds to Capitalist Pigs: A Study in Animal Cultures and Evolutionary History," *Environmental History* 16, no. 1 (2011): 94–120. The study of tsarist and Soviet technology has been a very productive vein of history. Paul Josephson, *Would Trotsky Wear a Bluetooth? Technological Utopianism under Socialism, 1917–1989* (Johns Hopkins University Press, 2010); Kendall Bailes, *Technology and Society under Lenin and Stalin: Origins of the Soviet Technical Intelligentsia, 1917–1941* (Princeton University Press, 1978); Loren Graham, *What Have We Learned about Science and Technology from the Russian Experience?* (Stanford University Press, 1978).

33. Nancy Langston, *Toxic Bodies: Hormone Disruptors and the Legacy of DES* (Yale University Press, 2011); Brett Walker, *Toxic Archipelago: A History of Industrial Disease in Japan* (University of Washington Press, 2011); Chris Otter, Nicholas Breyfogle, and John Brooke, eds., *Technology, Ecology, and Human Health since 1850*, thematic forum in *Environmental History* 20, no. 4 (October 2015): 710–804.

34. Toby Craig Jones, *Desert Kingdom: How Oil and Water Forged Modern Saudi Arabia* (Harvard University Press, 2010); Christopher F. Jones, *Routes of Power: Energy and Modern America* (Harvard University Press, 2014); William F. Ruddiman, *Plows, Plagues and Petroleum: How Humans Took Control of Climate* (Princeton University Press, 2005); Rolf Sieferle, *The Subterranean Forest: Energy Systems and the Industrial Revolution* (White Horse Press, 2001); Vaclav Smil, *Energy and Civilization: A History* (MIT Press, 2017); Sonja Schmid, *Producing Power: The Pre-Chernobyl History of the Soviet Nuclear Industry* (MIT Press, 2015); Per Högselius, *Red Gas: Russia and the Origins of European Energy Dependence* (Palgrave Macmillan, 2013); Anne Rassweiler, *The Generation of Power: The History of Dneprostroi* (Oxford University Press, 1988); Paul Josephson, *Red Atom: Russia's Nuclear Power Program from Stalin to Today* (W.H. Freeman, 2000); Dorothy Zeisler-Vralsted, *Rivers, Memory, and Nation-Building: A History of the Volga and Mississippi Rivers* (Berghahn Books, 2014).

35. On zapovedniki and nature protection, see Douglas Weiner, *A Little Corner of Freedom: Russian Nature Protection from Stalin to Gorbachev* (University of California Press, 2002); Douglas Weiner, *Models of Nature: Ecology, Conservation, and Cultural Revolving in South Russia* (Indiana University Press, 1988); Feliks Shtil'mark, *History of the Russian Zapovedniks, 1895–1995* (Russian Nature Press, 2003). On national parks, see Alan Roe, *Into Russian Nature: National Parks and Environmental Protection* (Oxford University Press, forthcoming).

36. LeCain, *Matter of History*; John L. Brooke and Clark Spencer Larson, "The Nurture of Nature: Genetics, Epigenetics, and Environment in Human Biohistory," *AHR* 119, no. 5 (December 2014): 1500–513.

37. J. R. McNeill, *Mosquito Empires: Ecology and War in the Greater Caribbean, 1620–1914*

(Cambridge University Press, 2010); Otter, Breyfogle, and Brooke, *Technology*; Daniel Schneider, *Hybrid Nature: Sewage Treatment and the Contradictions of the Industrial Ecosystem* (MIT Press, 2011).

38. Will Steffen, Paul Crutzen, and John McNeill, "The Anthropocene: Are Humans Now Overwhelming the Great Forces of Nature?" *Ambio: A Journal of the Human Environment* 36 (2007): 614–21; McNeill, *Something New*; McNeill and Engelke, *Great Acceleration*.

39. John L. Brooke, *Climate Change and the Course of Global History: A Rough Journey* (Cambridge University Press, 2014); David Christian, *Maps of Time: An Introduction to Big History*, 2nd ed. (University of California Press, 2011); David Christian, Cynthia Brown, and Craig Benjamin, *Big History: Between Nothing and Everything* (McGraw-Hill Education, 2013); Cynthia Stokes Brown, *Big History: From the Big Bang to the Present* (New Press, 2012); Valentina B. Bazarova et al., "Late Holocene Paleoclimatic Events and Evolution of Environments in Southeastern Transbaikalia," *Quaternary International* 355 (January 2015): 44–51.

40. Frank Uekoetter, ed., *The Turning Points of Environmental History* (University of Pittsburgh Press, 2010); Scott, *Against the Grain*.

41. On spatial history, see Jeffrey S. Smith, *Explorations in Place Attachment* (Routledge, 2018); N. P. Baron, "New Spatial Histories of Twentieth Century Russia and the Soviet Union: Surveying the Landscape," *Jahrbücher für Geschichte Osteuropas* 55, no. 3 (2007): 374–400; Kelly O'Neill, *Claiming Crimea: A History of Catherine the Great's Southern Empire* (Yale University Press, 2017).

42. J. W. Powell, *Report on the Lands of the Arid Region of the United States*, online at http://pubs.usgs.gov/unnumbered/70039240/report.pdf. See also Donald Worster, *A River Running West: The Life of John Wesley Powell* (Oxford University Press, 2002); Mark Cioc, *The Rhine: An Eco-Biography, 1815–2000* (University of Washington Press, 2006); Daniel Haines, *Rivers Divided: Indus Basin Waters in the Making of India and Pakistan* (Oxford University Press, 2016).

43. LeCain, *Matter of History*; Russell, *Evolutionary History*, esp. 25–30; Edmund Russell, *Greyhound Nation: A Coevolutionary History of England, 1200–1900* (Cambridge University Press, 2018); Joseph E. Taylor III, *Making Salmon: An Environmental History of the Northwest Fisheries Crisis* (University of Washington Press, 1999), esp. 203–36; B. Lynn Ingram and Frances Malamud-Roam, *The West without Water: What Past Floods, Droughts, and Other Climatic Clues Tell Us about Tomorrow* (University of California Press, 2013); Brooke, *Climate Change*; White, *Cold Welcome;* Degroot, *Frigid Golden Age*; *AHR* Roundtable, "History Meets Biology."

44. Kate Brown, *Dispatches from Dystopia: Histories of Places Not Yet Forgotten* (University of Chicago Press, 2015); Yi-Fu Tuan, *Space and Place: The Perspective of Experience* (University of Minnesota Press, 1977); Peter Coates, David Moon, and Paul Warde, eds., *Local Places, Global Processes: Histories of Environmental Change in Britain and Beyond* (Windgather Press, 2016); David Moon, Alexandra Bekasova, and Nicholas Breyfogle, "Exploration and Place in Studying Russia's Environmental History," The White Horse Press blog, Feb. 21, 2017, online at https://whitehorsepress.blog/2017/02/21/exploration-and-place-in-studying-russias-environmental-history/.

45. Josephson, *Conquest of the Russian Arctic*; Bruno, *Nature of Soviet Power*; John McCannon, *A History of the Arctic: Nature, Exploration and Exploitation* (Reaktion Books, 2012); Nicholas Breyfogle, "Russia and the Race for the Arctic," *Origins: Current Events in Historical Perspective* 5, no. 11 (August 2012), online at http://origins.osu.edu/article/russia-and-race-arctic/.

Chapter 2: Planting Trees in Unsuitable Places

The author acknowledges the financial support of the UK Arts and Humanities Research Council, the British Academy, and the Leverhulme Trust. He is grateful to the editor, the participants at the conference, and two anonymous readers for comments on earlier drafts.

1. See Alfred J. Rieber, *The Struggle for the Eurasian Borderlands: From the Rise of Early Modern Empires to the End of the First World War* (Cambridge University Press, 2014); David

A. Bello, *Across Forest, Steppe, and Mountain: Environment, Identity, and Empire in Qing China's Borderlands* (Cambridge University Press, 2016).

2. Quoted in N. Sredinskii, "Kratkii istoricheskii ocherk lesorazvedeniia v iuzhno-russkihk stepiakh," *Lesnoi zhurnal* 6 (1887): 743–44.

3. See "Skarzhinskii, Viktor Petrovich," *Bol'shaia biograficheskaia entsiklopediia*, online at http://dic.academic.ru/dic.nsf/enc_biography/113523/; S. V. Postashkov, "Skarzhinskii, Viktor Petrovich," and "Skarzhinsogo eskadron," in *Otechestvennaia voina 1812 goda: Entsiklopediia*, ed. V. M. Bezotosnyi (ROSSPEN, 2004), 658.

4. For brief histories of the botanical garden, see "Ogród Zoobotaniczny w Toruniu," online at http://encyklopedia.naukowy.pl/Ogr%C3%B3d_Zoobotaniczny_w_Toruniu/; "Ogród Zoobotaniczny w Toruniu, Historia," online at http://www.zoo.torun.pl/pl/historia/.

5. On the steppe environment, see A. A. Chibilev, *Stepi Severnoi Evrazii: ekologo-geograficheskii ocherk i bibliografiia* (UrO RAN, 1998). On debates concerning reasons for the lack of trees, see David Moon, *The Plough that Broke the Steppes: Agriculture and Environment on Russia's Grasslands, 1700–1914* (Oxford University Press, 2013), 59–62, 98–111.

6. M. P. Borovskii, *Istoricheskii obzor piatidesiatiletnei deiatel'nosti Imperatorskogo Obshchestva sel'skogo khoziaistva Iuzhnoi Rossii s 1828 po 1878 god* (P. Frantsov, 1878). On Palimpsestov, see Moon, *Plough*, 52.

7. See Michael Khodarkovsky, *Russia's Steppe Frontier: The Making of a Colonial Empire, 1500–1800* (Indiana University Press, 2002); Willard Sunderland, *Taming the Wild Field: Colonization and Empire on the Russian Steppe* (Cornell University Press, 2004); Roger P. Bartlett, *Human Capital: The Settlement of Foreigners in Russia, 1762–1804* (Cambridge University Press, 1979).

8. See David Christian, *A History of Russia, Central Asia and Mongolia, Vol.1, Inner Eurasia from Prehistory to the Mongol Empire* (Blackwell, 1998); David Morgan, "The Decline and Fall of the Mongol Empire," *Journal of the Royal Asiatic Society* 19, no. 4 (2009): 427–37.

9. See Khodarkovsky, *Russia's Steppe Frontier*; V. P. Zagorovskii, *Belgorodskaia cherta* (Izd-vo Voronezhskogo Universiteta, 1969); Brian J. Boeck, *Imperial Boundaries: Cossack Communities and Empire-Building in the Age of Peter the Great* (Cambridge University Press, 2009).

10. See François Hartog, *The Mirror of Herodotus: The Representation of the Other in the Writing of History*, trans. Janet Lloyd (University of California Press, 1988); Carl Thompson, *Travel Writing* (Routledge, 2011), 132–33.

11. On differing cultural perspectives, see Arash Khazeni, "Across the Black Sands and the Red: Travel Writing, Nature, and the Reclamation of the Eurasian Steppe circa 1850," *International Journal of Middle Eastern Studies* 42 (2010): 591–614.

12. Ibn Fadlan, *Ibn Fadlan and the Land of Darkness: Arab Travellers in the Far North*, trans. Paul Lunde and Caroline Stone (Penguin, 2012), xvi–xxiv, 3, 8–11, 32, 34, 41 (the quotes are from 11 and 41). I am grateful to Dr. Harry Munt for drawing my attention to this book.

13. Ibn Fadlan, *Ibn Fadlan and the Land of Darkness*, 116–17, 124–25 (quotations from 124 and 125).

14. [Tulisen], *Narrative of the Chinese Embassy to the Khan of the Tourgouth Tartars, in the years 1712, 13, 14, and 15*, trans. Sir George Thomas Staunton (John Murray, 1821), https://archive.org/details/narrativechines00tuligoog/, xiv, 16, 20, 25–27, 32, 43–45, 49, 72, 82, 112, 116, 122, 123, 128, 151. (Readers interested in locating the precise page references for the individual short quotations may search the electronic version of the source.) I am grateful to Professor Peter Perdue for drawing my attention to this source.

15. I. I. Lepekhin, "Zapiski puteshestviia Akademika Lepekhina," in *Polnoe sobranie uchenykh puteshestvii po Rossii* 3 (Imperatorskaia akademiia nauk, 1821), 90, 98–99, 126.

16. Lepekhin, "Zapiski," 228.

17. Lepekhin, "Zapiski," 310, 345, 352, 371–73.

18. Lepekhin, "Zapiski," 233–57.

19. See David Moon, "The Russian Academy of Sciences Expeditions to the Steppes in the Late-Eighteenth Century," *Slavonic and East European Review* 88 (2010): 204–36.

20. August von Haxthausen, *The Russian Empire: Its People, Institutions and Resources*, trans. Robert Farie (Chapman and Hall, 1856), 2:70.

21. Haxthausen, *Russian Empire*, 1:415. See also Moon, *Plough*, 2–3.

22. Harvey L. Dyck and John R. Staples, eds., *Transformation on the Southern Ukrainian Steppe: Letters and Papers of Johann Cornies, Vol. I: 1812–1835* (Toronto University Press, 2015), 129, 135, 151. The "Guardianship Committee for Foreign Colonists in New Russia" was the governmental body that had oversight of the settlers from abroad. John R. Staples, *Cross-Cultural Encounters on the Ukrainian Steppe: Settling the Molochna Basin, 1783–1861* (University of Toronto Press, 2003), 24, 114–21.

23. Stephen Brain, *Song of the Forest: Russian Forestry and Stalinist Environmentalism, 1905–1953* (University of Pittsburgh Press, 2011), 4.

24. See Moon, *Plough*, 4–6; David Moon, "Land and Environment," in *The Oxford Handbook of Modern Russian History*, ed. Simon Dixon (Oxford University Press, 2015), chapter 1, available online at DOI:10.1093/oxfordhb/9780199236701.013.001/.

25. See Christopher Ely, *This Meager Nature: Landscape and National Identity in Imperial Russia* (Northern Illinois University Press, 2002); Dorothy Zeisler-Vralsted, *Rivers, Memory, and Nation-Building: A History of the Volga and Mississippi Rivers* (Berghahn Books, 2014).

26. See Ely, *Meager Nature*, 165–92. The locations where Shishkin painted some of his landscapes are protected in the Lower Kama National Park, Tatarstan. See "Nizhniaia Kama Natsional'nyi Park," *FGBU Natsional'nyi Park Nizhniaia Kama*, online at http://nkama-park.ru/index/0-3/.

27. I. U. Palimpsestov, "Odin iz otvetov na vopros: 'byli li lesa na iuge Rossii?'" *Izvestiia imperatorskogo Obshchestva liubitelei estestvoznaniia, antropologii i etnografii* 41, no. 1 (1881): 17.

28. I. Palimpsestov, *Stepi iuga Rossii byli-li iskoni vekov stepami i vozmozhno-li oblesit' ikh?* revised ed. (Odessa: Tip. L. Nitche, 1890), 264.

29. Jane Costlow, *Heart-Pine Russia: Walking and Writing the Nineteenth-Century Forest* (Cornell University Press, 2012).

30. "Komar & Melamid: The Most Wanted Paintings on the Web," online at http://awp.diaart.org/km/.

31. Valerie Kivelson, *Cartographies of Tsardom: The Land and Its Meanings in Seventeenth-Century Russia* (Cornell University Press, 2006), 110–12.

32. Quoted in Jeffrey K. Wilson, *The German Forest: Nature, Identity, and the Contestation of a National Symbol, 1871–1914* (University of Toronto Press, 2012), 1. See also Simon Schama, *Landscape and Memory* (Fontana Press, 1996), 116. On the significance of forests and forestry across northern Europe, see *Managing Northern Europe's Forests*

Histories from the Age of Improvement to the Age of Ecology, ed. Richard Hölzl and Jan Oosthoek (Berghahn Books, 2018).

33. See M. V. Loskutova and A. A. Fedotova, *Stanovlenie prikladnykh biologicheskikh issledovanii v Rossii: Vzaimodeistvie nauki i praktiki v XIX–nachale XX vekakh.*

(Nestor-Istoriia, 2014), 7–39. The Ministry of State Domains was set up in 1837 to manage state lands.

34. Ledebur, "O razvedenii lesov po beregam Chernogo Moria dlia preduprezhdeniia zasukh," *Lesnoi zhurnal* 2, no. 1 (1834): 173–77.

35. See Ia. Veinberg, *Les: znachenie ego v prirode i mery k ego sokhraneniiu* (Tip. E. Lissner i Iu. Roman, 1884); Moon, *Plough*, 176–79.

36. [F. L.] Breitenbakh, "O pol'ze lesov v prirode," *Lesnoi zhurnal* 1, no. 3 (1835): 383–91 (quotation from 384). The passage in my text summarizes the article in its entirety. The author's name is spelled correctly in German in the text but transliterated from Cyrillic in the reference as the article was published in Russian.

37. See Richard Grove, *Green Imperialism: Colonial Scientists, Ecological Crises, and the History of Environmental Concern, 1600–1860* (Cambridge University Press, 1994); Berthold

Ribbentrop, *Forestry in British India* (1900; reis., Indus, 1989); Antti Erkkilä and Harri Siiskonen, *Forestry in Namibia, 1850–1990* (University of Joensuu, 1992), 66, 73–74, 85–104, 181.

38. Thomas Dunlap, *Nature and the English Diaspora: Environment and History in the United States, Canada, Australia, and New Zealand* (Cambridge University Press, 1999), 17.

39. Alfred Crosby, *Ecological Imperialism: The Biological Expansion of Europe, 900–1900* (Cambridge University Press, 1986). "Neo-Europes" are explained on 2–7.

40. For one version of the story, see Sredinskii, "Kratkii istoricheskii ocherk," 741.

41. *Istoricheskoe obozrenie piatidesiatiletnei deiatel'nosti Ministerstva Gosudarstvennykh Imushchestv, 1837–1887* (St. Petersburg, 1888), 3:172–73.

42. Aleksei Karimov, "State, Navy and the Origin of the Petrine Forest Cadastral Survey," in *Peter the Great and the West: New Perspectives*, ed. Lindsey Hughes (Palgrave, 2001), 118–29.

43. N. I. Rubtsov, *Kratkaia istoriia stepnogo i zashchitnogo lesonasazhdeniia* (Leningrad, 1970), 9.

44. A. I. Khodnev, *Istoriia Imperatorskogo Vol'nogo Ekonomicheskogo Obshchestva s 1765 do 1865 goda* (Tip. Obshchestvennaia pol'za, 1865); A.T. Bolotov, "O rublenii, popravlenii i zavedenii lesov," *Trudy Vol'nogo ekonomicheskogo obshchestva* 4 (1766): 68–158, and 5 (1767): 78–130.

45. Ol'ga Eliseeva, *Grigorii Potemkin* (Molodaia Gvardiia, 2006), 362, 367.

46. Vasilii Zuev, *Puteshestvennye zapiski Vasiliia Zueva ot S. Peterburga do Khersona, v 1781 i 1782 godu* (Imperatorskaia Akademiia Nauk, 1787), 226.

47. A. V. Khrapovitskii, *Zhurnal Vysochaishogo puteshestviia ee Velichestva Gosudarnyni Ekateriny II samoderzhavitsy vserossiiskoi v Poludennye Strany Rossii v 1787 godu* (Universitetskaia tipografiia, 1787), 71; *Polnoe sobranie zakonov Rossiiskoi imperii* (Complete collection of the laws of the Russian Empire) (Tip. II Otdeleniia Sobstvennoi Ego Imperatorskago Velichestva Kantseliarii, 1830; hereafter cited as *PSZ*), ser. 1, 22:859–60, no. 16548 (June 8, 1787).

48. I. I. Veber, "Primechaniia o razlichnykh predmetakh khoziaistva v Ekaterinoslavskom Namestnichestve," *Trudy Vol'nogo Ekonomicheskogo Obshchestva* 50 (1795): 179–81.

49. *PSZ*, ser. 1, 26:635–49, no. 19873 (May 16, 1801).

50. *PSZ*, ser. 1, 27:726–31, no. 20841 (July 7, 1803).

51. James Urry, *None but Saints: The Transformation of Mennonite Life in Russia, 1789–1889* (Hyperion Press, 1989), 98.

52. G. N. Vysotskii, "Stepnoe lesorazvedenie," *Polnaia entsiklopediia russkogo sel'skogo khoziaistva i soprikasaiushchikhsia s nimi nauk* (Devrien, 1905), vol. 9, collection 445.

53. See K. V. Rudchenko, "The Mennonite Role in the Development of Intensive Steppe Forest Cultivation in the First Half of the Nineteenth Century," in *Molochna—2004: Mennonites and Their Neighbours (1804–2004)* (Zaporizhia, 2004), 165–67.

54. Staples, *Cross-Cultural Encounters on the Ukrainian Steppe*, 73–74, 114–18; Dyck and Staples, *Transformation*, 227–35, 373–74.

55. I. Kornis (Johann Cornies), "O sostoianii khoziaistva v Molochanskikh Menonistskikh koloniiakh v 1842 g.," *Zhurnal Ministerstva Gosudarstvennikh Imushchestv* 8 (1843), 2nd pagn., 70–72.

56. "O sostoianii raznykh otraslei sel'skogo khoziaistva v iuzhnoi Rossii v 1848 godu," *Zapiski Imperatorskogo Obshchestva Sel'skogo Khoziaistva Iuzhnoi Rossii* 11 (1848): 708. See also Rossiiskii Gosudarstvennyi Istoricheskii Arkhiv (Russian State Historical Archive; hereafter cited as RGIA) f. 383, op. 29, 1837–1838, d. 609.

57. Haxthausen, *Russian Empire*, 2:72.

58. Haxthausen, *Russian Empire*, 2:75. On Haxthausen's visit to the Mennonites, see Haxthausen, *Russian Empire*, 1:420–30. For comments on Mennonite tree planting by visitors later in the century, see N. K. Kalageorgi and B. M. Borisov, *Ekskursiia na reku Molochnuiu: opyt sel'skokhoziaistvennogo i ekonomicheskogo issledovaniia iuzhnorusskikh khoziaistv* (Tip. V. F. Demakova, 1878), part 1, 73–75.

59. E. I. Druzhinina, *Iuzhnaia Ukraina v period krizisa feodalizma 1825–1860 gg.* (Nauka, 1981), 73–74.

60. A. Nordman, "O gorode Odesse v estestvoispytatel'nom otnoshenii," *Listki Imperatorskogo Obshchestva Sel'skogo-Khoziaistva Iuzhnoi Rossii* 4 (1837): 209–10.

61. Khodnev, *Istoriia*, 199.

62. Nordman, "O gorode Odesse," 208; A. Nordman, "Opisanie imperatorskogo Odesskogo sada i vzgliad na rastitel'nye i klimatologicheskie otnosheniia okrestnostei g. Odessy," *Zapiski Imperatorskogo Obshchestva Sel'skogo-Khoziaistva Iuzhnoi Rossii* 6 (1847): 108–11; Palimpsestov, *Stepi iuga Rossii*, 262. On Nordman, see N. G. Pavlov, "Aleksandr Davidovich Nordman (1803–1866)," online at http://jugan2.narod.ru/nordman.html.

63. J. G. Kohl, *Reisen in Südrussland* (Arnold, 1841), 1:74–78. The quotations are from an abridged translation: J. G. Kohl, *Russia: St. Petersburg, Moscow, Kharkoff, Riga, Odessa, the German provinces on the Baltic, the Steppes, the Crimea, and the Interior of the Empire* (Chapman and Hall, 1842), 425–26.

64. Haxthausen, *Russian Empire*, 2:69–71.

65. Haxthausen, *Studien*, 1:456.

66. Haxthausen, *Russian Empire*, 2:71–72.

67. Haxthausen, *Russian Empire*, 2:73.

68. *PSZ*, ser. 2, 10:453–639, no. 8163 (May 26, 1835), Prilozhenie I, pp. 78–79, 88–90.

69. GARO f. 243, op. 1, d. 19, ll. 1–18, 21ob.; GARO f. 301, op. 14, d. 7, 1849–1852, ll. 13–25, 143; GARO f. 301, op. 14, dd. 7, 10, 22; GARO f. 301, op. 27, d. 2. See also Moon, *Plough*, 182–84.

70. A. Bitnyi, "O popytkakh lesnogo upravleniia k obleseniiu stepei," *Zhurnal Ministerstva Gosudarstvennykh Imushchestv* 84 (1863), 2nd pagn., 274–83; P. Nevol'sin, "Otchet o puteshestvii v Orenburgskii i Astrakhanskii krai," *Vestnik Imperatorskogo Russkogo geograficheskogo obschestva* 4 (1852), 5th pagn., 1–34.

71. Frants Khristianovich Meier, "Stepnoe lesovodstvo," *Trudy Imperatorskogo Vol'nogo Ekonomicheskogo Obshchestva*, no. 3 (1860): 385–430; no. 4 (1860): 16–47. See also A. Sovetov, "Maier, Frantz Khristianovich," *Entsiklopedicheskii Slovar'* (Brokgauz i Efron, 1896), 35:369–70; M. D. Merzlenko, "Pioner stepnogo lesorazvedeniia (k 220-letiiu so dnia rozhdeniia Frantsa Khristianovicha Maiera)," *Ustoichivoe lesopol'zovanie* 1 (2003): 44–45, online at http://www.wwf.ru/data/publ_period/forest_mag1/10.pdf.

72. [Vasilii Iakovlevich] Lomikovskii, *Razvedenie lesa v sel'tse Trudoliubov* (Tip. Departamenta Vneshnei Torgovli, 1837).

73. Nikolai Gogol, *Dead Souls*, trans. David Magarshack (Penguin, 1961), 312–13.

74. Palimpsestov, *Stepi iuga Rossii*, 261.

75. See Ely, *Meager Nature*, 91–94; Edyta M. Bojanowska, *Nikolai Gogol: Between Ukrainian and Russian Nationalism* (Harvard University Press, 2007). See also Inga Vidugirite, *Geograficheskoe voobrazhenie. Gogol'* (Vilniaus universitetas, 2015).

76. See A. A. Fedotova and M. V. Loskutova, "Veliko-Anadol'skoe lesnichestvo i sozdanie lesnogo opytnogo delo v Rossii (1840–nachalo 1890-kh gg.)," *Izvestiia Russkogo geograficheskogo obshchestva* 145, no. 6 (2013): 52–65; Moon, *Plough*, 185–205.

77. Denis J. B. Shaw, "Mastering Nature through Science: Soviet Geographers and the Great Stalin Plan for the Transformation of Nature, 1948–53," *Slavonic and East European Review* 93 (2015): 120–46.

78. Vysotskii, "Stepnoe lesorazvedenie," cols. 443–99; G. N. Vyssotsky, "Shelterbelts in the Steppes of Russia," *Journal of Forestry* 33 (1935): 781–88.

79. See Grove, *Green Imperialism*, 108–25, 153–67, 266–84, 309–11.

80. See C. V. Kremenetski, "Human Impact on the Holocene Vegetation of the South Russian Plain," in *Landscapes in Flux: Central and Eastern Europe in Antiquity*, ed. John Chapman and Pavel Dolukhanov (Oxbow Books, 1997), 275–87.

81. Palimpsestov, *Stepi iuga Rossii.*

82. Sredinskii, "Kratkii istoricheskii ocherk," 741, 744–45; D. Strukov, "O lesakh Novorossi-

iskogo kraia i Bessarabii," *Zhurnal Ministerstva Gosudarstvennikh Imushchestv* 46, no. 1 (1853), 1st pagn., 224–25.

83. "V Trikraty, k Skarzhinskomu," online at http://www.voskres.ru/army/publicist/hristova.htm/.

84. See, for example "Trikrati—Trikrats'kii lis," *KoloKrai*, online at http://kolokray.com/uk/f/trikraty-trikratskiy-les-1360926105.html and at http://angatravel.com.ua/ru/trikratskiy-labirint.html. The forest can be located on Google Earth at 47° 42' 23" N and 31° 24' 42" E.

Chapter 3: "People Arrive but the Land Does Not Move"

In addition to the workshop at The Ohio State University where this essay originated, versions of this essay were presented at a working group on Russian history at the Library of Congress, at a workshop on comparative famines held at Nazarbayev University in Astana, Kazakhstan, and at a workshop entitled "New Directions in the History of Inner and Central Asia," held at Harvard University. I would like to thank all the participants at these workshops and the reviewers of this volume for their thoughtful comments on earlier drafts.

1. See, among others, Francine Hirsch, *Empire of Nations: Ethnographic Knowledge and the Making of the Soviet Union* (Cornell University Press, 2005); Peter Holquist, *Making War, Forging Revolution: Russia's Continuum of Crisis, 1914–1921* (Harvard University Press, 2002).

2. On the Kazakh famine, see Sarah Cameron, *The Hungry Steppe: Famine, Violence, and the Making of Soviet Kazakhstan* (Cornell University Press, 2018).

3. This question has received little attention from scholars. The one exception is Niccolò Pianciola, *Stalinismo di frontiera: colonizzazione agricola, sterminio dei nomadi e costruzione statale in Asia centrale, 1905–1936* (Viella, 2009), chapters 1 and 2. In this essay I seek to build upon Pianciola's observations, showing how the tools of environmental history might augment the economic approach he uses in his study.

4. Amartya Sen, *Poverty and Famines: An Essay on Entitlement and Deprivation* (Oxford University Press, 1981).

5. There are numerous examples of this approach. The classic account is Robert Conquest, *The Harvest of Sorrow: Soviet Collectivization and the Terror-Famine* (Oxford University Press, 1986). For an approach that combines political and economic history to analyze the Soviet collectivization famines, see R. W. Davies and Stephen G. Wheatcroft, *The Years of Hunger: Soviet Agriculture, 1931–1933* (Palgrave Macmillan, 2004).

6. For an overview of this debate, see John-Paul Himka, "Encumbered Memory: The Ukrainian Famine of 1932–33," *Kritika: Explorations in Russian and Eurasian History* 14, no. 2 (2013): 411–36.

7. For an alternate perspective, see Mark Tauger, "The 1932 Harvest and the Famine of 1933," *Slavic Review* 50, no. 1 (1991): 70–89. Tauger rejects the idea that the Ukrainian famine was man-made, and he argues that hunger was because of actual grain shortages brought about by a poor harvest in 1932.

8. Historians of the Russian Empire have devoted little attention to the ways in which borderland colonization changed native subsistence systems. These questions, however, have been explored extensively by US environmental historians focusing on Native American subsistence systems. Two classic studies are William Cronon, *Changes in the Land: Indians, Colonists and the Ecology of New England* (Hill and Wang, 1983) and Richard White, *The Roots of Dependency: Subsistence, Environment and Social Change among the Choctaws, Pawnees and Navajos* (University of Nebraska Press, 1983).

9. For examples of this approach, see Mike Davis, *Late Victorian Holocausts: El Niño Famines and the Making of the Third World* (Verso, 2002); Steven Serels, *Starvation and the State: Famine, Slavery, and Power in Sudan, 1883–1956* (Palgrave MacMillan, 2013); Michael Watts, *Silent Violence: Food, Famine, and Peasantry in Northern Nigeria* (University of California Press, 1983).

10. I take this figure from George J. Demko, in *The Russian Colonization of Kazakhstan, 1896–1916* (Indiana University Publications, 1969), who focuses on the steppe oblasti of Akmolinsk,

Semipalatinsk, Ural'sk, and Turgai and parts of Semirech'e and Syr Darya, all of which would eventually become part of Soviet Kazakhstan. On peasant settlement of the Kazakh Steppe, see Ian W. Campbell, *Knowledge and the Ends of Empire: Kazak Intermediaries and Russian Rule on the Steppe, 1731–1917* (Cornell University Press, 2017), 125–56; Gulnar Kendirbai, *Land and People: The Russian Colonization of the Kazak Steppe* (Klaus Schwarz Verlag, 2002); C. N. Maltusynov, *Agrarnyi vopros v Kazakhstane i Gosudarstvennaia duma Rossii 1906–1917 gg. (sotsiokul'turnyi podkhod)* (Daik-Press, 2006); Virginia Martin, *Law and Custom in the Steppe: The Kazakhs of the Middle Horde and Russian Colonialism in the Nineteenth Century* (Curzon: 2001), 60–86; Ihor Stebelsky, "The Frontier in Central Asia," in *Studies in Russian Historical Geography*, ed. James H. Bater and R. A. French (Academic Press, 1983), 1:149–52. On the settlement of Russian Turkestan (parts of which would be joined to the steppe to form Soviet Kazakhstan), see Daniel Brower, *Turkestan and the Fate of the Russian Empire* (RoutledgeCurzon, 2003); Alexander Morrison, "Peasant Settlers and the Civilizing Mission in Russian Turkestan, 1865–1917," *Journal of Imperial and Commonwealth History* 43, no. 3 (2015), 387–417. The classic Soviet account is A. B. Tursunbaev, *Iz istorii krest'ianskogo pereseleniia v Kazakhstan* (Akademiia nauk Kazakhskoi SSR, 1950).

11. The first station to measure precipitation and temperature in the Kazakh Steppe was Vernyi (present-day Almaty), where temperature and precipitation measurements began in 1879. Complete data for other stations in the Kazakh Steppe, such as Fort Alexandrovsk (present-day Fort Shevchenko) and Kazalinsk followed only in 1891. Thus, it is very difficult to assess change over time, as well as regional differences in drought patterns in the Kazakh Steppe for this period. For temperature and precipitation data for Central Asia, see M. W. Williams and V. G. Konovalov, *Central Asia Temperature and Precipitation Data, 1879–2003* (USA National Snow and Ice Data Center, 2008).

12. Michael D. Frachetti, *Pastoralist Landscapes and Social Interaction in Bronze Age Eurasia* (University of California Press, 2008).

13. Robert N. Taaffe, "The Geographic Setting," in *The Cambridge History of Early Inner Asia*, ed. Denis Sinor (Cambridge University Press 1990), 19–40.

14. For the argument that environmental and political changes during the Iron Age led pastoral nomads in Semirech'e to move from a "more pastoral orientation to a more intensive agricultural focus," see Arlene Miller Rosen, Claudia Chang, and Fedor Pavlovich Grigoriev, "Paleoenvironments and Economy of Iron Age Saka-Wusun Agro-Pastoralists in Southeastern Kazakhstan," *Antiquity* 70, no. 285 (2000): 611–23.

15. For a discussion of Kazakh nomads' practice of burning, see A. N. Zarudnyi, *Strana svobodnykh zemel'* (St. Petersburg, 1908), 16. On the construction of wells, see F. Fiel'strur, "Skotovodstvo i kochevanie v chasti stepei zapadnogo Kazakstana," in *Kazaki: Antropologicheskie ocherki, sbornik II*, ed. S. I. Rudenko (Izd-vo. Akademii Nauk SSSR, 1927), 104.

16. I thank Nicholas Breyfogle for suggesting this wording.

17. In part because of the lack of sources from this period, it is difficult to date the formation of the three Kazakh hordes. The first mention of the three hordes by name appears in 1731, but most scholars date their formation to the late sixteenth century. See the discussion in Allen J. Frank, "The Qazaqs and the Russians," in *The Cambridge History of Inner Asia: The Chinggisid Age*, ed. Nicola Di Cosmo, Allen J. Frank, and Peter B. Golden (Cambridge University Press, 2009), 364–65.

18. I have adopted the classifications found in Taaffe, "Geographic Setting."

19. Occasionally, commentators referred to this plateau as the *severnaia golodnaia step'*, or the northern Hungry Steppe, to differentiate it from another, smaller plain with the same name. The second Hungry Steppe, part of present-day Uzbekistan, is located between the city of Dzhizak and the Syr Darya River.

20. S. P. Suslov, *The Physical Geography of Asiatic Russia*, trans. Noah D. Gershevsky (W. H. Freeman, 1961), 447–51.

21. Taaffe, "Geographic Setting," 30–31.

22. Taaffe, "Geographic Setting," 26.

23. Fahu Chen, Xiaozhong Huang, Jianwu Zhang, J. A. Holmes, and Jianhui Chen, "Humid Little Ice Age in Arid Central Asia Documented by Bosten Lake, Xinjiang China," *Science in China Series D: Earth Sciences* 49, no. 12 (2006): 1280–90. I thank John Brooke for pointing me to this reference.

24. Taaffe, "Geographic Setting," 28, 35–37.

25. Demko, *Russian Colonization*, 14.

26. Suslov, *Physical Geography*, 427.

27. Martin, *Law and Custom*, 22.

28. Nurbulat Masanov, *Kochevaia tsivilizatsiia kazakhov: Osnovy zhiznedeiatel'nosti nomadnogo obshchestva* (Sotsinvest, 2011), 277–80.

29. Martin, *Law and Custom*, 21.

30. On how Russian imperial officials and Kazakh intellectuals came to understand *zhŭt* as a disaster, see Ian Campbell (this volume); on droughts in the steppe in the 1960s, see Marc Elie (this volume).

31. On this particular period of migration, see Francois-Xavier Coquin, *La Sibérie: Peuplement et immigration paysanne au 19e siècle* (Institut d'études slaves, 1969); Willard Sunderland, *Taming the Wild Field: Colonization and Empire on the Russian Steppe* (Cornell University Press, 2004), 177–221; Donald W. Treadgold, *The Great Siberian Migration: Government and Peasant in Resettlement from Emancipation to the First World War* (Princeton University Press, 1957). For a broader overview of colonization and its place in Eurasian history, see the introduction to Nicholas B. Breyfogle, Abby Schrader, and Willard Sunderland, eds., *Peopling the Russian Periphery: Borderland Colonization in Eurasian History* (Routledge, 2007).

32. In 1801 Russian authorities had removed a large group of Little Horde Kazakhs from the authority of their khan, renaming them the Inner Horde, or Bukei Horde. Members of the Inner Horde were then encouraged to migrate in the territory between the Volga and Ural Rivers, behind the Siberian Line. This territory was part of Astrakhan province. See Frank, "The Qazaqs," 371.

33. Demko, *Russian Colonization*, 114.

34. Stebelsky, "Frontier," 167–68.

35. Demko, *Russian Colonization*, 63.

36. Demko, *Russian Colonization*, 13.

37. *Kratkii istoricheskii ocherk Semipalatinskogo kraia (do 1917 goda)* (Semipalatinsk, 1929), 33–34.

38. Orazgul Mŭkhatova, *Qazaqstandaghï XX-ghasïrdïng alghashqï onzhïldïq—tarïndaghï agrarlïq reformalar tarikhnamasï (1920–1929 zhïldar)* (Taldïqorghan universitetï, 1998), 3.

39. Lewis Siegelbaum, "Those Elusive Scouts: Pioneering Peasants and the Russian State, 1870s–1950," *Kritika: Explorations in Russian and Eurasian History* 14, no. 1 (2013): 31–58.

40. Demko, *Russian Colonization*, 120.

41. For a petition by the vice-governor of Akmolinsk oblast to close his region to colonization in 1906 and the refusal, see Rossiiskii gosudarstvennyi istoricheskii arkhiv (Russian State Historical Archive) f. 391, op. 3, d. 105, ll. 5–6, 36, reprinted in S. N. Maltusynov, ed., *Agrarnaia istoriia Kazakhstana (konets XIX–nachalo XX v.): Sbornik dokumenov i materialov* (Daik-Press, 2006), 99–100. In 1902 in Turkestan, the governor-general begged the Internal Affairs Ministry to cut off the flow of illegal settlers. See Brower, *Turkestan*, 135.

42. Stebelsky, "Frontier," 158.

43. Letter to the Committee for Assistance to the Starving from the Head of the Settlers' Affairs in Semirech'e oblast, April 11, 1907, Tsentral'nyi gosudarstvennyi arkhiv Respubliki Kazakhstan (Central State Archive of the Republic of Kazakhstan; hereafter cited as TsGARK) f. 19, op. 1, d. 57, ll. 12–12ob).

44. Letter to the Steppe administration of the Russian society of the Red Cross from a detachment of the Red Cross, April 6, 1907 (TsGARK f. 64, op. 1, d. 4701, l. 27).

45. Telegram from the secretary of the financial committee in Omsk to the minister of trade and industry, St. Petersburg. July 2, 1911 (TsGARK f. 64, op. 1, d. 4752, ll. 3–4).

46. George Demko estimates that more than 22 percent of all settlers went back to European Russia. See Demko, *Russian Colonization*, 203.

47. Demko, *Russian Colonization*, 174–75. The situation was similar further south in Turkestan. There, peasants in Semirech'e expanded their practice of stock raising to protect themselves in case of a bad harvest. Brower, *Turkestan*, 132–33. On the efforts of settlers to adapt their practice of farming to the steppes of European Russia, see David Moon, *The Plough that Broke the Steppes: Agriculture and Environment on Russia's Grasslands, 1700–1914* (Oxford University Press, 2013), 18–20.

48. A. S. Bezkhovich, "Zemledelie ukraintsev-pereselentsev iuzhnoi chasti Semipalatinskoi gubernii," in *Ukraintsy-pereselentsy Semipalatinskoi gubernii*, ed. S. I. Rudenko (Izd-vo. AN SSSR, 1930), 16.

49. Bezkhovich, "Zemledelie," 103. Unlike the settlers that David Moon discusses (this volume), those who came to the Kazakh Steppe do not appear to have made widespread efforts to cultivate trees, perhaps a reflection of the steppe's more arid conditions.

50. Martin, *Law and Custom*, 74–83.

51. K. A. Chuvelev, "O reorganizatsii kochevogo i polukochevogo khoziaistva," *Narodnoe khoziaistvo Kazakstana* 2–3 (1928): 47.

52. Anatoly M. Khazanov, *Nomads and the Outside World*, trans. Julia Crookenden (Cambridge University Press, 1984), 46–49.

53. In Arakaraisk *volost* (part of Kustanai *uezd*), the number of cattle in Kazakhs' herds increased by 20 percent from 1898 to 1905. *Vliianie kolonizatsii na kirgizskoe khoziaistvo* (Izdanie pereselenchskago upravleniia, 1907), 11. See also A. Iu. Bykov, *Istoki modernizatsii Kazakhstana* (Izd-vo. Azbuka, 2003), 208.

54. P. S. Shulkov, "Ob"em rynochnykh sviazei Kazakskogo khoziaistva," *Narodnoe khoziaistvo Kazakstana* 11–12 (1928): 64–74.

55. Martin, *Law and Custom*, 66–67; Bykov, *Istoki modernizatsii Kazakhstana*, 200.

56. For further discussion of this issue, see Campbell (this volume).

57. TsGARK f. 25, op. 1, d. 2822, ll. 213–214ob., reprinted in Maltusynov, *Agrarnaia istoriia Kazakhstana*, 109–11. Although the rental of pasturelands was technically illegal (St. Petersburg claimed to own nomads' pasturelands), nomads in areas of intense peasant settlement such as the northern reaches of Turgai oblast began to rent pasturelands to supplement their practice of pastoralism.

58. Anthropologists have shown that pastoral nomadism is a flexible and adaptive system. In an important study, *The Kirghiz and Wakhi of Afghanistan: Adaptation to Closed Frontiers* (University of Washington Press, 1979), the anthropologist M. Nazif Mohib Shahrani explores how Kyrgyz nomads adapted their practice of pastoral nomadism to the closure of the Soviet and Chinese frontiers.

59. The construction of Cossack settlements in the northern Kazakh Steppe in the eighteenth century had led to an increase in this trade. On the extensive contact between Siberian Cossacks and Kazakhs, see Yuriy Malikov, *Tsars, Cossacks, and Nomads: The Formation of a Borderland Culture in Northern Kazakhstan in the 18th and 19th Centuries* (Klaus Schwarz Verlag, 2011).

60. On the steppe trade see Frank, "The Qazaqs," 374–75; T. K. Shcheglova, *Iarmarki Zapadnoi sibiri i stepnykh oblastei vo vtoroi polovine XIX veka: iz istorii rossiisko-aziatskoi torgovli* (Izd-vo. Barnaul'skogo gosudarstvennogo pedagogicheskogo universiteta, 2002).

61. Demko, *Russian Colonization*, 186.

62. Kh. A. Argynbaev, *Istoriko-kul'turnye sviazi russkogo i kazakhskogo narodov* (Altyn kitap, 2005), 66. Shahrani finds that Kyrgyz pastoralists of the Afghan Pamirs made similar adaptations after the closure of the Soviet and Chinese frontiers. The region's animal population increased, as the Kyrgyz responded to the growing possibilities for trade with settled groups. The Kyrgyz then began to increase the numbers of sheep and yaks in their herds, while

decreasing the number of camels and horses. See Shahrani, *Kirghiz and Wakhi*, 224. Kazakhs' production of cattle for Russian markets was part of a broader shift. During the late nineteenth and twentieth centuries, as consumers began to eat more meat and distribution systems improved, the world's cattle population rose dramatically. See Chris Otter, "Planet of Meat: A Biological History," in *Challenging (the) Humanities*, ed. Tony Bennett (Australian Academy of the Humanities, 2013), 33–49.

63. Michael Khodarkovsky, *Russia's Steppe Frontier: The Making of a Colonial Empire, 1500–1800* (Indiana University Press, 2002), 161.

64. V. K. Kuznetsov, ed., *Kirgizskoe khoziaistvo v Akmolinskoi oblasti, t. 1, Kokchetavskii uezd. Povtornoe issledovanie 1907 g.* (Pereselencheskoe upravlenie, 1910), v.

65. Kuznetsov, *Kirgizskoe khoziaistvo*, 144.

66. Nurila Z. Shakanova, "The System of Nourishment among the Eurasian Nomads: The Kazakh Example," in *Ecology and Empire: Nomads in the Cultural Evolution of the Old World*, ed. Gary Seaman (University of Southern California: 1989), 111–19; G. E. Taizhanova, ed., *Kazakhi: Istoriko-etnograficheskoe issledovanie* (Kazakhstan, 1995), 163–76.

67. On the Russian state's increasing interest in human nutritional needs, see Alison K. Smith, *Recipes for Russia: Food and Nationhood under the Tsars* (Northern Illinois University Press, 2008). On the development of food consumption surveys in the 1880s, see Stephen G. Wheatcroft, "Famine and Food Consumption Records in Early Soviet History, 1917–25," in *Food, Diet, and Economic Change Past and Present*, ed. Catherine Geissler and Derek J. Oddy (St. Martin's Press, 1993), 151–74. On the eighteenth-century grain trade in Russia, see Robert E. Jones, *Bread upon the Waters: The St. Petersburg Grain Trade and the Russian Economy, 1703–1811* (University of Pittsburgh Press, 2013).

68. Kuznetsov, *Kirgizskoe khoziaistvo*, 139. I thank Ian Campbell for pointing me to this reference. See Campbell, *Knowledge*, 125–56.

69. Kuznetsov, *Kirgizskoe khoziaistvo*, 145.

70. Demko, *Russian Colonization*, 200.

71. Demko, *Russian Colonization*, 174.

72. On the relative absence of wars and its role in human and animal population growth, see Martin, *Law and Custom*, 191n.

73. Georgii Safarov, *Kolonial'naia revoliutsiia (opyt Turkestana)* (Zhalyn, 1996), 76. Over the course of the nineteenth century, settlers in European Russia also cut back the time that fields were left fallow. David Moon attributes this to several factors, including the demand for grain on domestic and international markets and population increases. See Moon, *Plough*, 113.

74. Kuznetsov, *Kirgizskoe khoziaistvo*, 12–13.

75. Letter from the Akmolinsk governor to the steppe governor-general, June 30, 1890 (TsGARK f. 64, op. 1, d. 4324, l. 34).

76. Telegram from the secretary of the financial committee in Omsk to the minister of trade and industry, St. Petersburg, July 2, 1911 (TsGARK f. 64, op. 1, d. 4752, l. 121).

77. Several factors may explain this. Orazgul Mŭkhatova suggests that settlement was a concerted strategy by Kazakh nomads. Kazakhs settled to protect their lands from further peasant encroachment. See *Qazaqstandaghï XX-ghasïrdïng alghashqï onzhïldïq*. Bykov speculates that nomads' growing consumption of grain could have prompted them to settle. See *Istoki modernizatsii Kazakhstana*, 215. The large increases in the steppe's human and animal populations during the late Russian Empire may have also resulted in the overgrazing of pasturelands, prompting some Kazakhs to settle.

78. Kuznetsov, *Kirgizskoe khoziaistvo*, 48.

79. Argynbaev, *Istoriko-kul'turnye sviazi*, 58.

80. By the 1920s most Soviet sources classify the majority of Kazakhs as "semi-nomadic," 20–30 percent of Kazakhs as "settled," and a minority, usually less than 10 percent, of Kazakhs as "nomadic." For an analysis based upon data from 1929, see B. Rodnevich, *Ot kolonial'nogo*

vyrozhdeniia k sotsialisticheskomu rastsvetu: O Kazakstane (Vlast' sovetov' pri prezidiume Vtsik, 1931), 15.

81. Khazanov, *Nomads*, 21; Caroline Humphrey and David Sneath, *The End of Nomadism? Society, State, and the Environment in Inner Asia* (Duke University Press, 1999), 196.

82. Numbers for Dzehtisu (Semirech'e) give some sense of the enormity of these losses. From 1916 to 1920, the province's sown field area fell from 640,000 desiatiny to 300,000 desiatiny, while livestock levels fell from 9.2 million to 2.6 million. V. L. Genis, "Deportatsiia russkikh iz Turkestana v 1921 gody ("Delo Safarova")," *Voprosy istorii* 1 (1998): 57.

83. On the Pampas, see Adrián Gustavo Zarrilli, "Capitalism, Ecology, and Agrarian Expansion in the Pampean Region, 1890–1950," *Environmental History* 6 (2001), 561–83. On the Great Plains, see Andrew C. Isenberg, *The Destruction of the Bison: An Environmental History, 1750-1920* (Cambridge University Press, 2000). Indeed, Soviet planners consciously compared these landscapes, seeking to learn from the experiences of planners in the Great Plains or the Pampas. See N. I. Maslov, "Perspektivy ratsionalizatsii i rekonstruktsii skotovodstva zasushlivogo tsentral'nogo Kazakstana," *Narodnoe khoziaistvo Kazakstana* 1–2 (1929): 3–22.

Chapter 4: "The Scourge of Stock Raising"

Research for this paper was completed with the support of a Fulbright-Hays DDRA grant, a postdoctoral fellowship from the Davis Center for Russian and Eurasian Studies at Harvard University, and graduate fellowships from the History Department of the University of Michigan. Earlier versions of the essay were presented at a conference entitled "Frost, Ice, and Snow: Cold Climate in Russian History," organized by the Rachel Carson Center (Munich) and held at the German Historical Institute (Moscow) and at the UC-Berkeley Russian history *kruzhok*. I thank all of these institutions for their support.

1. *Kirgizskaia stepnaia gazeta* (hereafter cited as *KSG*), January 16, 1894. "Kirgiz" as an ethnonym referred, in the nineteenth century, to the ethnic groups known today as Kazakhs and Kyrgyz (also sometimes called "kara-kirgiz" or "dikokamennyi kirgiz" at the time).

2. The other common Kazakh term for a *zhŭt*, *qoian-zhïlï*—year of the hare—refers to a year of the Zodiac cycle, implying a period of twelve years for its worst manifestations; the last three major zhŭts of the nineteenth century occurred in the winters of 1867–1868, 1879–1880, and 1891–1892.

3. N. I. Krasovskii, ed., *Materialy dlia geografii Rossiiskoi imperii, sobrannye ofitserami General'nogo shtaba: oblast' Sibirskikh kirgizov*, part 2 (Transhel'; Rotger i Shneider, 1868), 10; O. A. Shkapskii, "Nekotoryia dannyia dlia osveshcheniia kirgizskogo voprosa," *Russkaia mysl'* 7 (1897): 36.

4. The classic statement of this thesis with respect to famines is Amartya Sen, *Poverty and Famines: An Essay on Entitlement and Deprivation* (Oxford University Press, 1981). Mike Davis offers a devastating critique of "the malign interaction between climatic and economic processes" in nineteenth-century famines in *Late Victorian Holocausts: El Niño Famines and the Making of the Third World* (Verso, 2001), 12. Also see Andy Bruno, "Tumbling Snow: Vulnerability to Avalanches in the Soviet North," *Environmental History* 18, no. 4 (2013): 683–709.

5. For the connection between the ruling ideology of Confucianism and famine relief in China, see Lillian Li, *Fighting Famine in North China: State, Market, and Environmental Decline, 1690s–1990s* (Stanford University Press, 2007), 3. For nineteenth-century conflicts about the priority to be given to famine relief, see Kathryn Edgerton-Tarpley, *Tears from Iron: Cultural Responses to Famine in Nineteenth-Century China* (University of California Press, 2008), esp. 90–113. The literature on the Irish famine of the 1840s is vast; I have been influenced by the argument in Peter Gray, *Famine, Land, and Politics: British Government and Irish Society, 1843–1850* (Irish Academic Press, 1999).

6. James Vernon, *Hunger: A Modern History* (Belknap Press, 2007), 17–40.

7. Michel Foucault, "Governmentality," in *The Foucault Effect: Studies in Governmentality*,

ed. Graham Burchell et al. (University of Chicago Press, 1991), 87–104. A few examples will suffice. Between 1866 and 1876, all zemstvo provinces of the Russian Empire developed compulsory fire insurance programs to address the widespread destruction of peasant property in fires; see Cathy A. Frierson, *All Russia Is Burning! A Cultural History of Fire and Arson in Late Imperial Russia* (University of Washington Press, 2002), 191–92. An 1866 law created the most comprehensive food aid program that had hitherto existed in the Russian Empire; see Richard Robbins, *Famine in Russia, 1891–1892: The Imperial Government Responds to a Crisis* (Columbia University Press, 1975), 20–21. An 1865 law regularized and brought smallpox inoculation under state control; see *Polnoe sobranie zakonov Rossiiskoi imperii* (Complete collection of the laws of the Russian Empire), ser. 2, 40:886–87, no. 42375 (August 6, 1865).

8. See, for example, "Ob ustroistve upravleniia v Kirgizskikh stepiakh," *Turkestanskii sbornik* 6:277.

9. Compare Alice Conklin's emphasis on *mise en valeur* (rational development) of natural and human resources in French West Africa, in Conklin, *A Mission to Civilize: The Republican Idea of Empire in France and West Africa, 1895–1930* (Stanford University Press, 1997), 6–7. On the Russian Empire's civilizing mission, see for example, Ricarda Vulpius, "The Russian Empire's Civilizing Mission in the Eighteenth Century: A Comparative Perspective," in *Asiatic Russia: Imperial Power in Regional and International Contexts*, ed. Tomohiko Uyama (Routledge, 2011), 13–31.

10. Mark Carey, *In the Shadow of Melting Glaciers: Climate Change and Andean Society* (Oxford University Press, 2010), 12. For transformative agendas in other settings, see for example, Tamara Whited, *Forests and Peasant Politics in Modern France* (Yale University Press, 2000); Darren C. Zook, "Famine in the Landscape: Imagining Hunger in South Asian History, 1860–1990," in *Agrarian Environments: Resources, Representation, and Rule in India*, ed. Arun Agrawal and K. Sivaramakrishnan (Duke University Press, 2000), 107–31.

11. Compare the arguments of Sergey Glebov, "Siberian Middle Ground: Languages of Rule and Accommodation on the Siberian Frontier," in *Empire Speaks Out: Languages of Rationalization and Self-Description in the Russian Empire*, ed. Ilya Gerasimov et al. (Brill, 2009), 121–51; Tomohiko Uyama, "A Strategic Alliance between Kazakh Intellectuals and Russian Administrators: Imagined Communities in *Dala Walayatïnïng Gazeti*," in *The Construction and Deconstruction of National Histories in Slavic Eurasia*, ed. Tadayuki Hayashi (Slavic-Eurasian Research Center, 2003), 237–59.

12. Compare Osama Abi-Mershed, in *Apostles of Modernity: Saint-Simonians and the Civilizing Mission in Algeria* (Stanford University Press, 2010), 11, who argues that French administrators holding otherwise very different views of imperialism shared principles including "French subjugation of indigenous society and the dissolution of the latter's retrograde features."

13. See Sarah Cameron (this volume).

14. On these diverse forms, see Carole Ferret, "Discontinuités spatiales et pastoralisme nomade en Asie intérieure au tournant des XIXe et XXe siècles," *Annales. Histoire, Sciences Sociales* 2014, no. 4 (2014): 957–96.

15. Musa Chormanov, "Kazakhskie narodnye obychai," republished in *Musa Chormanov: kazakhskie narodnye obychai*, ed. E. M. Aryn (EKO, 2005), 9–10. For a classic statement of the conflict between state borders and transhumance, see M. Nazif Shahrani, *The Kirghiz and Wakhi of Afghanistan: Adaptation to Closed Frontiers* (University of Washington Press, 1979).

16. See the discussions of reserve pastures, for example, in Älikhan Bökeikhanov, "Ovtsevodstvo v stepnom krae," in *A. N. Bokeikhanov: Kazakhi: istoriko-etnograficheskie trudy*, ed. Zh. O. Artykbaev (Altïn-kïtap, 2007), 202.

17. B. D. Daulbaev, "Rasskaz o zhizni Kirgiz Nikolaevskogo uezda Turgaiskoi oblasti," *Zapiski Orenburgskogo otdela Imperatorskogo russkogo geograficheskogo obshchestva* (hereafter cited as *IRGO*) 4 (1881): 106–7.

18. "Prostoi sposob bor'by s dzhutom," *KSG*, November 23, 1897.

19. "O bedstviiakh preterpennykh kirgizami ot neurozhaia i padezha skota" (1851), Rossiiskii

gosudarstvennyi istoricheskii arkhiv (Russian State Historical Archive; hereafter RGIA) f. 1291, op. 81, d. 219, ll. 20–21; "Otchet Orenburgskoi pogranichnoi kommissii po upravleniiu Zauralskimi Kirgizami Orenburgskogo vedomstva i shkoloi dlia kirgizskikh detei" (1856), RGIA f. 1291, op. 81, d. 233, ll. 15–16ob.

20. N. Ia. Konshin, "Zametka ob odnom istoricheskom dzhute," *Zapiski Semipalatinskogo pod"otdela Zapadno-sibirskogo otdela IRGO* 2 (1905): 1–17.

21. A. O. Diugamel' to P. A. Valuev, "Ob upadke u Kirgiz rogatogo skota i loshadei, i o vydache posobiia nuzhdaiushchimsia iz nikh," May 30, 1862 (RGIA f. 1291, op. 82, d. 9, ll. 2–3ob.).

22. David A. Bello, "Relieving Mongols of Their Pastoral Identity: Disaster Management on the Eighteenth-Century Qing China Steppe," *Environmental History* 19, no. 3 (2014): 482.

23. The classic work of P. I. Rychkov indicates that Kazakhs of the Small Horde were successfully growing grain in the mid-1700s. See Rychkov, *Topografiia Orenburgskoi gubernii* (1762; Breslin, 1887), 100–101.

24. A. A. Katenin to A. M. Gorchakov, March 4, 1859 (RGIA f. 1291, op. 82, d. 9, ll. 5–6ob.).

25. Reşat Kasaba, *A Moveable Empire: Ottoman Nomads, Migrants, and Refugees* (University of Washington Press, 2009), 86–89.

26. On shortfalls, see the report from V. A. Perovskii, April 14, 1856 (RGIA f. 1291, op. 81, d. 219, l. 30); on unpredictability, see the report from M. A. Taube (1899), Tsentral'nyi gosudarstvennyi arkhiv respubliki Kazakhstan (Central State Archive of the Republic of Kazakhstan; hereafter cited as TsGARK) f. 64, op. 1, d. 647, ll. 199ob.–200. This lack of information about taxable assets led to a flat tax on each Kazakh tent, a situation similar to that described for rural communes in Yanni Kotsonis, *States of Obligation: Taxes and Citizenship in the Russian Empire and Early Soviet Republic* (University of Toronto Press, 2014).

27. A. K. Geins, "Kharakter upravleniia kirgizami russkoi vlast'iu," in *Sobranie literaturnykh trudov A. K. Geinsa* (Stasiulevich, 1897), 1:112.

28. A. I. Maksheev, *Opisanie nizov'ev Syr-dar'i* (Imperatorskaia Akademiia nauk, 1856), 53.

29. Shkapskii, "Nekotoryia dannyia," 46, 48–49. For more on Shkapskii, see Daniel Brower, *Turkestan and the Fate of the Russian Empire* (RoutledgeCurzon, 2003), 136–38.

30. Krasovskii, *Materialy*, 72–74.

31. For maintenance of the commune post-emancipation as a means of maintaining social control and supervision over peasants, see for example, P. A. Zaionchkovskii, *The Abolition of Serfdom in Russia*, trans. and ed. Susan Wobst (Academic International Press, 1978), 101; Daniel Field, *The End of Serfdom: Nobility and Bureaucracy in Russia, 1855–1861* (Harvard University Press, 1976), 347–48.

32. The concern here was interference with the practice of mobile pastoralism caused by private plots of land. See RGIA f. 1291, op. 82, d. 5c, ll. 92–94ob. Similar rhetoric rejecting private property in land for Russian settlers in Semirech'e is found in TsGARK f. 64, op. 1, d. 3, l. 7ob. (proposal of late 1869, rejection in 1870).

33. A. P. Bezak to P. A. Valuev, July 12, 1863 (RGIA f. 1291, op. 82, d. 22, l. 1). Bezak cites the example of a ssudnaia kassa that had existed for the Kazakh Inner Horde since 1854 (ll. 1ob.–2).

34. N. G. Kaznakov to L. S. Makov, November 26, 1879 (RGIA f. 1291, op. 82, d. 26, ll. 1–2).

35. "Proekt polozheniia o ssudnoi kasse pri oblastnom pravlenii Orenburgskimi kirgizami," 1863 (RGIA f. 1291, op. 82, d. 22, ll. 4–4ob.).

36. "Proekt ustava ssudnoi kassy kirgizov Semipalatinskoi oblasti," 1873 (RGIA f. 1291, op. 82, d. 1, ll. 5ob.–6).

37. Report from A. P. Konstantinovich, December 10, 1881 (RGIA f. 1291, op. 82, d. 39, ll. 7–7ob.).

38. Report from Ia. F. Barabash, September 20, 1898 (RGIA f. 1291, op. 84, d. 14, ll. 6–6ob.).

39. Report from the Council of the Main Administration of Western Siberia, October 20, 1866 (RGIA f. 1291, op. 82, d. 5c, ll. 101–8, and d. 23, l. 2). The governor-general of Western Siberia, A. O. Diugamel', forwarded this report to the Ministry of Internal Affairs the following day (RGIA f. 1291, op. 82, d. 23, l. 1).

40. N. G. Kaznakov to A. E. Timashev, January 20, 1876, citing a report from A. P. Khrushchov, November 30, 1874, Rossiiskii gosudarstvennyi voenno-istoricheskii arkhiv (Russian State Military History Archive), f. 400, op. 1, d. 498, ll. 2, 5ob.–6ob.

41. G. A. Kolpakovskii to N. N. Obruchev, May 2, 1886, Rossiiskii gosudarstvennyi voenno-istoricheskii arkhiv f. 400, op. 1, d. 1077, ll. 2–9.

42. TsGARK f. 64, op. 1, d. 4230, ll. 37ob.–38.

43. Anatolii Remnev, "Colonization and 'Russification' in the Imperial Geography of Asiatic Russia: From the Nineteenth to the Early Twentieth Centuries," trans. Paul Werth, in *Asiatic Russia: Imperial Power in Regional and International Contexts*, ed. Tomohiko Uyama (Routledge, 2011), 102–28.

44. For the panoply of views around Slavic peasants as a civilizing force, see A. V. Remnev and N. G. Suvorova, "'Obrusenie' aziatskikh okrain Rossiiskoi imperii: optimizm i pessimizm russkoi kolonizatsii," *Istoricheskie zapiski* 11 (2008): 132–79.

45. "Po voprosu o khlebopashestve v kirgizskikh stepiakh," *KSG*, January 9, 1894.

46. *KSG*, January 6, 1895.

47. M. Imshenetskii, "Nesmetnoe bogatstvo, sokrytoe v kirgizsksoi stepi," *KSG*, January 1, 1891.

48. Amre Nogerbekov, "Moi zaniatiia khlebopashestvom," *KSG*, December 11, 1894; "Eshche raz o preimushchestvakh zemledel'cheskogo khoziaistva," *KSG*, June 5, 1894.

49. Meiram Ïsqaqov, "Qala bolushïlar," in *"Aiqap,"* ed. U. Subkhanberdina and S. Däuitov (Qazaq entsiklopediiasï, 1995), 201–2.

50. Pey-Yi Chu (this volume) defines adaptation as referring to "to a process, as the term is understood in evolutionary biology, rather than to any specific practice." This is a useful lens for understanding the position of committed agriculturalists on the Kazakh steppe.

51. A. Kurmanbaev, "K voprosu o chul-bidae," *KSG*, June 5, 1894; correspondence selected from other newspapers reported on its successful importation and use in other dry regions of the empire, for example, *KSG*, March 26, 1895. The term *chul* (*cho'l*, in modern Uzbek) refers to a desert or barren place. Thanks to Maya Peterson for the reminder.

52. St. Artemenko, "Ob iskusstvennom oblesenii stepei," *KSG*, January 23, 1894; David Moon (this volume).

53. Shkapskii, "Nekotoryia dannyia," 45, 48. The most comprehensive study of the politics and praxis of irrigation in Central Asia is Maya K. Peterson, "Technologies of Rule: Empire, Water, and the Modernization of Central Asia, 1867–1941" (PhD diss., Harvard University, 2011).

54. Donald Treadgold, *The Great Siberian Migration: Government and Peasant in Resettlement from Emancipation to the First World War* (Princeton University Press, 1957), 116–19. George Demko estimates that two-thirds of settlers in the northern steppe between 1894 and 1903 had arrived without permission. See George Demko, *The Russian Colonization of Kazakhstan, 1896–1916* (Indiana University Press, 1969), 59.

55. Willard Sunderland, "The 'Colonization Question': Visions of Colonization in Late Imperial Russia," *Jahrbücher für Geschichte Osteuropas* 48, no. 2 (2000): 224. There is an extensive bibliography of works endorsing this viewpoint on the same page.

56. "O merakh obespecheniia narodnogo prodovol'stviia kochevogo naseleniia stepnykh oblastei," September 13, 1894 (RGIA f. 1291, op. 82, d. 34 [1893], ll. 22ob.–23).

57. *KSG*, February 12–19, 1893.

58. *KSG*, February 12–19, 1893. On attempts to improve Central Asian horse keeping (for the imperial cavalry, in particular) and the resulting intersections between metropolitan and local knowledge practices, see Carole Ferret, "Des chevaux pour l'empire," *Cahiers d'Asie central* 17/18 (2009): 211–53.

59. *KSG*, February 12–19, 1893.

60. For example, *KSG*, January 2, 1894.

61. V. Mikhailov, "Kirgizskie stepi Akmolinskoi oblasti (po obsledovaniiam veterinarnykh vrachei)," *Zapiski Zapadno-Sibirskogo otdela IRGO* 16, no. 1 (1893): 20.

62. *KSG*, January 23, February 6, March 13, 1894.

63. Sannikov, *KSG*, February 12–19, 1893; Mikhailov, "Besedy o skotovodstve (V)," *KSG*, March 27, 1894; "V chem spasenie ot 'dzhuta,'" *KSG*, January 30, 1894.

64. "O kormovykh surrogatakh dlia skota," *KSG*, October 24, 1892.

65. "Broshiura 'Obshchii vyvod o skotovodstve kirgiz Turgaiskoi oblasti' i perepiska po etomu voprosu," 1895 (TsGARK f. 64, op. 1, d. 4437, l. 46ob.).

66. A. I. Dobrosmyslov, "Turgaiskaia oblast': istoricheskii ocherk," *Zapiski Orenburgskogo otdela IRGO* 17 (1902): 488. Reporting on the assemblies leading to this decision can be found at *KSG*, March 19, 1893.

67. A similar claim by a Kazakh, Makash Tulemysov, that both "agriculture and stock raising depend on God" called forth a strident defense of preferences for agriculture in *KSG*, June 5, 1894.

68. I. Altynsarin, "Po povodu goloda v Kirgizskoi stepi," in *Ibrahim Altynsarin: sobranie sochinenii v trekh tomakh*, ed. B. Suleimenov (Nauka, 1976), 2:100. Contrast the endorsement of peasant settlement in the same newspaper; "Zaselenie kirgizskikh stepei," in *Natsional'nye men'shinstva v Orenburzh'e (khronika obshchestvennoi i kul'turnoi zhizni po materialam periodicheskoi pechati kontsa XIX–nachala XX vv.)*, ed. A. V. Amelin (OGAU, 2003), 18.

69. "Zapiska na imia voennogo gubernatora Turgaiskoi obl., 'O vvedenii professialno-tekhnicheskogo obucheniia v dvukhklassnykh russko-kazakhskikh shkolakh,'" in Suleimenov, *Ibrahim Altynsarin*, 2:190–200.

70. Peter Rottier, "The Kazakness of Sedentarization: Promoting Progress as Tradition in Response to the Land problem," *Central Asian Survey* 22, no. 1 (2003): 67–81; Mïrzhaqïp Dŭlatov, "Öner mektebï," in *"Qazaq" gazetï: Alash azamattarïnïng rukhïna baghyshtaladï*, ed. U. Subkhanberdina et al. (Qazaq entsiklopediiasï, 1998), 222.

71. B. Benkevich, "Kirgizskoe stepnoe skotovodstvo i mery k ego uluchsheniia," *Zapiski Semipalatinskogo pod"otdela Zapadno-sibirskogo otdela IRGO* 1 (1903): 1–24.

72. I. P. Poddubnyi, "Etnograficheskii ocherk," in *Aziatskaia Rossiia t. 1—liudi i poriadki za Uralom* (1914; repr. Oriental Research Partners, 1974), 160; F. A. Shcherbina and E. Dobrovol'skii, *Materialy po kirgizskomu zemlepol'zovaniiu, t. 3—Akmolinskaia oblast,' Akmolinskii uezd* (Veseloi, 1909), 129–30.

73. G. Chirkin, "Zemleustroistvo Kirgiz v sviazi s kolonizatsiei stepi," *Voprosy kolonizatsii* 2 (1907): 54. For the one specific case, see A. Uspenskii, "Vliianie kolonizatsii na kirgizskoe khoziaistvo (po dannym povtornogo izsledovaniia Arakaragaiskoi volosti, Kustanaiskogo uezda, Turgaiskoi oblasti, proizvedennogo L. N. Tsabelem)," *Voprosy kolonizatsii* 2 (1907): 29–43 (esp. 40).

74. New regulations in early 1906 granted all peasants the right to migrate wherever they chose and to receive state assistance in doing so. On the hopes and anxieties associated with this measure, see Charles Steinwedel, "Resettling People, Unsettling the Empire: Migration and the Challenge of Governance," in *Peopling the Russian Periphery: Borderland Colonization in Eurasian History*, ed. Nicholas Breyfogle, Abby Schrader, and Willard Sunderland (Routledge, 2007), 128–47.

75. Ä. Bökeikhanov, "Zhauap khat," in *Alash qozghalïsï: qŭzhattar men materialdar zhinaghï*, ed. M. Q. Qoigeldiev (Alash, 2004), 151. According to Bökeikhanov, this new norm "cut off the Kazakhs' nose [*qazaqtï pŭshtitïp*]."

76. V. Kuznetsov, *Kirgizskoe khoziaistvo v Akmolinskoi oblasti, t. 1—Kokchetavskii uezd. Povtornoe issledovanie 1907 g.* (St. Petersburg, 1909), 111.

77. Frierson, *All Russia Is Burning*, 7.

78. For claims that some settlers to the steppe were fleeing agrarian crisis at home, see Ihor Stebelsky, "The Frontier in Central Asia," in *Studies in Russian Historical Geography*, ed. James H. Bater and R. A. French (Academic Press, 1983), 1:159, 165.

79. Compare Lewis Siegelbaum and Leslie Page Moch, *Broad Is My Native Land: Repertoires and Regimes of Migration in Russia's Twentieth Century* (Cornell University Press, 2014), 8.

Chapter 5: Desiccated Steppes

Research for this article has been made possible by a grant from the French National Research Agency (ANR) and the German Research Foundation (DFG) for the project "Contemporary environmental history of the Soviet Union and the successor states, 1970–2000. Ecological globalization and regional dynamics" (ANR-13-FRAL-0003, EcoGlobReg, 2014–2018).

1. Martin McCauley, *Khrushchev and the Development of Soviet Agriculture: The Virgin Land Programme, 1953–1964* (Holmes and Meier, 1976); Nikolai Dronin and Edward G. Bellinger, *Climate Dependence and Food Problems in Russia, 1900–1990: The Interaction of Climate and Agricultural Policy and Their Effect on Food Problems* (Central European University Press, 2005), 171–217; Zauresh G. Saktaganova, *Istoriya osushchestvleniya sovetskogo opyta ekonomicheskoy modernizatsii v Kazakhstane (1946–1970)* (Izdatel'stvo KarGU, 2004), 143–76.

2. These regions made up 65 percent of plow land, constituted two-thirds of the rich black soils *chernozem*, and produced one-half of the overall agricultural production of the USSS. Aleksandr Aleksandrovich Nikonov, *Spiral' mnogovekovoy dramy: agrarnaya nauka i politika Rossii (XVIII—XX vv.)* (Entsiklopediya rossiyskikh dereven,' 1995), 389.

3. I. E. Zelenin, "Agrarnaya politika N. S. Khrushcheva i sel'skoe khozyaystvo strany," *Otechestvennaya istoriya* 1 (2000): 84; Dronin and Bellinger, *Climate Dependence*, 176, 257, 278.

4. Stephan Merl, "Why Did the Attempt under Stalin to Increase Agricultural Productivity Prove to Be Such a Fundamental Failure? On Blocking the Implementation of Progress in Agrarian Technology (1929–1941)", *Cahiers du monde russe* 57, no. 1 (2016): 191–220.

5. Dronin and Bellinger, *Climate Dependence*, 6, 9.

6. Dronin and Bellinger, *Climate Dependence*, 108.

7. For overviews of works centered on vulnerability, see Phil O'Keefe, Ken Westgate, and Ben Wisner, "Taking the Naturalness out of Natural Disasters," *Nature* 260, no. 5552 (April 15, 1976): 566–67; Anthony Oliver-Smith, "Anthropological Research on Hazards and Disasters," *Annual Review of Anthropology* 25 (1996): 303–28; Sandrine Revet, "Penser et affronter les désastres: un panorama des recherches en sciences sociales et des politiques internationales," *Critique Internationale* 52, no. 3 (2011): 157–73.

8. For a recent manifesto, see Jean-Christophe Gaillard et al., "Alternatives pour une réduction durable des risques de catastrophe. Alternatives for Sustained Disaster Risk Reduction. Alternativas para la reducción sustentable de peligrosos de riesgo," *Human Geography* 3, no. 1 (2010): 66–88.

9. One classical study on historically constructed vulnerability is Anthony Oliver-Smith, "Peru's 500-Year-Earthquake: Vulnerability in Historical Perspective," in *The Angry Earth: Disaster in Anthropological Perspective*, ed. Anthony Oliver-Smith and Susanna Hoffmann (Routledge, 1999).

10. Claude Meillassoux, "Development or Exploitation: Is the Sahel Famine Good Business?" *Review of African Political Economy* 1, no. 1 (1974): 27–33, provides a Marxist view of the Sahel famine as a neocolonial disaster. The landmark in the transformation of disaster studies is Rolando García's project "Drought and Man: The 1972 Case History," with three volumes published in the 1980s: *Nature Pleads Not Guilty: The 1972 Case History* (Pergamon Press, 1981); *The Constant Catastrophe: Malnutrition, Famines and Drought* (Pergamon Press, 1982); Rolando Victor García and Pierre Spitz, *The Roots of Catastrophe* (Pergamon Press, 1986).

11. García, *Nature*, xii.

12. "What Is Drought?" *National Drought Mitigation Center*, online at http://drought.unl.edu/DroughtBasics/WhatisDrought.aspx/.

13. Donald A. Wilhite and Michael H. Glantz, "Understanding the Drought Phenomenon: The Role of Definitions," *Water International* 10, no. 3 (1985): 111–20.

14. García, *Nature*, 169.

15. Marcel Mazoyer and Laurence Roudart, *Histoire des agricultures du monde: du néolitique à la crise contemporaine* (Éditions du Seuil, 2002).

16. Christophe Bonneuil and Jean-Baptise Fressoz, *The Shock of the Anthropocene: The Earth, History and Us* (Verso, 2016).

17. Paul R. Josephson, *Industrialized Nature: Brute Force Technology and the Transformation of the Natural World* (Island Press, 2002).

18. The classic argument for how technology and nature are intertwined is made by Richard White, *The Organic Machine: The Remaking of the Columbia River* (Hill and Wang, 1996).

19. John L. Brooke, *Climate Change and the Course of Global History: A Rough Journey* (Cambridge University Press, 2014).

20. Donald Worster, *Dust Bowl : The Southern Plains in the 1930s*, twenty-fifth anniversary edition (Oxford University Press, 2004).

21. Dronin and Bellinger, *Climate Dependence*, 177.

22. On high modernism, see James Scott, *Seeing like a State: How Certain Schemes to Improve the Human Condition Have Failed* (Yale University Press, 1998). On nature's agency, see Andy Bruno (this volume).

23. Marc Elie, "The Soviet Dust Bowl and the Canadian Erosion Experience in the New Lands of Kazakhstan, 1950s–1960s," *Global Environment* 8, no. 2 (January 1, 2015): 259–92.

24. Vladimir Fedorovich Loginov, A. I. Neushkin, and E. V. Rocheva, *Zasukhi, ikh vozmozhnye prichiny i predposylki predskazaniya* (VNIIGMI-MTsD, 1975), 4.

25. Anatoliy Sergeevich Uteshev, "K probleme issledovaniya i prognozirovaniya zasukh," in *Zemel'nye resursy mira, ikh ispol'zovanie i okhrana* (Nauka, 1978), 145.

26. Feofan Farneevich Davitaia, "Zasukhi v SSSR i nauchnoe obosnovanie mer bor'by s nimi po prirodnym zonam," *VASKhNIL* 1 (1959): 57.

27. Quoted by Georgiy Timofeevich Selianinov, "Proiskhozhdenie i dinamika zasukh," in *Zasukhi v SSSR, ikh proiskhozhdenie, povtoriaemost' i vliianie na urozhai*, ed. A. I. Rudenko (Gidrometeoizdat, 1958), 9.

28. Oleg Alekseevich Drozdov, *Zasukha i dinamika uvlazhneniya* (Gidrometeoizdat, 1980), 5.

29. Davitaia, "Zasukhi," 57.

30. Drozdov, *Zasukha*, 5–6.

31. Ted Steinberg, *Acts of God: The Unnatural History of Natural Disaster in America* (Oxford University Press, 2006).

32. Richard R. Heim, "Drought Indices. A Review," in *Drought: A Global Assessment*, ed. Donald Wilhite (Routledge, 2000), 1:164; E. I. Khlebnikova, T. V. Pavlova, and N. A. Speranskaya, "Zasukhi," in *Metody otsenki posledstviy izmeneniy klimata dlya fizicheskikh i biologicheskikh sistem* (Rosgidromet, 2012), 144.

33. Dronin and Bellinger, *Climate Dependence*. See also Stephen G. Wheatcroft, "The Significance of Climatic and Weather Change on Soviet Agriculture (with Particular Reference to the 1920s and the 1930s)," CREES Discussion Paper Soviet Industrialisation Project Series (SIPs) no. 11 (Centre for Russian and East European Studies, University of Birmingham, 1977).

34. Worster, *Dust Bowl*.

35. Vasiliy Vasil'evich Dokuchaev, *Nashi stepi prezhde i teper'* (OGIZ-SEL'KhOZGIZ, 1936 [1892]); David Moon, "The Environmental History of the Russian Steppes: Vasilii Dokuchaev and the Harvest Failure of 1891," *Transactions of the Royal Historical Society* 15 (December 2005): 149–74.

36. I. G. Gringof, V. V. Popova, and V. N. Strashnyi, *Agrometeorologiya* (Gidrometizdat, 1987).

37. Selianinov, "Proiskhozhdenie," 16; Khlebnikova, Pavlova, and Speranskaya, "Zasukhi," 128.

38. The increase was from 150 stations in 1915 to 1,500 in 1928. A. D. Kleshchenko and I. G. Gringof, eds., *Razvitie sel'skokhozyaystvennoy meteorologii v Rossii*, 2nd ed. (Rosgidromet, 2009), 57–58, 60.

39. The Russian Drought Monitoring Center uses the HTC for its ten-day summary evaluations of drought conditions. Tsentr monitoringa zasukhi, Vserossiiskii nauchno-issledovatel'nyi institut sel'sko-khoziaistvennoi meteorologii (VNIICXM), "Dekadnyi Byul-

leten,'" *VNIICXM*, online at http://cxm.obninsk.ru/index.php?id=200/. Dronin and Bellinger, *Climate Dependence*, uses the HTC index to draw maps showing which regions were affected for every severe drought in Russia and the USSR since 1891.

40. G. T. Selianinov, "O sel'skokhozyaystvennoy otsenke klimata," *Trudy po sel'skokhozyaystvennoy meteorologii*, no. 20 (1928): 165–77. The formula is written:

$$HTC = 10 \times \frac{\sum Q}{\sum T}$$

41. Feofan Farneevich Davitaia, "Itogi diskussii po agrometeorologii," *Meteorologiia i gidrologiia* 10 (1952): 1–8.

42. Kleshchenko and Gringof, *Razvitie*, 75.

43. M. I. Budyko et al., *Izmenenie klimata v svyazi s planom preobrazovaniya prirody* (Gidrometeorlogicheskoe izdatel'stvo, 1952). On the history of the plan, see Paul Josephson et al., *An Environmental History of Russia* (Cambridge University Press, 2013), 121–28; Stephen Brain, "The Great Stalin Plan for the Transformation of Nature," *Environmental History* 15, no. 4 (October 2010): 670–700.

44. Viktor Abramovich Kovda, *Velikiy plan preobrazovaniya prirody* (Izdatel'stvo Akademii nauk SSSR, 1952), 62–73.

45. Davitaia, "Zasukhi."

46. Mariia Vasil'evna Zavarina, *Zasukha i bor'ba s nei* (Gosudarstvennoe izdatel'stvo geograficheskoi literatury, 1954), 18–22. This misconception has proved persistent. For Wheatcroft and Davies, the sukhovei "blow from Central Asia across the Volga region, North Caucasus and Ukraine." R. W. Davies, Mark Harrison, and S. G. Wheatcroft, eds., *The Economic Transformation of the Soviet Union, 1913–1945* (Cambridge University Press, 1994), 124.

47. Feofan Farneevich Davitaia, *Zasukhi v SSSR i nauchnoe obosnovanie mer bor'by s nimi po prirodnym zonam* (Izdatel'stvo Ministerstva sel'skogo khoziaistva SSSR, 1958).

48. On the debate around the planting of trees in the steppe in tsarist Russia, see David Moon, *The Plough that Broke the Steppes Agriculture and Environment on Russia's Grasslands, 1700–1914* (Oxford University Press, 2013).

49. Feofan Farneevich Davitaia, ed., *Agroklimaticheskie i vodnye resursy rayonov osvoeniia tselinnykh i zalezhnykh zemel'* (Gidrometeorologicheskoe izdatel'stvo, 1955), 275–79.

50. T. F. Yakubov, "Vetrovaia erosiia pochvy v raionakh osvoeniia tselinnykh i zalezhnykh zemel' Kazakhstana," Rossiikii Gosudarstvennyi Arkhiv Ekonomiki Russian State Economic Archive f. 401, op. 2, d. 420, ll. 1–59 (1956).

51. Elie, "Soviet Dust Bowl."

52. Selianinov, "Proiskhozhdenie i dinamika zasukh," in A. I. Rudenko, *Zasukhi v SSSR, ikh proiskhozhdenie, povtoriaemost' i vliianie na urozhai* (Gidrometeorologicheskoe izdatel'stvo, 1958), 14.

53. Anatoliy Sergeevich Uteshev, ed., *Klimat Kazakhstana* (Gidrometeorologicheskoe izdatel'stvo, 1959), 117–38.

54. Anatoliy Sergeevich Uteshev, *Atmosfernye zasukhi i ikh vliianie na prirodnye iavleniia* (Nauka, 1972).

55. Khrushchev had taken many fatal decisions regarding agriculture in the first half of the 1960s.

56. McCauley, *Soviet Agriculture*, 82, 99; Dronin and Bellinger, *Climate Dependence*, 176.

57. Dronin and Bellinger, *Climate Dependence*, 175.

58. Elie, "Soviet Dust Bowl." On Khrushchev's errors, see Dronin and Bellinger, *Climate Dependence*, 197–99, 208–13.

59. Dronin and Bellinger, *Climate Dependence*, 251, 295–300, 304.

60. A. V. Meshcherskaya and V. G. Blazhevich identify 1981 as the largest and 1975 as the third-largest drought in the European USSR. See A. V. Meshcherskaya and V. G. Blazhevich, "The Drought and Excessive Moisture Indices in a Historical Perspective in the Principal Grain-Producing Regions of the Former Soviet Union," *Journal of Climate* 10, no. 10 (October 1, 1997): 2670–82.

61. Iu. A. Izrael', "Ekologiia, klimat i vliianie vozmozhnykh ego izmenenii na sel'skoe khoziaistvo strany," *Vestnik sel'sko-khoziaistvennoi nauki* 10 (1987): 27.

62. Zelenin, "Agrarnaia politika," 83–84.

63. Dronin and Bellinger, *Climate Dependence*, 213.

64. "Protokoly sobraniy partiynogo aktiva TsK KP Kazakhstana," Otdel organizatsionno-partiynoy raboty TsK KPSS, 1973 in Rossiiskii gosudarstvennyi arkhiv sotsial'no-politicheskoi istorii (Russian State Archive of Social and Political History) f. 17, op. 141, d. 1551, l. 93.

65. Dronin and Bellinger, *Climate Dependence*, 303.

66. "O nauchnykh osnovakh zemledeliia v stepnykh raionakh SSSR v desiatoi piatiletke i rasshirenii proizvodstva protivoerozionnoi tekhniki," 1975 in Gosudarstvennyi arkhiv Rossiiskoi Federatsii (State Archive of the Russian Federation; hereafter cited as GARF) f. R-5446, op. 110, d. 932, l. 13).

67. GARF f. R-5446, op. 110, d. 932, ll. 9–11.

68. GARF f. R-5446, op. 110, d. 932, ll. 18–19.

69. GARF f. R-5446, op. 110, d. 932, l. 13.

70. Drozdov, *Zasukha*, 61–62.

71. Drozdov, *Zasukha*, 63–67. He calculated a reduction in precipitation of 17.4 percent.

72. M. I. Budyko, *Vliianie cheloveka na klimat* (Gidrometizdat, 1972), 4. In an earlier article, Budyko established that the "process of climate change under the economic activity of humankind accelerates," and this could lead to "dramatic changes of climate on large territories." Mikhail Ivanovich Budyko, "Nekotorye puti vozdeistviia na klimat," *Meteorologiia i Gidrologiia* 2 (1962): 3–8.

73. On Budyko's general contribution to climate studies, see Jonathan D. Oldfield, "Mikhail Budyko's (1920–2001) Contributions to Global Climate Science: From Heat Balances to Climate Change and Global Ecology," *Wiley Interdisciplinary Reviews: Climate Change* 7, no. 5 (June 2016): 682–92.

74. Mikhail Ivanovich Budyko, *Sovremennye izmeneniia klimata* (Gidrometizdat, 1977), 5, 43.

75. Budyko, *Sovremennye izmeneniia klimata*, 39, 40, 43.

76. Budyko, *Sovremennye izmeneniia klimata*, 40.

77. G. V. Menzhulin and S. P. Savvataev, "Vliianie sovremennykh izmeneniy klimata na urozhaynost' sel'skokhoziaistvennykh kul'tur," in *Problemy atmosfernogo uglekislogo gaza. Trudy sovetsko-amerikanskogo simpoziuma. Dushanbe, 12–20 oktiabria 1978g.* (Gidrometizdat, 1980), 186–97.

78. In 1975–1988 the dryness index rose in all regions of the Soviet Union. Budyko calculated this index as the difference between temperature anomalies and precipitation anomalies, each one divided by its standard deviation. Mikhail Ivanovich Budyko, ed., *Predstoyashchie izmeneniia klimata: sovmestnyy sovetsko-amerikanskii otchet o klimate i ego izmeneniiakh* (Gidrometizdat, 1991), 235.

79. M. I. Budyko and Iu. A. Izrael', *Antropogennye izmeneniya klimata* (Gidrometizdat, 1987), 331, 333, 334.

80. Viktor A. Kovda, *Land Aridization and Drought Control* (Westview Press, 1980), 49; Marc Elie, "Formulating the Global Environment: Soviet Soil Scientists and the International Desertification Discussion, 1968–91," *Slavonic and East European Review* 93, no. 1 (January 2015): 198.

81. Viktor Abramovich Kovda, *Aridizatsiya sushi i bor'ba s zasukhoy* (Nauka, 1977), 25–54; Kovda, *Land Aridization*, 21–55.

82. Kovda, *Land Aridization*, 26–27, 46–48. Lamb matched climate cycles with the sun's activity.

83. Elie, "Global Environment," 18–19.

84. Kovda, *Land Aridization*, 49.

85. David Moon, "The Debate over Climate Change in the Steppe Region in Nineteenth-Century Russia," *Russian Review* 69, no. 2 (April 2010): 251–75; Jonathan D. Oldfield, "Climate

Modification and Climate Change Debates among Soviet Physical Geographers, 1940s–1960s," *Wiley Interdisciplinary Reviews: Climate Change* 4, no. 6 (November 2013): 513–24; Elie, "Global Environment."

86. Oleg Dmitrievich Sirotenko and Vera Nikolaevna Pavlova, "Metody otsenki vliianiia izmenenii klimata na produktivnost' sel'skogo khoziaistva," in *Metody otsenki posledstvii izmeneniia klimata dlia fizicheskikh i biologicheskikh sistem* (Rosgidromet, 2012), 171.

87. Sirotenko and Pavlova, "Metody," 170.

88. Sirotenko and Pavlova, "Metody," 176.

89. O. D. Sirotenko and E. V. Abashina, "Vliianie global'nogo potepleniia na agroklimaticheskie resursy i produktivnost' sel'skogo khoziaistva Rossii," *Meteorologiya i gidrologiya* 4 (1994): 101; S. M. Semenov, V. V. Yasiukevich, and E. S. Gel'ver, *Vyiavlenie klimatogennykh izmenenii* (Izdatel'skii tsentr Meteorologiya i gidrologiya, 2006), 160–64. For more on the actual climatic situation in Kazakhstan, see Marc Elie, "Les steppes bouleversées," *Etudes rurales* 200 (2017): 80–105.

90. Oldfield, "Climate Modification"; Oldfield, "Mikhail Budyko"; Moon, "The Debate."

91. Donald Worster, "Transformations of the Earth: Toward an Agroecological Perspective in History," *Journal of American History* 76, no. 4 (March 1, 1990): 1087.

Chapter 6: Leviathan on the Oxus

Research for this paper was made possible by generous support from Volkswagen Foundation, Hannover, Germany, and the German Academic Exchange Service.

1. Vladimir Lenin, Letter to Caucasian communists, dated April 14, 1921, in *Polnoe sobranie sochinenii* (Izd-vo politicheskoi literatury, 1963), 43:200; Kaganovich at Orgbiuro meeting, May 8, 1929 in Rossiiskii gosudarstvennyi arkhiv sotsial'no-politicheskoi istorii (Russian State Archive of Social and Political History; hereafter cited as RGASPI) f. 17, op. 113, d. 725, ll. 55–56); Usman Iusupov, "K novym pobedam kommunizma!" *Pravda Vostoka*, February 4, 1940, 3.

2. Joachim Radkau, *Nature and Power: A Global History of the Environment* (Cambridge University Press, 2008), 86.

3. There is, however, a growing body of academic work addressing this caveat. See, for example, Nikolai Ssorin-Chaikov, *The Social Life of the State in the Subarctic Siberia* (Stanford University Press, 2003); Nick Baron, *Soviet Karelia: Politics, Planning and Terror in Stalin's Russia, 1920–1939* (Taylor and Francis, 2009); Andy Bruno, *The Nature of Soviet Power: An Arctic Environmental History* (Cambridge University Press, 2016).

4. Vladimir Kaganskii, *Kul'turnyi landshaft i sovetskoe obitaemoe prostranstvo* (Novoe literaturnoe obozrenie, 2001), 137–49; Douglas Weiner, "The Predatory Tribute-Taking State: A Framework for Understanding Russian Environmental History," in *The Environment and World History*, ed. Edmund Burke III and Kenneth Pomeranz (University of California Press, 2009), 276–336; Alexander Etkind, *Internal Colonization: Russia's Imperial Experience* (Polity Press, 2011), 72–73.

5. Stephen Kotkin, "The State—Is It Us? Memoirs, Archives, and Kremlinologists," in *Russian Review* 61, no. 1 (2002): 35–51.

6. Karl August Wittfogel, *Oriental Despotism: A Comparative Study of Total Power* (Yale University Press, 1957). Wittfogel's hypothesis was controversially discussed in *Slavic Review* 22, no. 4 (1963), 627–55; James C. Scott, *Seeing like a State: How Certain Schemes to Improve the Human Condition Have Failed* (Yale University Press, 1998), 3–5.

7. Christopher A. Bayly, *The Birth of the Modern World, 1780–1914: Global Connections and Comparisons* (Blackwell, 2004), 252. Bayly's argument continues: "In methodological terms, the emphasis on the state often involves a problematic reliance on certain categories of sources. Historians generally work with state documents. The argument becomes circular. It is very easy to assume that states are therefore responsible for any and all social change."

8. Frederick Cooper, *Colonialism in Question: Theory, Knowledge, History* (University of California Press, 2005). Cooper is to the point when criticizing historians' use of the concept

of modernity: "The use by historians and others of the concept of colonial modernity flattens history, elevating messy histories into a consistent project and underplaying the efforts of colonized people to deflect and appropriate elements of colonizing policies, taking apart the packaging that the critics of modernity leave intact" (117).

9. Water, irrigation, and the cotton economy in tsarist and Soviet Central Asia have been addressed in a number of recent book-length studies such as J. Michael Thurman, "Modes of Organization in Central Asian Irrigation: The Ferghana Valley, 1876 to Present" (PhD diss., Indiana University, 1999); Maya Peterson, "Technologies of Rule: Empire, Water, and the Modernization of Central Asia, 1867–1941" (PhD diss., Harvard University, 2011); Sergey Abashin, *Sovetskii kishlak:Mezhdu kolonializmom i modernizatsiei* (Novoe literaturnoe obozrenie, 2015); Christian Teichmann, *Macht der Unordnung: Stalins Herrschaft in Zentralasien, 1920–1950* (Hamburger Edition, 2016); Julia Obertreis, *Imperial Desert Dreams: Cotton Growing and Irrigation in Central Asia, 1860–1991* (V&R unipress, 2017).

10. Solotskii to Karakalpak *obkom* in Qaraqalpaqstan Respublikasi Orayliq Ma'mleketlik Arxivi (State Archive of the Autonomous Republic of Karakalpakstan; hereafter cited as QROMA) f. 17, op. 1, d. 77, l. 44).

11. Botakoz Kassymbekova and Christian Teichmann, "The Red Man's Burden: Soviet European Officials in Central Asia in the 1920s and 1930s," in *Helpless Imperialists: Imperial Failure, Fear and Radicalization*, ed. Marus Reinkowski and Gregor Thum (Vandenhoeck and Ruprecht, 2013), 163–86.

12. "Kharakteristika zemel'nykh otnoshenii v oazisakh nizovoi Amu-Dar'i," 1927 (RGASPI f. 62, op. 2, d. 1660, ll. 2–4).

13. "Karakalpakiia," *Sovetskaia step'*, December 10, 1927, 2.

14. For detailed descriptions of irrigation techniques, see for example, V. V. Tsinzerling, *Oroshenie na Amudar'e*, 2nd ed. (Izdanie upravleniia vodnogo khoziaistva Srednei Azii, 1927), 85–224; Ia. G. Guliamov, *Istoriia orosheniia Khorezma s drevneishikh vremen do nashikh dnei* (Izd-vo AN Uzbekskoi SSR, 1957), 268–96; M. Annanepsov, *Khoziaistvo Turkmen v XVIII—XIX vv.* (Ylym, 1972), 87–103, 133–73; M. V. Sazonova, *Traditsionnoe khoziaistvo uzbekov iuzhnogo Khorezma* (Nauka, 1978), 12–25; Ktaibek Sarybaev, *Istoriia orosheniia Karakalpakstana: s kontsa XIX veka do nashikh dnei* (Karakalpakstan, 1995), 80–124.

15. "Doklad inspektora-inzhenera B. P. Selivanovskogo," November 16, 1926 in Gosudarstvennyi arkhiv Rossiiskoi Federatsii (State Archive of the Russian Federation; hereafter cited as GARF) f. 3292, op. 1, d. 14, l. 252).

16. "Kharakteristika zemel'nykh otnoshenii v oazisakh nizovoi Amu-Dar'i," 1927 (RGASPI f. 62, op. 2, d. 1660, l. 4).

17. B. B. Karp and I. E. Suslov, eds., *Sovremennyi kishlak Srednei Azii (sotsial'no-ekonomicheskii ocherk): Khankinskaia volost' (Khorezmskoi oblasti Uzbekskoi SSR)* (Izdanie sredne-aziatskogo biuro TsK VKP(b), 1926), 125–28.

18. N. Subbota, "K voprosam irrigatsii," *Vestnik agitatsii i propagandy* (*Tashkent*) 2 (1922): 25–26.

19. For the collection of salguta tax, see M. Iu. Iuldashev, *K istorii krest'ian Khivy XIX veka* (Fan, 1966), 61–65. For zekat tax registers and kazu (seasonal work obligations) registers, see Iu. E. Bregel', *Dokumenty arkhiva Khivinskikh khanov po istorii i etnografii Karakalpakov* (Glavnaia redaktsiia vostochnoi literatury, 1967), 129–292, and 293–96, respectively.

20. Seymour Becker, *Russia's Protectorates in Central Asia: Bukhara and Khiva, 1865–1924* (1968; repr. RoutledgeCurzon, 2004).

21. Ekatarina Pravilova, "Reka imperii: Amudar'ia v geopoliticheskikh i irrigatsionnykh proektakh vtoroi poloviny XIX veka," in N. G. Suvorova, ed., *Aziatskaia Rossiia: Liudi i struktury imperii* (Izd-vo OmGU, 2005), 425–57.

22. Shoshana Keller, "The Central Asian Bureau, an Essential Tool in Governing Soviet Turkestan," in *Central Asian Survey* 22, no. 2–3 (2003): 281–97. See also Vladmir Genis, "'Butaforskaia revoliutsiia' ili rossiiskoe polpredstvo v Khive v 1920 godu," *Vostok* 2 (2000): 5–26.

23. N. V. Briullova-Shaskol'skaia, "Na Amudar'e: etnograficheskaia ekspeditsiia v Kerkinskii okrug TSSR," *Novyi Vostok* 16–17 (1927): 290–303.

24. "Doklad inspektora-inzhenera B. P. Selivanovskogo," November 16, 1926 (GARF f. 3292, op. 1, d. 14, ll. 250ob.–251). Similar conflicts evolved all over Soviet Central Asia; see Teichmann, *Macht der Unordnung*, 57–69.

25. Uzakbay Kulumbetov, "Pervye, no tverdye shagi," *Sovetskaia step'*, December 10, 1927, 2; Faizulla Khodzhaev, "Khlopok," *Revoliutsiia i natsional'nosti* 6 (1930): 27–32; Annakuli Artykov, *Pervye gody bor'by: vospominaniia partiinogo rabotnika o stroitel'stve sotsializma v Tashauzkom okruge* (Turkmenistan, 1969), 63.

26. V. I. Masal'skii, *Turkestanskii krai*, vol. 19 of *Rossiia: Polnoe geograficheskoe opisanie nashego otechetstva*, ed. V. P. Semenov-Tian-Shanskii (Izdanie A. F. Devrien, 1913), 415.

27. N. Subbota, "K voprosam irrigatsii," *Vestnik agitatsii i propagandy* (*Tashkent*) 2 (1922): 25–26.

28. For more evidence, see Teichmann, *Macht der Unordnung*, 85–90.

29. K. Zubrek, "Zemledelie i irrigatsiia v Khorezme," *Novyi Vostok* 3 (1923): 208–19.

30. "Perepiska oblvodkhoza Amu-Darinskoi oblasti s mestnymi organami," 1925 (QROMA f. 17, op. 1, d. 46).

31. "Doklad inspektora-inzhenera B. P. Selivanovskogo" (GARF f.3292, op. 1, d. 14, l. 346ob.).

32. Petrachuk to Water Managerment Authority of Kazakhstan (UVKh Kazakstana) (GARF f. 3292, op. 1, d. 5, l. 64ob.).

33. Solotskii to Karakalpak *obkom* (QROMA f. 17, op. 1, d. 77, ll. 7, 13–14).

34. On the distinction of bureaucratic, intermediary, and despotic forms of rule, see Gerd Spittler, "Administrative Despotism in Peasant Societies," in *The Foundations of Bureaucracy in Economic and Social Thought*, ed. Bill Jenkins and Edward C. Page (Elgar, 2004), 1:339–50.

35. "Stenogramma kustnogo soveshchaniia predstavitelei partiinykh i sovetskikh organizatsii Barnaul'skogo, Biiskogo i Rubtsovskogo okrugov po voprosam khlebozagotovok s uchastiem general'nogo sekretaria TsK VPK(b) tov. Stalina," January 22, 1928 (RGASPI f. 558, op. 11, ll. 118, 81–82).

36. Gotfrid to Khorezm *obkom*, March 1930 (RGASPI f. 62, op. 2, d. 2137, l. 60).

37. "Stenograficheskii otchet soveshchaniia komissii po voprosam o meropriiatiiach po provedeniiu plana khlopkovych posevov i vytesneniiu zernovykh kul'tur," September 29, 1929, O'zbekiston respublikasi markaziy davlat arxivi (State Archive of the Republic of Uzbekistan; hereafter cited as ORMDA) f. 837, op. 26, d. 289, ll. 26–26ob.

38. "Dokladnaia zapiska po voprosu o kolkhoznom stroitel'stve v KKASSR," June 19, 1934 (QROMA f. 322, op. 1, d. 209, ll. 3–13). The version of the report that was sent to Moscow (RGASPI f. 62, op. 2, d. 3274) did not contain the cited passages.

39. "Dokladnaia zapiska po voprosu o kolkhoznom stroitel'stve v KKASSR," June 19, 1934 (QROMA f. 322, op. 1, d. 209, l. 7).

40. This was the case in other regions of the Soviet Union. See Lewis H. Siegelbaum, "Dear Comrade, You Ask What We Need: Socialist Paternalism and Soviet Rural 'Notables' in the Mid-1930s," in *Stalinism: New Directions*, ed. Sheila Fitzpatrick (Routledge, 2000), 231–55, esp. 235–48.

41. The following passages rely on Gerd Spittler, *Administrative Despotism*, 339–45. Spittler developed the concept of administrative despotism in a number of articles, most notably in Gerd Spittler, "Administration in a Peasant State," *Sociologia Ruralis* 23, no. 2 (1983): 130–44; Gerd Spittler, "Passivität statt sozialer Bewegung. Familiäre Subsistenzwirtschaft als Basis für defensive Strategien der Bauern und Passivität der Verwaltung," in *Soziale Bewegungen in Entwicklungsländern*, ed. Rolf Hanisch (Nomos, 1983), 45–73; Gerd Spittler, "Volkszählung und bürokratische Herrschaft in Bauernstaaten," in *Bauerngesellschaften im Industriezeitalter. Zur Rekonstruktion ländlicher Lebensformen*, ed. Christian Giordano and Robert Hettlage (Reimer, 1989), 97–108.

42. Gerd Spittler, "Administrative Despotism," 340.

43. Christian Teichmann, "Canals, Cotton, and the Limits of De-colonization in Soviet Uzbekistan, 1924–1941," in *Central Asian Survey* 26, no. 4 (2007): 499–519, esp. 508–9, 512.

44. This restriction also applied to the large-scale "narodnye stroiki" construction projects in Uzbekistan during the 1940s; see Teichmann, *Macht der Unordnung*, 220–24.

45. "Istoriia voprosa bor'by s deigishem u goroda Turtkulia za period 1932–1937 gg." (ORMDA f. 837, op. 32, d. 357, l. 2).

46. Narkomzem KKASSR Kudabaev to Sovnarkom KKASSR Kurbanov, July 5, 1936 (QROMA f. 322, op. 1, d. 342, ll. 46–47); Mambet'iarov to Kurbanov and Kudabaev (QROMA f. 322, op. 1, d. 342, l. 60).

47. "Postanovlenie No. 291 SNK SSSR," February 17, 1937 (GARF f. 5446, op. 19, d. 4, ll. 199–200).

48. Chernov to Molotov, September 7, 1937 (ORMDA f. 837, op. 32, d. 577, l. 18); Syromiatnikov to Uzvodchoz, November 25, 1937 (QROMA f. 322, op. 1, d. 509, ll. 35–37).

49. Besednov to Karakalpak SNK, January 13, 1938 (QROMA f. 322, op. 1, d. 509, ll. 6–7).

50. Syromiatnikov to Glavvodchoz NKZem SSSR, June 29, 1937, and Khodzhaev to Chernov, March 25, 1937 (ORMDA f. 837, op. 32, d. 46, ll. 75, 76, 159, 168).

51. Pershin to Segizbaev, December 2, 1937 (ORMDA f. 837, op. 32, d. 577, l. 43).

52. Shevelev to Segizbaev, December 7, 1937, Pershin to Segizbaev, December 2, 1937, Segizbaev to Rakhmatullaev, December 21, 1937 (ORMDA f. 837, op. 32, d. 577, ll. 50, 63, 64).

Chapter 7: Soviet Irrigation Policies under Fire

1. To name just two examples of the Western treatment of the Aral Sea siltation, see René Létolle and Monique Mainguet, *Aral* (Springer, 1993) and the more recent Philip Micklin, "The Aral Sea Crisis and Its Future: An Assessment in 2006," *Eurasian Geography and Economics* 47, no. 5 (2006): 546–67. The numerous instructive and well-informed publications by Philip Micklin have contributed to the rise of the subject of the Aral Sea's fate. The Aral Sea siltation has become a major topic in books on environmental problems in the Soviet Union and has been characterized as a major event in global environmental history as well. An early example of the former is Marshall Irwin Goldman, *The Spoils of Progress: Environmental Pollution in the Soviet Union* (MIT Press, 1972). An example of the latter is J. R. McNeill, *Something New under the Sun: An Environmental History of the Twentieth-Century World* (Global Century Series, 2001), 162–66.

2. Two prominent examples are Edward Allworth, ed., *Central Asia: 130 Years of Russian Dominance, a Historical Overview* (Duke University Press, 1994), and Adeeb Khalid, *Islam after Communism: Religion and Politics in Central Asia* (University of California Press, 2007). For a more recent study on environmental history, see Maya Peterson, *Technologies of Rule: Empire, Water, and the Modernization of Central Asia, 1867–1941* (PhD diss., Harvard University, 2011).

3. Influential works include Douglas R. Weiner, *A Little Corner of Freedom: Russian Nature Protection from Stalin to Gorbachëv* (University of California Press, 1999); Stephen Brain, *Song of the Forest: Russian Forestry and Stalinist Environmentalism, 1905–1953* (Pittsburgh University Press, 2011). A German-language contribution focusing on the period 1948–1967 and including aspects of environmental history is Klaus Gestwa, *Die Stalinschen Großbauten des Kommunismus: Sowjetische Technik- und Umweltgeschichte, 1948–1967* (De Gruyter, 2010).

4. McNeill, *Something New*, 350.

5. Jens Ivo Engels, "Modern Environmentalism," in *The Turning Points of Environmental History*, ed. Frank Uekötter (Pittsburgh University Press, 2010), 119–31. See also Dieter Rucht, "Environmental Movement Organizations in West Germany and France: Structure and Interorganizational Relations," *International Social Movement Research* 2 (1989): 61–94; Angela G. Mertig and Riley E. Dunlap, "Environmentalism, New Social Movements, and the New Class: A Cross-National Investigation," *Rural Sociology* 66 (2001): 113–36; Jens Ivo Engels, *Naturpolitik in der Bundesrepublik: Ideenwelt und politische Verhaltensstile in Naturschutz und Umweltbewegung 1950–1980* (Schoeningh, 2006), 322–38.

6. Joachim Radkau, *The Age of Ecology: A Global History* (Cambridge University Press, 2014).

7. McNeill, *Something New*, 355.

8. The most prominent contribution on "ecocide" is Murray Feshbach and Alfred Friendly Jr., *Ecocide in the USSR: Health and Nature under Siege* (Basic Books, 1992).

9. To be sure, in certain spheres indigenous cadres were well represented in the early Soviet period already, especially in highly visible party positions and in the educational and cultural spheres. A seminal general work on the early Soviet period is Terry Martin, *The Affirmative Action Empire: Nations and Nationalism in the Soviet Union, 1923–1939* (Cornell University Press, 2001). A detailed study on the professional advance of Central Asians for the late Soviet period is Nancy Lubin, *Labour and Nationality in Soviet Central Asia* (Palgrave, 1984).

10. In 1970, for example, it delivered 4.5 million of the 6.9 million total tons of cotton produced. *Narodnoe khoziaistvo Uzbekskoi SSR v 1971 g.: Statisticheskii ezhegodnik* (Gosudarstvennyi statisticheskoe izd-vo, 1972), 114. Cotton production in Uzbekistan in the 1970s and 1980s has been outlined by Boris Z. Rumer, "Central Asia's Cotton: The Picture Now," *Central Asian Survey* 6 (1987): 75–88. See also Boris Z. Rumer, *Soviet Central Asia: "A Tragic Experiment"* (Unwin Hyman, 1989).

11. See also Julia Obertreis, "Der 'Angriff auf die Wüste' in Zentralasien: Zur Umweltgeschichte der Sowjetunion," *Osteuropa* 58 (2008): 4–5, 37–56.

12. Stephan Merl, "III.C Wirtschaftsrefomen, 2. Agrarpolitik," in *Handbuch der Geschichte Russlands, vol. 5.1: 1945–1991: Vom Ende des Zweiten Weltkriegs bis zum Zusammenbruch der Sowjetunion*, ed. Stefan Plaggenborg (Hiersemann, 2003), 212–25.

13. For an overview on the development history of the Hungry Steppe, see Ian M. Matley, "The Golodnaya Steppe: A Russian Irrigation Venture in Central Asia," *Geographical Review* 60, no. 3 (1970): 328–46. One example of the high hopes set for the Hungry Steppe during the tsarist period is N. Petrov, *Ob irrigatsii v Turkestanskom krae* (S. I. Petukhov, 1894), 15.

14. Details of the official (but not necessarily uncritical) Soviet history of the Hungry Steppe can be found in publications such as R. S. Igamberdyev and A. A. Razzakov, *Istoriia melioratsii v Uzbekistane (na materialakh Golodnoi stepi)* (FAN, 1978). The central Soviet survey is *Irrigatsiia Uzbekistana: v chetyrekh tomakh*, 4 vols. (FAN, 1975).

15. On the continuity beyond Stalinism of the nature domination discourse, see Gestwa, *Stalinschen Großbauten*, 563. See also Julia Obertreis, *Imperial Desert Dreams: Cotton Growing and Irrigation in Central Asia, 1860–1991* (V&R unipress, 2017), esp. 465–66.

16. For an overview, see *Irrigatsiia Uzbekistana*, 1:208–11.

17. For a voice from the tsarist period when engineers, officials, and scholars claimed that the feeding of the Aral Sea was useless and the rivers' water should be used for irrigation instead, see Petrov, *Ob irrigatsii*, 57. On the discovery of the rapid shrinking of the Aral Sea beginning in the 1960s, see Rumer, *Soviet Central Asia*, 82–83. For concerned voices in the Uzbek Society of Nature Protection in 1975, see Tsentral'nyi gosudarstvennyi arkhiv Respubliki Uzbekistan (Central State Archive of the Republic of Uzbekistan; hereafter cited as TsGARUz) f. 2472, op. 1, d. 153, l. 81.

18. A vivid illustration is in Viktor Yaroshenko, *Ekspeditsiia "Zhivaia voda"* (Molodaia gvardiia, 1989), 350, 352.

19. On the effect of water scarcity on the cotton harvests, see F. Amanov, "Problemy povysheniia proizvodstva tonkovoloknistogo khlopka," *Sel'skoe khoziaistvo Uzbekistana* (Agriculture in Uzbekistan; hereafter cited as *SKhU*) 12 (1980): 2–4.

20. For example, in Khorezm province in 1981, still only 41 percent of water was treated sufficiently to remove biological and chemical wastes (TsGARUz f. 2483, op. 2, d. 4234, ll. 46, 90). The same year, 72 percent of the rural population in Uzbekistan used surface water as drinking water. People drank water with toxic substances that was polluted up to ten times the allowed concentrations (TsGARUz f. 2483, op. 2, d. 4239, l. 97).

21. For a survey on the history of the river diversion up to the mid-1980s, see Philip P.

Micklin, "The Fate of 'Sibaral': Soviet Water Politics in the Gorbachev Era," *Central Asian Survey* 6, no. 2 (1987): 67–88. On the broad support for the diversion of Siberian rivers in Central Asia, see Yaroshenko, *Ekspeditsiia*, 332.

22. The Uzbek republic received its own Academy of Sciences in November 1943; before it had only a branch of the Soviet Academy. Eric Sievers, "Academy Science in Central Asia, 1922–1998," *Central Asian Survey* 22, no. 2 (2003): 264–66.

23. According to Joan DeBardeleben, the most open and controversial debate on environmental issues on the USSR level had come up in the 1970s in the Moscow Academy's journal *Voprosy filosofii* (Questions of philosophy). She wrote in the 1980s that the Moscow Academy of Sciences "has come to enjoy considerable autonomy from direct political intrusion," at least in regard to its scientists and philosophers. Joan DeBardeleben, *The Environment and Marxism-Leninism: The Soviet and East German Experience* (Westview Press, 1985), 77.

24. For example, see McNeill, *Something New*, 350.

25. Even before and in different contexts, the labeling "rational/irrational usage of natural resources" was common in the Soviet context. See, among many examples, Mark Sokolsky (this volume).

26. Stephan Merl, "The Soviet Economy in the 1970s: Reflections on the Relationship between Socialist Modernity, Crisis and the Administrative Command Economy," in *The Crisis of Socialist Modernity: The Soviet Union and Yugoslavia in the 1970s*, ed. Marie-Janine Calic, Dietmar Neutatz, and Julia Obertreis (Vandenhoeck and Ruprecht, 2011), 28–65, especially 37.

27. For one example, see B. Ovezmuradov, "K osvoeniiu tselinnykh i zalezhnykh zemel' v zone Karakumskogo kanala," *Sel'skoe khoziaistvo Turkmenistana* (Agriculture in Turkmenistan; hereafter cited as *SKhT*) 4 (1962): 49–53.

28. A water codex of the Turkmen Republic was enacted in 1973. M. Ataev, "Bol'she vnimaniia okhrane vod," *SKhT* 9 (1974): 36.

29. For example, see S. Mamatov, "O sisteme organov gosudarstvennogo upravleniia UzSSR po okhrane prirody," *Obshchestvennye Nauki Uzbekistana* (Social Sciences of Uzbekistan; hereafter cited as *ONU*) 10 (1973), 39–44.

30. Yu. I. Alekseev, "Biologicheskii metod bor'by s vrediteliami khlopchatnika," *SKhT* 11 (1971): 29–30.

31. A. E. Ishmukhamedov, "Sotsial'no-ekonomicheskie problemy prirodopol'zovaniia v usloviiakh nauchno-tekhnicheskogo progressa," *ONU* 4 (1975): 32–37.

32. Z. M. Kuzham'iarova, "Ekologicheskaia tekhnologiia kak faktor optimizatsii vzaimootnoshenii sub-ekta i ob-ekta," *ONU* 4 (1975): 39, 41.

33. Ishmukhamedov, "Sotsial'no-ekonomicheskie problemy," 37.

34. For one typical example, see N. Aksarin, "Ekonomicheskie i sotsial'nye problemy okhrany okruzhaiushchei sredy. Konsul'tatsii po kursu 'Agrarnaia teoriia i politika KPSS' (dlia propagandistov i slushatelei sistemy politicheskogo i ekonomicheskogo obrazovaniia)," *SKhU* 2 (1980): 49–51.

35. The program was presented in the news block of *Gidrotekhnika i Melioratsiia* (Hydroengineering and melioration) in 1972 as a new program, but it is not clear when exactly the Soviet Union joined it. "Chelovek i biosfera," *Gidrotekhnika i Melioratsiia* 9 (1972): 97; Martin W. Holdgate, *The Green Web: A Union for World Conservation* (Earthscan, 1999), 97–98.

36. Barbara Ward Jackson and René J. Dubos, *Only One Earth* (W. W. Norton, 1972). A Russian translation of this work appeared in 1975. See Barbara Uord and Rene Dubos, *Zemlia tol'ko odna* (Progress Publishers, 1975).

37. For example, see *Vsestoronnyi analiz okruzhaiushchei sredy. Trudy sovetsko-amerikanskogo simpoziuma, Tbilisi 25–29 marta 1974 g.* (Gidrometeoizdat, 1975). See also DeBardeleben, *Environment and Marxism-Leninism*, 124–25, where she says that Soviet experts used the offers for an exchange of ideas with Western colleagues more actively than the experts from the GDR.

38. Jonathan D. Oldfield and Denis J. B. Shaw, "V.I. Vernadskii and the Development of Biogeochemical Understandings of the Biosphere, c. 1880s–1968," *British Journal for the History of Science* 46, no. 2 (2013): 287–310.

39. For example, in 1972 the Council of Ministers of the Uzbek Republic issued a decree entitled "O sostoianii i meropriiatiiakh po uluchsheniiu okhrany prirody v Uzbekskoi SSR" (On the state and measures for improvement of nature protection in the Uzbek SSR) (TsGARUz f. 2742, op. 1, d. 153, l. 88).

40. It was entitled "O dopolnitel'nykh merakh po usileniiu okhrany prirody i uluchsheniiu ispol'zovaniia prirodnykh resursov," *SKhT* 3 (1979): 2–3.

41. Cf. Gestwa, *Stalinschen Großbauten*, 531–34. For an example from the Ferghana Valley from the late 1970s, see TsGARUz f. 2483, op. 2, d. 4239, ll. 33–35, 41.

42. Gestwa, *Stalinschen Großbauten*, 541.

43. Article 18 of the constitution spoke of "safeguarding the reproduction of natural resources" and "the improvement of the human environment," and Article 67 ran: "Citizens of the USSR are bound to protect nature and safeguard its riches." See "The Constitution of 1977," trans. F. J. M. Feldbrugge and William B. Simons, in *The Constitutions of the USSR and the Union Republics: Analysis, Texts, Reports*, ed. Ferdinand J Feldbrugge (Springer, 1979), 85, 107. Such aspects were absent in the 1936 "Stalin" Constitution.

44. Michael Cotey Morgan, "The Seventies and the Rebirth of Human Rights," in *The Shock of the Global: The 1970s in Perspective*, ed. Niall Ferguson et al. (Harvard University Press, 2010), 237–50, esp. 247–49. Matthias Peter and Hermann Wentker, "'Helsinki-Mythos' oder 'Helsinki-Effekt'? Der KSZE-Prozess zwischen internationaler Politik und gesellschaftlicher Transformation. Zur Einleitung," in *Die KSZE im Ost-West-Konflikt. Internationale Politik und gesellschaftliche Transformation 1975–1990*, ed. Peter and H. Wentker (De Gruyter, 2012), 1–14.

45. A. Khakyev, "Okhraniat' prirodu—nakaz partii," *SKhT* 5 (1979): 31–32.

46. This evaluation is based on an analysis of material of the Uzbek Society of Nature Protection from the years 1975, 1976, and 1985 in TsGARUz f. 2742, op. 1, dd. 153, 164, 329, 330.

47. Khakyev, "Okhraniat' prirodu," 31, 32.

48. Yaroshenko, *Ekspeditsiia*.

49. For data on Karakalpakstan for 1981, see TsGARUz f. 2483, op. 2, d. 4222, ll. 1–2, 13.

50. On the record harvest and how it was praised in poems by Uzbek poets, see S. I. Vorob'eva, "Obychai i traditsii v sisteme sotsial'nykh norm sotsialisticheskogo obshchestva," *ONU* 6 (1981): 14–20.

51. D. Sheripov, "Liutserna—kak melioriruiushchaia kul'tura," *SKhT* 8 (1980): 34–35; also M. Sorokin and N. Safiulin, "Khlopkovye sevooboroty na lugovykh poliakh," *SKhU* 11 (1980): 24–25.

52. K. Redzhepbaev, I. I. Iurashkevich, and A. A. Ibatulina, "Bol'she vnimaniia bor'be s zasoleniem oroshaemykh pochv," *SKhT* 6 (1981): 22–23.

53. For example, see R. Turaev, "Biometodu—zelenuiu ulitsu," *SKhU* 6 (1980): 50–51.

54. Aksarin, "Ekonomicheskie."

55. I. Khashimova, "Berech' prirodnye resursy," *SKhU* 3 (1980): 62.

56. James Critchlow, *Nationalism in Uzbekistan: A Soviet Republic's Road to Sovereignty* (Westview Press, 1991), 43–44.

57. On the cotton scandal and its implications for Uzbek nationalism, see Critchlow, *Nationalism*, 39–54. This subject needs further exploration. The best account of the political landscape and changes in the Perestroika years avoids the subject. William Fierman, "Political Development in Uzbekistan: Democratization?" in *Conflict, Cleavage, and Change in Central Asia and the Caucasus*, ed. Karen Dawisha (Cambridge University Press, 1997), 360–408.

58. Obertreis, *Imperial Desert Dreams*, 365.

59. Khalid, *Islam after Communism*, 154; Michael Ochs, "Turkmenistan: The Quest for Stability and Control," in Dawisha, *Conflict, Cleavage, and Change*, 312–59, especially 313, 324–26.

60. Fierman, "Political Development," 367; Kathleen Collins, *Clan Politics and Regime*

Transition in Central Asia (Cambridge University Press, 2006), 147; Laura L. Adams, *The Spectacular State: Culture and National Identity in Uzbekistan* (Duke University Press, 2010), 46–48.

61. There had been an earlier expedition to Central Asia from Moscow in 1978. It was one of a series of expeditions under the title Zhivaia voda (Living water) expedition. Yaroshenko, *Ekspeditsiia.*

62. Zalygin's appointment was one of several appointments of chief editors or major journals initiated by the propaganda department of the Central Committee. Archie Brown, *The Gorbachev Factor* (Oxford University Press, 1996), 163–64.

63. Grigorii Reznichenko, *Aral'skaia katastrofa: Dnevnik ekspeditsii (s otstupleniiami i kommentariiami)* (Novosti, 1992), 7–9.

64. *Mirza-Ali Valievich Mukhamedzhanov. Materialy k biobibliografii uchenykh Uzbekistana* (FAN, 1985).

65. M. Mukhamedzhanov, *Berech' zemliu, umnozhat' ee plodorodie* (Mekhnat, 1985), 8.

66. Patricia M. Carley, "The Price of the Plan: Perceptions of Cotton and Health in Uzbekistan and Turkmenistan," *Central Asian Survey* 8, no. 4 (1989): 1–38, especially 2, 10–11.

67. The reasons for the ban of the project have been discussed by various authors. In the end it seems that neither environmental concerns nor national resentments played a decisive role; it was simply a matter of money. Nevertheless, the downturn of "the diversion" had an enormous significance in regard to infrastructural politics and the relation of politics and the public sphere. It was the first time that one of the big transformative projects had been stopped. Marshall I. Goldman, "Umweltverschmutzung in der Sowjetunion: Die Abwesenheit einer aktiven Umweltbewegung und die Folgen," in *Umweltprobleme in Mittel- und Osteuropa*, ed. Helmut Schreiber (Springer, 1989), 162–83; Micklin, "The Fate of 'Sibaral,'" 78–79, 81; Weiner, *A Little Corner of Freedom*, 427.

68. Weiner, *A Little Corner of Freedom*, 419–26; Gestwa, *Stalinschen Großbauten*, 548. Village Prose was a Russian literary movement that helped shape national Russian consciousness by addressing the threat of losing (idealized) rural heritage by Soviet modernization.

69. TsGARUz f. 837, op. 41, d. 7209, ll. 132–36.

70. TsGARUz f. 837, op. 41, d. 6826, ll. 88–92.

71. TsGARUz f. 837, op. 41, d. 7209, ll. 181–82.

72. Reznichenko, *Aral'skaia katastrofa*, 8.

73. Irina Aliab'eva, "'I poka ia na zemle, ia speshu priniat' mery . . . ,'" *Zvezda Vostoka* 5 (1989): 75–76, 87.

74. Reznichenko, *Aral'skaia katastrofa*, 5.

75. For an early account of the protests against the pollution of Lake Baikal, see Thane Gustafson, *Reform in Soviet Politics: Lessons of Recent Policies on Land and Water* (Cambridge University Press, 1981), 40–46.

76. "Vystuplenie Tulepbergena Kaipbergenova na s"ezde Soiuza Pisatelei," *Literaturnaia gazeta* 19 (May 6, 1987): 6.

77. Aliab'eva, "I poka ia na zemle," 75.

78. Aliab'eva, "I poka ia na zemle," 75, 80. Note that the Russian word for "nature [*priroda*]" is feminine.

79. I. Chervanev, "Otkaz ot stikhiinogo—vo imia razumnogo," *Zvezda Vostoka* 2 (1990): 25.

80. B. Atchabarov and T. Sharmanov, "Voda dlia regiona. Sibirskaia voda dlia Srednei Azii—absurd ili nasushchnaia neobkhodimost'?" *Zvezda Vostoka* 4 (1990): 117–22.

81. TsGARUz f. 837, op. 41, d. 7417, ll. 62–65.

82. For example, see TsGARUz f. 837, op. 41, d. 7417, ll. 83, 89.

83. A joint initiative by the USSR Minvodkhoz and Gosagroprom along with other authorities suggested the founding of a new structure of administrations of the existing irrigation systems in March 1987. It is unknown if there were attempts to implement this initiative (TsGARUz f. 837, op. 41, d. 7209, ll. 71–72).

84. See TsGARUz f. 837, op. 41, d. 7417, ll. 86–87.

85. Ulrich Weißenburger, "Der Umweltschutz in der Sowjetunion: Zwang zum Handeln," in *Umweltprobleme in Mittel- und Osteuropa*, ed. Helmut Schreiber (Springer, 1989), 184–96.

86. Bo Libert, *The Environmental Heritage of Soviet Agriculture* (Cabi, 1995), 49.

87. Obertreis, *Imperial Desert Dreams*, chapter 4.

88. On a reevaluation of the major periods of Soviet political history, see Susanne Schattenberg, "Von Chruščev zu Gorbačev—Die Sowjetunion zwischen Reform und Zusammenbruch," *Neue Politische Literatur* 55 (2010): 255–84.

89. Besides studies from Western authors, see also the contributions by the Russian sociologist Oleg Yanitskii—among them, Oleg Yanitsky, "Environmental Initiatives in Russia: East-West Comparisons," in *Environment and Democratic Transition: Policy and Politics in Central and Eastern Europe*, ed. A. Vari (Springer, 1993), 120–45.

90. See, for example, the ecological section of the Biological Society of the CSSR Academy of Sciences whose members had access to environmental data often classified as top secret. See Petr Jehlička, Philip Sarre, and Juraj Podoba, *Czech Environmental Discourse after a Decade of Western Influence: Transformation beyond Recognition or Continuity of the Pre-1989 Perspective?* EUI Working Papers, RSC No 2002/24 (European University Institute, 2002), 9–10.

91. Gestwa, *Stalinschen Großbauten*, 550–51.

92. Radkau, *Age of Ecology*, 356.

93. D. J. Peterson, *Troubled Lands: The Legacy of Soviet Environmental Destruction* (Westview Press, 1993), 215; Weiner, *A Little Corner of Freedom*, 9, 21.

94. Feshbach and Friendly, *Ecocide*, 231–34; Peterson, *Troubled Lands*, 211–15.

95. On rising interethnic tensions in the southern districts of Karakalpakstan, which were caused by water scarcity, see A. Zholdasov, "Ekologiia . . . minus etnos?" *Zvezda Vostoka* 12 (1990): 9–16. On the interethnic clashes of the late 1980s in Central Asia, see Valery Tishkov, *Ethnicity, Nationalism and Conflict in and after the Soviet Union: The Mind Aflame* (Sage, 1997), 136–37.

96. To my knowledge, there is no study on African or other Asian countries devoted exactly to this phenomenon. However, on the role of indigenous representatives in planning processes in Africa, see Andreas Eckert: "'We Are All Planners Now.' Planung und Dekolonisation in Afrika," *Geschichte und Gesellschaft* 34, no. 3 (2008), 375–97. For highlighting the agendas of Europeans and Africans in a (post-)colonial irrigation and cotton-growing project in Sudan, see Maurits W. Ertsen, *Improvising Planned Development on the Gezira Plain, Sudan, 1900–1980* (Palgrave Macmillan, 2016). See also the ongoing Indian French research project "Engineers and Society in India from c. 1850 to Present Times," *Engineers and Society in Colonial and Post-Colonial India*, online at http://engind.hypotheses.org/2177#more-2177/.

97. For example, see S. Usmanov, "Gordost' i bol' Uzbekistana," *Zvezda Vostoka* 3 (1988): 123–35.

98. See Jane I. Dawson, *Eco-nationalism: Anti-nuclear Activism and National Identity in Russia, Lithuania, and Ukraine* (Duke University Press, 1996).

Chapter 8: Models of Soil and Society

1. "An Obituary for Justus Liebig," *The Chemical News*, April 25, 1873.

2. Karl Marx and Frederick Engels, *Selected Correspondence, 1846–1895* (International Publishers, 1942), 204.

3. William H. Brock, *Justus von Liebig: The Chemical Gatekeeper* (Cambridge University Press, 2002), 98.

4. For further context on Liebig, see Brock, *Justus von Liebig*; E. Patrick Munday, "Sturm und Dung: Justus von Liebig and the Chemistry of Agriculture" (PhD diss., Cornell University, 1990).

5. A. A. Rode, *Soil Science* (Israel Program for Scientific Translations, 1962), 5.

6. Justus von Liebig, *Chemistry in Its Application to Agriculture and Physiology*, 3rd ed. (Taylor and Walton, 1843), 178.

7. For more on the use of *Stoffwechsel* in German, see Robert U. Ayres and Leslie Ayres, *A Handbook of Industrial Ecology* (Edward Elgar, 2002), 16–17.

8. Justus von Liebig, "An Autobiographical Sketch," trans. J. Campbell Brown, *The Chemical News*, June 5, 1891.

9. Justus Liebig, *Familiar Letters on Chemistry*, 4th ed. (Walton and Maberly, 1859), 26.

10. Quoted in Brock, *Justus von Liebig*, 194.

11. Alexander Vucinich notes four major lineages of Russian chemistry: "It was virtually impossible to find a Russian chemist of the time who had not taken a course from Justus von Liebig at Giessen, Heinrich Rose at Berlin, R. W. Bunsen at Heidelberg, or Marcellin Berthelot at the College de France." See Alexander Vucinich, *Science in Russian Culture, 1861–1917* (Stanford University Press, 1970), 136. The term "chemist breeder" was coined by J. B. Morrell; see "The Chemist Breeders," *Ambix* 19 (1972): 1–46.

12. For more on Liebig's influence on the development of chemistry in nineteenth-century Russia, see Iu. S. Musabekov, *Iustus Libikh, 1803–1873* (Izd-vo AN SSSR, 1962), 169–77.

13. For more on Dokuchaev, see David Moon, *The Plough that Broke the Steppes: Agriculture and Environment on Russia's Grasslands, 1700–1914* (Oxford University Press, 2013); David Moon, "The Environmental History of the Russian Steppes: Vasilii Dokuchaev and the Harvest Failure of 1891," *Transactions of the Royal Historical Society* 15 (2005): 149–74; Catherine Evtuhov, "The Roots of Dokuchaev's Scientific Contributions: Cadastral Soil Mapping and Agro-environmental Issues," in *Footprints in the Soil: People and Ideas in Soil History*, ed. Benno P. Warkentin (Elsevier, 2006); Catherine Evtuhov, *Portrait of a Russian Province: Economy, Society, and Civilization in Nineteenth-Century Nizhnii Novgorod* (University of Pittsburgh Press, 2011).

14. Geert Vanpaemel notes that, although German scientific practices and institutions were looked to as an ideal model, they were rarely "copied" precisely but, rather, were adapted to local conditions at the "periphery." See Geert Vanpaemel, "The German Model of Laboratory Science and the European Periphery (1860–1914)," in *Sciences in the Universities of Europe, Nineteenth and Twentieth Centuries: Academic Landscapes*, ed. Ana Simões, Maria Paula Diogo, and Kōstas Gavroglou (Springer, 2015), 211–25.

15. A short work attributed to Liebig, *Iskusstvennye udobreniia ili tuki* (Artificial fertilizers), was published in Petersburg in 1850. This may be a translation of the 1845 pamphlet *An Address to the Agriculturists of Great Britain Explaining the Principles and Use of His Artificial Manures*, which was designed to explain and promote Liebig's patent fertilizers.

16. A. N. Engel'gardt, "Libikh v russkom perevode," *Sankt Peterburgskie vedomosti*, December 6, 1863, 1105–6.

17. Iustus Libikh, *Khimiia v prilozhenii k zemledeliu i fiziologii rastenii, sed'moe izdanie*, trans. P. A. Il'enkov (Fiveg and Son, 1864).

18. V. Kurilov, "Libikh" in *Entsiklopedicheskii slovar,'* ed. I. E. Andreevskii (St. Petersburg: Brokgauz i Efron, 1890–1907), 34:640. The author was echoing the assessment of German chemist August Wilhelm von Hofmann. Please note: all translations are mine unless otherwise indicated.

19. For a discussion of the role of the kruzhki in the formation of the Russian Chemical Society, see Michael D. Gordin, "The Heidelberg Circle: German Inflections on the Professionalization of Russian Chemistry in the 1860s," *Osiris* 23, no. 1 (2008): 23–49.

20. Dostoevskii parodies the Russian reader of Liebig: "This rival of Liebig, who maybe didn't even finish a course in high school, and who of course wouldn't get into arguing with Liebig about his superiority if someone pointed out that it was Liebig. . . . It would be another thing if, for example, he met Liebig not knowing that it was Liebig, say in a train car. And if they began a conversation about chemistry and our gentleman managed to stick with the conversation, then, no doubt, he would carry out the most learned debate, knowing of

chemistry just one word: 'chemistry.' Liebig would be surprised, of course, but—who knows—our gentleman might be considered the winner in the eyes of a spectator. For there are almost no limits to the brazenness of a Russian's scientific language." Dostoevskii, *Dnevnik pisatelia* (1873), in *Polnoe sobranie sochinenii* (Nauka, 1980), 21:121. Dostoevskii further mocks the reader of Liebig elsewhere.

21. The Slavophile Ivan Aksakov, son of Sergei Aksakov, wrote, "Beyond our native soil there is no basis, outside of the national nothing is real or living, and every thought that is good; any establishment not connected by its roots to our historical native soil, or not grown organically out of it, does not bear fruit and turns to rubbish." I. S. Aksakov, "O despotizme teorii nad zhizn'iu," in *Otchego tak nelegko zhivetsia v Rossii?* (ROSSPEN, 2002), 283.

22. Fedor Dostoevskii, *Polnoe sobranie sochinenii* (Nauka, 1972–1990), 18:36.

23. Apollon Grigor'ev, "Neskol'ko slov o zakonakh i terminakh organicheskoi kritiki," in *Apologiia pochvennichestva* (Institut russkoi tsivilizatsii, 2008), 138.

24. M. A. Antonovich, "Strizham" (Poslanie ober-strizhu, gospodinu Dostoevskomu, 1864), reprinted in *Literaturno-kriticheskie stat'i* (Gos. izd-vo khudozhestvennoi literatury, 1961), 182.

25. V. V. Vinogradov, *Istoriia slov* (Nauka, 1999), 527–28.

26. Vinogradov, *Istorii*, 529.

27. Gordin, "Heidelberg Circle," 29.

28. Nicholas Riasanovskii, *Nicholas I and Official Nationality in Russia, 1825–1855* (University of California Press, 1959), 218.

29. Victoria S. Frede, "Materialism and the Radical Intelligentsia: The 1860s," in *A History of Russian Philosophy, 1830–1930: Faith, Reason, and the Defense of Human Dignity*, ed. G. M. Hamburg and Randall A. Poole (Cambridge University Press, 2010), 71.

30. The Dutch physiologist Jacob Moleschott wrote several works polemicizing with Liebig, including *Der Kreislauf des Lebens: physiologische Antworten auf Liebig's Chemische Briefe*, 1852 (The cycle of life: Physiological replies to Liebig's "Chemical Letters"). For more on Liebig's differences with the young materialists, see Brock, *Justus von Liebig*, 311.

31. Robin Feuer Miller, "Fathers and Children," in *Ivan Turgenev* (Chelsea House Publishers, 2003), 23.

32. Feuer Miller, "Fathers and Children," 23.

33. Ivan Turgenev, *Polnoe sobranie sochinenii i pisem v dvadtsati vos'mi tomakh* (Nauka, 1964), 259–60.

34. Nikolai Chernyshevskii, *Chto delat'? Iz rasskazov o novykh liudiakh* (Nauka, 1975), 123. The English translation is taken from Nikolai Chernyshevsky, *What Is to Be Done?* trans. Michael R. Katz (Cornell University Press, 1989), 180.

35. Irina Paperno addresses the Christian symbolism of this scene in *Chernyshevskii and the Age of Realism: A Study in the Semiotics of Behavior* (Stanford University Press, 1988), 215–17.

36. For more on the rivalry between Lawes and Liebig, see Brock, *Justus von Liebig*, 121–22.

37. Marx and Engels, *Selected Correspondence*, 204.

38. See Joan Martinez Alier, "Marxism, Social Metabolism, and International Trade," in *Rethinking Environmental History: World-System History and Global Environmental Change*, ed. Alf Hornborg, J. R. McNeill, and Joan Martinez Alier (AltaMira Press, 2007), 221–37; John Bellamy Foster, "Liebig, Marx, and the Depletion of Soil Fertility," in *Ecology against Capitalism* (Monthly Review Press, 2002), 154–70; John Bellamy Foster, "Marx's Theory of Metabolic Rift: Classical Foundations for Environmental Sociology," *American Journal of Sociology* 105, no. 2 (1999): 66–405.

39. Justus Liebig, *Familiar Letters on Chemistry in Its Relations to Physiology, Dietetics, Agriculture, Commerce, and Political Economy*, 3rd ed. (Taylor, Walton, and Maberly, 1851), 473.

40. Liebig notes: "The importation of urine or of solid excrements from a foreign land, is quite equivalent to the importation of corn and cattle. All these matters, in a certain time, assume the form of corn, flesh, and bones; they pass into the bodies of men, and again assume the same form which they originally possessed. The only true loss that we experience, and

that we cannot prevent, on account of the habits of our times, is the loss of phosphates, which man carries in his bones to the grave." Liebig, *Chemistry in Its Application to Agriculture and Physiology*, 3rd ed. (Taylor and Walton, 1843), 178.

41. Karl Marx, *Capital: A Critique of Political Economy*, ed. Frederick Engels (International Publishers, 1976), 3:813. Translation adjusted.

42. Foster, "Marx's Theory," 366–405.

43. Friedrich Engels, "The Housing Question," in *On Historical Materialism: A Collection*, ed. T. Borodulina (International Publishers, 1974), 158.

44. D. I. Pisarev, "Ocherki iz istorii truda," in *Istoricheskie eskizy: Izbrannye stat'i* (Izdatel'stvo Pravda, 1989), 105.

45. Cathy Frierson, "Introduction," in *Aleksandr Nikolaevich Engelgardt's* Letters from the Country, *1872–1887*, trans. and ed. Cathy Frierson (Oxford University Press, 1993), 5.

46. A. N. Engel'gardt, *Iz derevni. 12 pisem. 1872–1887*, ed. A. V. Tikhonova (Nauka, 1999), 420–21.

47. See P. M. Luk'janov, ed., *Sechs unbekannte Briefe J. Liebigs an den russischen Chemiker P. A. Il'enkov* (Akademie-Verlag, 1960), 6–7.

48. Cathy Frierson, "Introduction," 6–7.

49. Engel'gardt, *Iz derevni*, 370. Engel'gardt's use of the term "soil particles" is idiosyncratic; he presumably means minerals.

50. Engel'gardt, *Iz derevni*, 371.

51. Engel'gardt, *Iz derevni*, 372.

52. See Vladimir Il'ich Lenin, "Istoriia khoziastva Engel'gardta," in *Polnoe sobranie sochinenii V. I. Lenina* (Izdatel'stvo politicheskoi literatury, 1971), 3:210–13.

53. Lenin directly alludes to Liebig in "Marxist Views on the Agrarian Question in Europe and in Russia," in which he states, "there is no doubt that capitalism has upset the equilibrium between the exploitation of the land and fertilisation [*sic*] of the land (the role of the separation of the town from the countryside)." Vladimir Lenin, "Marxist Views on the Agrarian Question in Europe and in Russia," in *Lenin's Collected Works* (Progress Publishers, 1964), 6:345.

54. V. G. Mineev, "Udobrenie," in *Bol'shaia sovetskaia entsiklopediia*, ed. A. M. Prokhorov (Sovetskaia entsiklopedia, 1970–1981), 26:479.

55. Ellen Stroud, "Does Nature Always Matter? Following Dirt through History," *History and Theory* 42, no. 4 (2003): 75–81. See Andy Bruno (this volume) for further discussion of Stroud's intervention.

56. Quoted in Brock, *Justus von Liebig*, 194.

57. On Dokuchaev's work in the distinctive late nineteenth-century Russian context, see Evtuhov, "Roots," and Evtuhov, *Portrait*, 15–42.

Chapter 9: How a Rock Remade the Soviet North

The research for this paper was made possible by a grant from the National Science Foundation (NSF ARC 0922651), an International Dissertation Research Fellowship from the Social Science Research Council with funds from the Andrew W. Mellon Foundation, and a fellowship from the US Student Fulbright Program.

1. Some works addressing this issue by social scientists are Marshall I. Goldman, *The Spoils of Progress: Environmental Pollution in the Soviet Union* (MIT Press, 1972); Charles E. Ziegler, *Environmental Policy in the USSR* (University of Massachusetts Press, 1987); Philip Pryde, *Environmental Management in the Soviet Union* (Cambridge University Press, 1991); Murray Feshbach and Alfred Friendly, *Ecocide in the USSR: Health and Nature under Siege* (Basic Books, 1992).

2. Douglas R. Weiner, "The Genealogy of the Soviet and Post-Soviet Landscape of Risk," in *Understanding Russian Nature: Representations, Values and Concepts*, ed. Arja Rosenholm and Sari Autio-Sarasmo (Aleksanteri Papers 4/2005), 209.

3. Douglas R. Weiner, "The Predatory Tribute-Taking State: A Framework for Understanding

Russian Environmental History," in *The Environment and World History*, ed. Edmund Burke III and Kenneth Pomeranz (University of California Press, 2009), 277.

4. Weiner, "Predatory," 293–94.

5. Paul R. Josephson, *Would Trotsky Wear a Bluetooth? Technological Utopianism under Socialism* (Johns Hopkins University Press, 2010), 193–99. Also see Paul R. Josephson, "Industrial Deserts: Industry, Science, and the Destruction of Nature in the Soviet Union," *Slavonic and East European Review* 85, no. 2 (April 2007): 294–321. Josephson elsewhere highlights the influence of the fetish of planning, the large bureaucracy, and the constant struggle against enemies in a consideration of Soviet environmental degradation specifically during the Cold War. Paul Josephson, "War on Nature as Part of the Cold War: The Strategic and Ideological Roots of Environmental Degradation in the Soviet Union," in *Environmental Histories of the Cold War*, ed. J. R. McNeill and Corinna R. Unger (Cambridge University Press, 2010), 21–50.

6. Josephson, *Trotsky*, 201.

7. Stephen Brain, "Stalin's Environmentalism," *Russian Review* 69, no. 1 (January 2010): 93; Stephen Brain, *Song of the Forest: Russian Forestry and Stalinist Environmentalism* (University of Pittsburgh Press, 2011).

8. Donald Filtzer, *The Hazards of Urban Life in Late Stalinist Russia: Health, Hygiene, and Living Standards, 1943–1953* (Cambridge University Press, 2010), 105–26.

9. Kate Brown, *Plutopia: Nuclear Families, Atomic Cities, and the Great Soviet and American Plutonium Disasters* (Oxford University Press, 2013); Andy Bruno, *The Nature of Soviet Power: An Arctic Environmental History* (Cambridge University Press, 2016), 170–219.

10. Natalia Mirovitskaya and Marvin S. Soros, "Socialism and the Tragedy of the Commons: Reflections on Environmental Practice in the Soviet Union," *Journal of Environmental Development* 4, no. 1 (Winter 1995): 77–110.

11. Jonathan D. Oldfield, *Russian Nature: Exploring the Environmental Consequences of Societal Change* (Ashgate, 2005).

12. J. R. McNeill, *Something New under the Sun: An Environmental History of the Twentieth-Century World* (W. W. Norton, 2000), 3–17.

13. These population statistics come from "Naselenie SSSR," *Wikipedia*, online at https://ru.wikipedia.org/wiki/Население_СССР/. The figures align with what I originally found using official data of the Federal State Statistics Service of the Russian Federation, online at http://www.gks.ru/, but these statistics have since been altered to match the current borders of the Russian Federation.

14. Data on electricity production come from Philip R. Pryde, *Environmental Management in the Soviet Union* (Cambridge University Press, 1991), 57.

15. I make the case for the importance of the late Soviet period in Bruno, *Nature*, 170–219.

16. For an overview of some of this theory, see Paul Robbins, *Political Ecology: A Critical Introduction* (Blackwell Publishing, 2004). For an essay that gives rocks a voice in Russian history, see Julia Fein, "Talking Rocks in the Irkutsk Museum: Networks of Science in Late Imperial Siberia," *Russian Review* 72, no. 3 (July 2013): 409–26.

17. Ellen Stroud, "Does Nature Always Matter? Following Dirt through History," *History and Theory* 42, no. 4 (December 2003): 75–81.

18. The connection between the cold and climate in Russian political history was made by Montesquieu during the Enlightenment and later echoed by Richard Pipes. See Richard Pipes, *Russia under the Old Regime* (Charles Scribner's Sons, 1974). Another example of a largely deterministic environmental history that does not concern Russia is Jared Diamond, *Guns, Germs, and Steel: The Fate of Human Societies* (W. W. Norton, 1997).

19. Chad Staddon's discussion of trees' complicity in their theft in the Razlog Basin in Bulgaria makes such an attribution. Chad Staddon, "The Complicity of Trees: The Socionatural Field of/for Tree Theft in Bulgaria," *Slavic Review* 68, no. 1 (Spring 2009): 70–94. On actor-network-theory, see Bruno Latour, *Reassembling the Social: An Introduction to Actor-Network-Theory* (Oxford University Press, 2005).

20. A further elaboration of a similar approach appears in Timothy R. LeCain, *The Matter of History: How Things Create the Past* (Cambridge University Press, 2017).

21. J. R. McNeill, *Mosquito Empires: Ecology and War in the Greater Caribbean, 1620–1914* (Cambridge University Press, 2010), 7.

22. Timothy Mitchell, *Rule of Experts: Egypt, Techno-Politics, Modernity* (University of California Press, 2002), 52.

23. B. Vishnevskii, *Kamen' plodorodiia* (Partiinoe izd-vo, 1932).

24. A. V. Barabanov et al., *Gigant v Khibinakh: Istoriia otkrytogo aktsionernogo obshchestva "Apatit" (1929–1999)* (Izdatel'skii dom "Ruda i metally," 1999), 44–66; M. D. Petrova, S. M. Salimova, and T. I. Podgorbunskaia, eds., *Kirovsk v dokumentakh i faktakh, 1920–1945 gg. Khrestomatiia* (Apatit-Media, 2006), 119, 190.

25. Prof. P. Borisov, "Khibinskie nefelinovye sienity i pervoe stelko iz nikh," *Karelo-Murmanksii krai* 4 (1927): 14–16.

26. A. E. Fersman and N. I. Vlodatsev, "Nefelin, ego mestorozhdeniia, zapasy, primenenie, i ekonomika," in *Khibinskie Apatity*, ed. A. E. Fersman (Gosudarstvennoe nauchno-tekhnicheskoe izd-vo Lenkhimsektor, 1931), 3:33–44; A. E. Fersman, ed., *Khibinskie Apatity i nefeliny: Nefelinovoi sbornik*, vol. 4 (Goskhimtekhizdat Leningradskoe otdelenie, 1932).

27. Gosudarstvennyi arkhiv Murmanskoi oblasti (State Archive of Murmansk oblast; hereafter cited as GAMO) f. 773 (individual *fond* of V. I. Kondrikova, 1929–1936), op. 1, d. 1, ll. 100–100ob.; I. L. Tartakovskaia, "Fersman i Kondrikov," *Zhivaia Arktika* 1 (October 2001): 34–37.

28. "Letopis' sobytii goroda Khibinogorska," *Zhivaia Arktika* 1 (October 2001): 31.

29. A. E. Fersman, "Predislovie," in *Khibinskie Apatity i nefeliny*, ed. A. E. Fersman (Goskhimtekhizdat Leningradskoe otdelenie, 1932), 4:1–2; Fersman and Vlodatsev, "Nefelin," 23–44.

30. N. Jarret, "Process Description," and L. K. Hudson, "Alumina Production," in *Production of Aluminium and Alumina*, ed. A. R. Burkin (John Wiley and Sons, 1987), 3–46.

31. Arkhiv Rossiiskoi Akademii Nauk f. 544 (Individual *fond* of academician A. E. Fersman), op. 1, d. 378, l. 9ob.; Theodore Shabad, *Basic Industrial Resources of the USSR* (Columbia University Press, 1969), 58–63, 115; Theodore Shabad, "The Soviet Aluminum Industry: Recent Developments," *Soviet Geography* 24, no. 2 (February 1983): 89–99.

32. Olli Salmi and Aino Toppinen, "Embedding Science in Politics: 'Complex Utilization' and Industrial Ecology as Models of Natural Resource Use," *Journal of Industrial Ecology* 11, no. 3 (2007): 93–111; GAMO f. 773, op. 1, d. 9, ll. 8–10; GAMO f. 773, op. 1, d. 5, l. 174.

33. A. E. Fersman, *Kompleksnoe ispol'zovanie iskopaemogo syr'ia* (Izd-vo Akademii Nauk SSSR, 1932), 1.

34. Fersman, *Kompleksnoe ispol'zovanie*, 11–12.

35. Fersman, *Kompleksnoe ispol'zovanie*, 19.

36. GAMO f. 773, op. 1, d. 1, ll. 107–8, 149–54; GAMO f. 773, op. 1, d. 51, ll. 209–10; Petrova, Salimova, and Podgorbunskaia, *Kirovsk*, 33–34; G. F. Smirnov, "Obogashchenie apatito-nefelinovoi porody Khibinskogo mestorozhdeniia," in *Khibinskie Apatity: Sbornik*, ed. A. E. Fersman (Izdanie Gostresta Apatit, 1930), 1:122–39; N. N. Vorontsov, "Khibinskoe stroitel'stvo," and O. R. Munts, "Gorod Khibinogorsk i ego planirovka," in *Khibinskie Apatity*, ed. A. E. Fersman (VSNKh SSSR Lenzkimsektor, 1932), 2:182–91, 192–207.

37. GAMO f. 773, op. 1, d. 2, ll. 238–44.

38. Rossiiskii gosudarstvennyi arkhiv ekonomiki (Russian State Archive of the Economy; hereafter cited as RGAE) f. 4372 (State planning committee of the Soviet of Ministers of the USSR, 1931, 1952, 1960, 1976–1991), op. 29, d. 832, l.1.

39. GAMO f. 773, op. 1, d. 9, l. 8.

40. GAMO f. 773, op. 1, d. 9, ll. 7–11; GAMO f. 773, op. 1, d. 15, ll. 102–4.

41. GAMO f. 773, op. 1, d. 9, l. 11; Barabanov et al., *Gigant*, 50.

42. GAMO f. 773, op. 1, d. 9, l. 283; M. M. Kossov and B. I. Kagan, "Severnyi gorno-khimicheskii trest 'Apatit' vo 2-m piatiletii," *Karelo-Murmanskii krai* 3–4 (1932): 14–22.

43. On "biting back," see Zsuzsa Gille, *From the Cult of Waste to the Trash Heap of History: The Politics of Waste in Socialist and Postsocialist Hungary* (Indiana University Press, 2007), 11–37.

44. Arkhiv Rossiiskoi Akademii Nauk f. 544, op. 1, d. 161, ll. 40–42.

45. L. D. Zverev, "Rasskaz o zhizni bogatoi," in *Spetspereselentsy v Khibinakh: Spetspereselentsy i zakliuchennye v istorii osvoeniia Khibin (Kniga vospominanii)* (Khibinskoe obshchestvo Memorial, 1997), 16.

46. I expand on this argument in Andy Bruno, "Industrial Life in a Limiting Landscape: An Environmental Interpretation of Stalinist Social Conditions in the Far North," *International Review of Social History* 55, supplement 18 (December 2010): 153–74.

47. GAMO f. 773, op. 1, d. 9, l. 283ob.

48. Arkhiv Rossiiskoi Akademii Nauk f. 544, op. 1, d. 378, l. 9ob.

49. GAMO f. 773, op. 1, d. 52, ll. 61–82, 114–15, 248–51; GAMO f. 773, op. 1, d. 51, ll. 402–9.

50. GAMO f. 773, op. 1, d. 59, ll. 16ob., 31ob.–32ob.

51. GAMO f. 773, op. 1, d. 9, ll. 157–66, 267–74, 309–26, 371; GAMO f. 773, op. 1, d. 59, ll. 39–47, 218–44.

52. GAMO f. 773, op. 1, d. 59, l. 47.

53. GAMO f. 773, op. 1, d. 55, ll. 237–65.

54. A. V. Barabanov and T. A. Kalinina, *"Apatit": vek iz veka* (Laplandia Minerals, 2004), 95; "Kol'skii khronograf," *Zhivaia Arktika* 3–4 (1999): 95; GAMO f. R-990, op. 1, d. 3, l. 19; GAMO f. 773, op. 1, d. 55, l. 242.

55. Shabad, *Basic Industrial Resources*, 59–62.

56. RGAE f. 4372, op. 52, d. 1179, l. 1.

57. It should be noted that although planners such as Fersman knew from the beginning that a substance like limestone was needed to make alumina out of nepheline, they had hoped to use processed apatite ore as a locally available replacement. This substitute ultimately proved ineffective. N. N. Markova, "Pervaia Poliarnaia konferentsiia," *Karelo-Murmanskii krai* 3–4 (1932): 24.

58. Theodore Shabad, "News Notes," *Soviet Geography* 15, no. 6 (June 1974): 386–87; Matthew J. Sagers, "The Soviet Aluminum Industry in the Former USSR in 1992," *Post-Soviet Geography* 33, no. 9 (September 1992): 591–601; Theodore Shabad and Matthew J. Sagers, "The Chlor-Alkali Industries" in *The Chemical Industry in the USSR: An Economic Geography*, ed. Matthew J. Sagers and Theodore Shabad (Westview Press, 1990), 151.

59. Shabad, *Basic Industrial Resources*, 62. By the 1970s this formula had changed slightly. Shabad, "News Notes," *Soviet Geography* 15, no. 6 (June 1974): 386–87.

60. GAMO f. 773, op. 1, d. 9, l. 310.

61. Arkhiv Rossiiskoi Akademii Nauk f. 544, op. 1, d. 378, ll. 9–9ob.; GAMO f. 773, op. 1, d. 9, ll. 53–77, 94–107, 252–74.

62. Gosudarstvennyi arkhiv Rossiiskoi Federatsii (State Archive of the Russian Federation; hereafter cited as GARF) f. A-482 (Ministry of Health of the RSFSR, op. 49, d. 7243, l. 130.

63. GARF f. A-482, op. 49, d. 7243, l. 125; GARF f. A-482, op. 50, d. 6178, ll. 47–52; GARF f. A-482, op. 54, d. 5009, l. 50.

64. GARF f. A-482, op. 54, d. 350, ll. 46–47, 52; GARF f. A-482, op. 50, d. 6178, l. 43.

65. Barabanov et al., *Gigant*, 82–129; R. Caron Cooper and Matthew J. Sagers, "The Fertilizer Complex and Other Agricultural Chemicals" in Sagers and Shabad, *Chemical Industry*, 81–84.

66. Barabanov et al., *Gigant*, 102.

67. Sagers, "The Soviet Aluminum Industry."

68. V. A. Fedoseev and A. V. Istomin, "Ekonomicheskie predposylki osvoeniia novykh mestorozhdenii i sozdaniia pererabatyvaiushchikh proizvodstv na Kol'skom poluostrove," in *Osvoenie mineral'nykh bogatstv Kol'skogo poluostrova*, ed. I. A. Turchaninov (Murmanskoe knizhnoe izd-vo, 1974), 230–31.

69. S. V. Petrov, "Economic Deposits Associated with Alkaline and Ultrabasic Complexes of

the Kola Peninsula," in *Phoscorites and Carbonatites from Mantle to Mine: The Key Example of the Kola Alkaline Province*, ed. F. Wall and A. N. Zaitsev (Mineralogical Society of Great Britain and Ireland, 2004), 486–87.

70. Olga Rigina, "Environmental Impact Assessment of the Mining and Concentration Activities in the Kola Peninsula, Russia by Multidate Remote Sensing," *Environmental Monitoring and Assessment* 75, no. 1 (April 2002): 11–31; T. I. Moisennko et al., "Anthropogenic Transformation of the Arctic Ecosystem of Lake Imandra: Tendencies for Recovery after Long Period of Pollution," *Water Resources* 36, no. 3 (May 2009): 296–309.

71. RGAE f. 4372, op. 67, d. 1961, ll. 81–82.

72. RGAE f. 4372, op. 67, d. 1961, ll. 81–83, 123, 163–66; N. Aleinikov and A. Smirov, "Problemy and mneniia: 'Apatit' i Imandra. Chto nuzhno dlia dobrogo sosedstva," *Poliarnaia pravda* (September 4, 1973), 2; P. A. Markov, "Prirodokhrannaia deiatel'nost' OAO Apatit," in *Kompleksnost' ispol'zovaniia mineral'no-syr'evyx resursov: osnova povysheniia ekologicheskoi besopasnosti regiona*, ed. A. I. Nikolaev (Kol'skii Nauchnyi Tsentr AN, 2005), 17–21.

73. Rigina, "Environmental Impact Assessment," 26, 23–27.

74. Boris Komarov (Ze'ev Wolfson), *The Destruction of Nature in the Soviet Union* (M. E. Sharpe, 1980), 93.

Chapter 10: Encounters with Permafrost

Funding for the research and writing of this chapter was generously granted by a Fulbright-Hays award from the US Department of Education, the Eurasia Program of the SSRC with Title VIII funds provided by the US Department of State, and the Davis Center for Russian and Eurasian Studies at Harvard University. My thinking benefited greatly from exchanges with participants of the Davis Center's "Informing Eurasia" seminar and the lunchtime colloquium of the Rachel Carson Center for Environment and Society, especially Tomasz Blusiewicz, Alan MacEachern, and Robert M. Wilson. I also thank Jonathan Oldfield, Andrew Stuhl, and the reviewers of this volume for their insights and incisive critiques.

1. Definitions of "the North" vary and do not correspond to a specific set of geographic coordinates. The Soviet pioneer of northern studies (*severovedenie*) S. V. Slavin noted that different authors meant different things by "the North." Slavin's own definition was revealing of the Soviet Union's modernizing ambitions: "regions located to the north of historically inhabited territories of the USSR requiring industrial development of their natural resources for the development of the country's economy." See S. V. Slavin, *Promyshlennoe i transportnoe osvoenie severa SSSR* (Izd-vo ekonomicheskoi literatury, 1961), 7. Definitions of "the North" in the North American context were also fluid. See Edward Jones-Imhotep, "Communicating the North: Scientific Practice and Canadian Postwar Identity," *Osiris* 24 (2009): 144–64.

2. M. I. Sumgin and B. N. Demchinskii, *Zavoevanie severa (v oblasti vechnoi merzloty)* (Izd-vo AN SSSR, 1938), 8, 81.

3. Sumgin and Demchinskii, *Zavoevanie*, 44.

4. On the potential negative consequences of permafrost thawing, see Gail Whiteman, Chris Hope, and Peter Wadhams, "Vast Costs of Arctic Change," *Nature* 499 (July 25, 2013): 401–3.

5. For a discussion of variations on the declension narrative in the literature on the environmental history of the Soviet Union, see Andy Bruno (this volume).

6. In this chapter, I use "frozen earth" and "permafrost" interchangeably. For a more detailed examination of how the concept of "permafrost" emerged out of the phenomenon of frozen earth, see Pey-Yi Chu, "Mapping Permafrost Country: Creating an Environmental Object in the Soviet Union, 1920s–1940s," *Environmental History* 20, no. 3 (July 1, 2015): 396–421.

7. On the importance of using the term "adaptation" to refer to a process rather than to specific features, see Richard M. Burian, "Adaptation: Historical Perspectives," in *Keywords in Evolutionary Biology* (Harvard University Press, 1992), 7–12.

8. John McNeill offers a critical view of the "Hegelian position" in environmental history that sees behavior as "fundamentally conditioned by ideas and culture" in J. R. McNeill, "Ob-

servations on the Nature and Culture of Environmental History," *History and Theory* 42 (December 2003): 7–8. For an elaboration on the non-Hegelian approach, see Yi-Fu Tuan, "Discrepancies between Environmental Attitude and Behaviour: Examples from Europe and China," *Canadian Geographer* 12, no. 3 (1968): 176–91.

9. For example, Andy Bruno has treated conquest as a unified set of perceptions and practices that "involved subjecting a natural terrain to purposefully aggressive manipulation. To conquer nature was to destroy any obstacles in the way of making it serve human-dictated purposes." See Andy Bruno, *The Nature of Soviet Power: An Arctic Environmental History* (Cambridge University Press, 2016), 30–31. Elsewhere, however, Bruno problematizes the relationship between the rhetoric of conquest and actual practice. See Andy Bruno, "Tumbling Snow: Vulnerability to Avalanches in the Soviet North," *Environmental History* 18, no. 4 (October 2013): 683–709. Paul Josephson has argued that "the path of Soviet Arctic conquest . . . was also exceedingly costly from the environmental and social points of view" and that Soviet actors "accelerated environmental degradation whenever they pursued conquest." See Paul R. Josephson, *The Conquest of the Russian Arctic* (Harvard University Press, 2014), 2, 11.

10. Alla Bolotova has also strived to complicate the rhetoric of conquest by showing how this "hegemonic discourse" did not match the perceptions of people who participated in industrialization. See Alla Bolotova, "Colonization of Nature in the Soviet Union: State Ideology, Public Discourse, and the Experience of Geologists," *Historical Social Research* 29, no. 3 (2004): 104–23; Alla Bolotova, "Loving and Conquering Nature: Shifting Perceptions of the Environment in the Industrialised Russian North," *Europe-Asia Studies* 64, no. 4 (June 2012): 645–71. Whereas Bolotova focuses on the diversity of views that existed alongside the rhetoric of conquest, I explore the political and social motivations that scientists had for using that rhetoric.

11. See, for example, Bruno, *Nature*; Stephen Brain, *Song of the Forest: Russian Forestry and Stalinist Environmentalism, 1905–1953* (University of Pittsburgh Press, 2011); Jenny Leigh Smith, *Works in Progress: Plans and Realities on Soviet Farms, 1930–1963* (Yale University Press, 2014).

12. V. A. Obruchev and A. M. Chekotillo, "M. I. Sumgin," *Izvestiia Akademii Nauk SSSR, seriia geologicheskaia* 2 (1943): 98; N. A. Vel'mina, *K tainam vechnoi merzloty: kniga ob osnovopolozhnike merzlotovedeniia Mikhaile Ivanoviche Sumgine* (Institut merzlotovedeniia SO RAN, 1994), 5.

13. David Blackbourn, *The Conquest of Nature: Water, Landscape, and the Making of Modern Germany* (Norton, 2006), 5–6. See also Clarence Glacken, *Traces on the Rhodian Shore: Nature and Culture in Western Thought from Ancient Times to the End of the Eighteenth Century* (University of California Press, 1976), 461–62.

14. Katerina Clark, *The Soviet Novel: History as Ritual* (University of Chicago Press, 1981), 100–106; John McCannon, "To Storm the Arctic: Soviet Polar Expeditions and Public Visions of Nature in the USSR, 1932–1939," *Ecumene* 2, no. 1 (January 1995): 15–31; John McCannon, *Red Arctic: Polar Exploration and the Myth of the North in the Soviet Union, 1932–1939* (Oxford University Press, 1998), 83–87; Emma Widdis, *Visions of a New Land: Soviet Film from Revolution to the Second World War* (Yale University Press, 2003), 146–55; Bolotova, "Colonization of Nature," 109–11; William B. Husband, "'Correcting Nature's Mistakes': Transforming the Environment and Soviet Children's Literature, 1928–1941," *Environmental History* 11, no. 2 (April 2006): 305–7, 312–13; Bruno, *Nature*, 69.

15. Widdis, *Visions*, 7; Emma Widdis, "Borders: The Aesthetic of Conquest in Soviet Cinema of the 1930s," *Journal of European Studies* (2000): 404–15. On the relationship between *osvoenie* and *zavoevanie*, I agree with Widdis's view that the two concepts became rhetorically conflated in the Soviet era. I agree less with Andy Bruno's view that the two concepts were rhetorically distinct and corresponded to different modes of interacting with the environment. Permafrost scientists such as Sumgin used the terms *osvoenie* and *zavoevanie* simultaneously. See Bruno, *Nature*, 30–32.

16. Clark, *Soviet Novel*, 98, 101–3; McCannon, "To Storm the Arctic," 19–20.

17. Rossiiskii gosudarstvennyi arkhiv ekonomiki (Russian State Archive of the Economy; hereafter RGAE) f. 662, op. 1, d. 8, ll. 17–20; Mordovskii respublikanskii ob"edinennyi kraevedcheskii muzei (Unified Local Studies Museum of the Mordvin Republic; hereafter cited as MROKM) f. R-5, osnovnoi fond (hereafter cited as "of.") 4961/49, l. 2; Vel'mina, *K tainam*, 37–39.

18. At 2,000 versty (roughly 2,134 kilometers), the Amur section of the Trans-Siberian Railroad extended from the town of Sretensk in the Trans-Baikal region to Khabarovsk on the Amur River. A. V. Khobta, *Stroitel'stvo Transsiba: ocherki istorii (konets xx–nachalo xx vv.)* (OOO NPF Zemlia Irkutskaia, 2009), 11, 16.

19. M. S., "Geograficheskoe rasprostranenie 'vechnoi merzloty' v Amurskoi oblasti s kartoi," *Izvestiia Meteorologicheskogo biuro Amurskogo raiona* 2 (1914), 5, 28–30; M. I. Sumgin, *Vechnaia merzlota pochvy v predelakh SSSR* (N.K.Z. Dal'ne-vostochnaia Geofizicheskaia Observatoriia, 1927), 1, 152–53, 252, 296–97, 314; Nikolay Shiklomanov, "From Exploration to Systematic Investigation: Development of Geocryology in 19th- and Early-20th-Century Russia," *Physical Geography* 26, no. 4 (January 1, 2005): 261.

20. Kendall Bailes, *Technology and Society under Lenin and Stalin: Origins of the Soviet Technical Intelligentsia, 1917–1941* (Princeton University Press, 1978), 44–49; Loren Graham, "The Formation of Soviet Research Institutes: A Combination of Revolutionary Innovation and International Borrowing," *Social Studies of Science* 5 (1975): 309–13; Alexei Kojevnikov, "The Phenomenon of Soviet Science," *Osiris* 23 (2008): 117–22.

21. M. I. Sumgin, "Sovremennoe polozhenie vechnoi merzloty v SSSR i zhelatel'naia postanovka etikh issledovanii v blizhaishem budushchem," in *Vechnaia merzlota: sbornik* (Izd-vo Akademii Nauk, 1930), 40–41.

22. Sumgin exercised de facto leadership of KIVM, but nominally its chairman was the geologist Vladimir Obruchev. Arkhiv Rossiiskoi Akademii Nauk (Archive of the Russian Academy of Sciences; hereafter cited as ARAN) f. 268, op. 3, d. 1, ll. 14, 22, 24, 78; ARAN f. 642, op. 3, d. 134, ll. 32–33; Obruchev and Chekotillo, "M. I. Sumgin," 99.

23. ARAN f. 268, op. 1, d. 3, ll. 3–5, and d. 48, l. 7; ARAN f. 268, op. 3, d. 4, ll. 16, 18, and d. 53, ll. 11, 77, 79–80, 82–87.

24. ARAN f. 268, op. 3, d. 3, l. 15.

25. Gosudarstvennyi arkhiv Rossiiskoi Federatsii (State Archive of the Russian Federation) f. R5446, op. 23, d. 1652, l. 1; RGAE f. 82, op. 1, d. 13, l. 24; L. A. Meister and N. I. Saltykov, *K istorii geokriologicheskikh issledovanii v SSSR* (Komi knizhnoe izd-vo, 1958), 9–10.

26. KIVM's laboratory was located on the premises of the Institute of Geology (Institut geologicheskikh nauk), whose director, Andrei Arkhangel'skii, attempted to absorb the commission and claim its laboratory in 1935. MROKM f. R-5, of. 4961/50, l. 3; MROKM f. R-5, of. 4961/35, l. 24; Vel'mina, *K tainam*, 87, 96.

27. RGAE f. 82, op. 1, d. 3, ll. 41–48.

28. V. Elenevskii, "Problemy stroitel'stva v raionakh merzloty," *Sotsialisticheskii transport* 8–9 (1937): 90.

29. Karl Schlögel, *Moscow, 1937*, trans. Rodney Livingstone (Polity Press, 2012), 267. Sumgin joined the Socialist Revolutionary Party in 1902 and served briefly on its Central Committee in 1918. RGAE f. 622, op. 1, d. 8, ll. 13, 26.

30. RGAE f. 82, op. 1, d. 3, l. 46.

31. Sumgin and Demchinskii, *Zavoevanie*, 34.

32. Sumgin and Demchinskii, *Zavoevanie*, 21, 73–74, 81.

33. Sumgin and Demchinskii, *Zavoevanie*, 6. The agency of permafrost as suggested by Soviet scientists' use of the rhetoric of conquest hewed more closely to the vitalism of Romantic aesthetics than to the sociological idea of objects as actors articulated in Bruno Latour's actor-network-theory. The rhetoric of conquest animated humans and nonhuman nature in epic narratives of conflict. By contrast, actor-network-theory embeds human and nonhuman actors together in networks of mutual influence. See Andy Bruno (this volume) for an example of granting agency to nonhuman actors in a Latourian sense.

34. Sumgin and Demchinskii, *Zavoevanie*, 58, 62, 92, 94, 100, 118.

35. MROKM f. R-5, of. 3355/58, l. 1.

36. Sumgin, *Vechnaia merzlota*, 20.

37. Sumgin, *Vechnaia merzlota*, 218.

38. Sumgin, *Vechnaia merzlota*, 9–10. Sumgin's mentioning of the "time of the mammoths" was meant to indicate that he considered *vechnaia merzlota* to have originated during the last glacial period, which ended roughly twelve thousand years ago. See Sumgin, *Vechnaia merzlota*, 242. For more on how Sumgin arrived at his definition of *vechnaia merzlota*, see Pey-Yi Chu, "Mapping Permafrost," 399–402.

39. ARAN f. 258, op. 3, d. 11, ll. 43–45; M. I. Sumgin, *Vechnaia merzlota pochvy v predelakh SSSR* (Izd-vo AN SSSR, 1937), 30; M. I. Sumgin et al., *Obshchee merzlotovedenie* (Izd-vo AN SSSR, 1940), 14–15.

40. The definition of "permafrost" used today also accords with Sumgin's—that is, "ground (soil or rock and included ice and organic material) that remains at or below 0°C for at least two consecutive years." See Robert O. van Everdingen, ed., *Multi-language Glossary of Permafrost and Related Ground-Ice Terms* (International Permafrost Association, 1998, 2005), 63; US Geological Survey, *Permafrost or Permanently Frozen Ground and Related Engineering Problems*, Strategic Engineering Study (Intelligence Branch Office, Chief of Engineers, 1943), 62:2–3; Victor V. Kabakchi and Ronald R. Butters, "Are Permafrost and Vernalization Loan Translations from Russian?" *American Speech* 64, no. 3 (Fall 1989): 287–88.

41. According to his colleagues, Sumgin was both a dedicated and a gifted science communicator. MROKM f. R-5, of. 3355/143, l. 2; MROKM f. R-5, of. 3355/147, l. 2; MROKM f. R-5, of. 4961/53, l. 21. The first use of *vechnaia merzlota* in print was by geographer Petr Kropotkin; the term was one of several used in the nineteenth century to refer to perennially frozen earth. See P. Kropotkin, "Otchet ob Olekminsko-Vitimskoi ekspeditsii," *Zapiski Imperatorskogo russkogo geograficheskogo obschchestva po obshchei geografii* 3 (1873): 240.

42. Sumgin and Demchinskii, *Zavoevanie*, 45. For Sumgin's defense of the term *vechnaia*, see Sumgin, *Vechnaia merzlota*, 13. The critical characterization of *vechnaia* was articulated by the soil scientist Robert Abolin. See R. I. Abolin, "Postoiannaia merzlota gruntov i iskopaemyi kamennyi led," *Zapiski Chitinskogo otdeleniia Priamurskogo otdela Imperatorskogo russkogo geograficheskogo obshchestva* 9 (1913): 50.

43. On frozen earth as a product of the contemporary climate, see A. V. Voznesenskii and V. B. Shostakovich, *Osnovnyia dannyia dlia izucheniia klimata Vostochnoi Sibiri s atlasom* (Tipo-litografiia P. I. Makushina i V. M. Posokhina, 1913), 142. Even the "time of the mammoths," other scientists argued, could hardly be considered "hoary antiquity," as Sumgin put it, but was merely the "contemporary (Quaternary) geological period." Sumgin's temporal framework was distinctly anthropocentric. See Sumgin, *Vechnaia merzlota*, 213, 220; ARAN f. 594, op. 1, d. 84, l. 27.

44. Stephen Hanson, *Time and Revolution: Marxism and the Design of Soviet Institutions* (University of North Carolina Press, 1997), 131–32.

45. Hanson, *Time and Revolution*, 13, 136, 149–53.

46. Sumgin and Demchinskii, *Zavoevanie*, 12. On Aleksandr Fersman and the "complex utilization" of resources, see Andy Bruno (this volume).

47. Sumgin and Demchinskii, *Zavoevanie*, 14–16.

48. V. V. Elenevskii and G. A. Nizovkin, *Zheleznodorozhnoe stroitel'stvo v usloviiakh merzloty: novye metody stroitel'stva v raionakh merzloty i glubokogo promerzaniia* (Gosudarstvennoe transportnoe zheleznodorozhnoe izd-vo, Transzheldorizdat, 1936), 122; "Protokol soveshchaniia, sozvannogo Komissiei po izucheniiu vechnoi merzloty po voprosu ustroistva avtoguzhevoi dorogi Dudnika-Noril'sk," *Trudy Komissii po izucheniiu vechnoi merzloty* 4 (1935): 146–50.

49. Elenevskii and Nizovkin, *Zheleznodorozhnoe*, 36.

50. Sumgin, *Vechnaia merzlota*, 283, 346.

51. P. N. Kapterev, *Po taige: putevye ocherki* (Moskovskii rabochii, 1948), 22.

52. N. I. Saltykov, "O fundamentakh zdanii g. Iakutska," *Trudy Instituta merzlotovedeniia im. V. A. Obrucheva* 1 (1946): 109, 120.

53. It could reach a thickness of between 1.5 and 1.75 meters. Saltykov, "O fundamentakh," 102; RGAE f. 82, op. 1, d. 13, l. 110.

54. Saltykov, "O fundamentakh," 103–4; RGAE f. 82, op. 1, d. 13, l. 110.

55. Saltykov, "O fundamentakh," 112–18.

56. Saltykov, "O fundamentakh," 110, 120.

57. RGAE f. 82, op. 1, d. 48, l. 8.

58. Saltykov, "O fundamentakh," 120–22.

59. "In Capitals of Autonomous Republics: Yakutsk," *Current Digest of the Soviet Press* 5, no. 23 (July 18, 1953), 44.

60. M. I. Sumgin, N. N. Geniev, and A. M. Chekotillo, eds., *Vodosnabzhenie zheleznykh dorog v raionakh vechnoi merzloty* (Gosudarsvennoe transportnoe zheleznodorozhnoe izd-vo, Transzheldorizdat, 1939), 6.

61. T. A. Kolpakova, "Epidemiologicheskie osobennosti Iakutskogo kraia," in *Kratkie otchety o rabotakh otriadov Iakutskoi ekspeditsii Akademii nauk SSSR, 1925–1926 gg.*, vol. 10, *Materialy Komissii po izucheniiu Iakutskoi Avtonomnoi Sovetskoi Sotsialisticheskoi Respubliki*, ed. V. Komarov (Izd-vo AN SSSR, 1929), 113–14.

62. ARAN f. 268, op. 1, d. 15, l. 50.

63. "Rezoliutsiia III soveshchaniia po izucheniiu vechnoi merzloty, sozvannogo KIVM Akademii Nauk SSSR, 10–16 ianvariia 1933 g.," *Trudy komissii po izucheniiu vechnoi merzloty* 3 (May 1934), 155.

64. I. M. Svetozarov, "Gidrogeologiia vechnoi merzloty po materialam issledovanii v raione Iakutska," *Problemy Sovetskoi geologii* 4, no. 10 (1934): 119–31; V. M. Maximov and N. I. Tolstikhin, "On Hydrogeological Conditions in the Vicinity of the Town of Yakutsk," *Comptes rendus (doklady) de l'Académie des Sciences de l'URSS* 28, no. 1 (1940): 93–96.

65. Svetozarov, "Gidrogeologiia," 125.

66. Rossiiskii gosudarstvennyi arkhiv nauchno-tekhnicheskoi dokumentasii (Russian State Archive of Scientific-Technical Documentation) f. R-571, op. 2–4, d. 3133, ll. 62, 62ob.; N. P. Anisimova, S. M. Fotiev, and V. V. Shepelev, "Otkrytie Iakutskogo artezianskogo basseina," *Kriosfera zemli* 2, no. 4 (December 1998): 19.

67. Maximov and Tolstikhin, "On Hydrogeological Conditions," 94–96; RGAE f. 82, op. 1, d. 13, l. 11.

68. A. I. Efimov and I. L. Iantsevich, "Predvaritel'noe soobshchenie o rezul'tatakh bureniia gidrogeologicheskoi skvazhiny v gorode Iakutska" (Biblioteka Instituta merzlotovedeniia im. Mel'nikova, 1941), 3.

69. Efimov and Iantsevich, "Predvaritel'noe," 33.

70. RGAE f. 82, op. 1, d. 13, ll. 49, 53ob.; Anisimova, Fotiev, and Shepelev, "Otkrytie," 21.

71. RGAE f. 662, op. 1, d. 6, l. 3.

72. ARAN f. 268, op. 1, d. 309, l. 19ob.; Anisimova, Fotiev, and Shepelev, "Otkrytie," 21; A. I. Efimov, "Iakutskii artezianskii bassein podmerzlotnykh vod," *Izvestiia Akademii Nauk SSSR, seriia geologicheskaia* 4 (1945): 133.

73. ARAN f. 268, op. 1, d. 309, l. 16–17.

74. RGAE f. 82, op. 2, d. 430, ll. 62–63.

75. RGAE f. 82, op. 2, d. 430, ll. 62–63; R. S. Kononova, "Istoriia gidrogeologicheskikh issledovanii po vodosnabzheniiu," in *Podzemnye vody Iakutii kak istochnik vodosnabzheniia* (Izd-vo Nauka, 1967), 6.

76. Sumgin and Demchinskii, *Zavoevaniae*, 6.

77. Elenevskii and Nizovkin, *Zheleznodorozhnoe*, 50.

78. M. Ia. Chernyshev, *Vodosnabzhenie v vechnoi merzlote* (Vsesoiuznyi nauchno-issledovatel'skii institut vodosnabzheniia i sanitarnoi tekhniki, 1933), 29.

79. Such instances are mentioned in Elenevskii and Nizovkin, *Zheleznodorozhnoe*, 198–99;

P. Strizhkov, *Aldanskie Priiski: Ocherki* (Molodaia gvardiia, 1931), 27–28; RGAE f. 82, op. 1, d. 1, l. 16; RGAE f. 82, op. 2, d. 1175, l. 6.

80. The purpose of the road was to improve access to gold mines in the Aldan River valley. The initial impulse was to build a road that was "direct and cheap." See I. F. Molodykh, "Puti soobshcheniia Iakutii," in *Iakutiia: sbornik statei*, ed. P. V. Vittenburg (Izd-vo AN SSSR, 1927), 655, 646; N. N. Zakharenko, *Aldan: ekonomicheskii ocherk* (Izdanie Iakutizdata, 1927), 19.

81. V. G. Petrov, "K voprosu o zashchite dorozhnykh sooruzhenii ot vrednogo vlianiia naledei," *Sovetskaia Aziia* 3–4 (1930): 72.

82. Petrov, "K voprosu," 72; Elenevskii and Nizovkin, *Zheleznodorozhnoe*, 200; RGAE f. 82, op. 1, d. 1, l. 17.

83. Petrov, "K voprosu," 73–74.

84. RGAE f. 82, op. 1, d. 1, l. 17; Elenevskii and Nizkovkin, *Zheleznodorozhnoe*, 200.

85. Elenevskii and Nizovkin, *Zheleznodorozhnoe*, 200–201.

86. RGAE f. 82, op. 2, d. 1175, l. 3.

87. See Timothy Mitchell, *Rule of Experts: Egypt, Techno-Politics, Modernity* (University of California Press, 2002); Kathryn Taylor Morse, *The Nature of Gold: An Environmental History of the Klondike Gold Rush* (University of Washington Press, 2003); Tyler Priest, "Extraction Not Creation: The History of Offshore Petroleum in the Gulf of Mexico," *Enterprise and Society* 8, no. 2 (2007): 227–67; Liza Piper, *The Industrial Transformation of Subarctic Canada* (UBC Press, 2009).

88. Notably, US scientists and engineers operating in Alaska sought to learn from Soviet practices. See Siemon William Muller, *Frozen in Time: Permafrost and Engineering Problems*, ed. Hugh M. French and Frederick E. Nelson (American Society of Civil Engineers, 2008), xxvii.

89. Alec Luhn, "Slow-Motion Wrecks: How Thawing Permafrost Is Destroying Arctic Cities," *The Guardian*, October 14, 2016, online at https://www.theguardian.com/cities/2016/oct/14/thawing-permafrost-destroying-arctic-cities-norilsk-russia/.

Chapter 11: The Christian Environmental Ethic of the Russian Pomor

This essay is a revised version of an article that appeared in the journal *Ecozon@* in 2011. See Stephen Brain, "The Christian Environmental Ethic of the Russian Pomor," *Ecozon@* 2, no. 2 (Fall 2011): 60–82.

1. P. S. Efimenko, *Materialy po etnografii russkogo naseleniia Arkhangel'skoi gubernii* (Tip. F. V. Miller, 1878), 69.

2. N. Vekhov, "Starinnye pomorskie promysly," *Okhota i okhotnich'e khoziaistvo* 6 (June 2000): 9. A note on terminology: Throughout this volume, we use "fishermen" rather than the gender neutral term "fisher" because, during the periods discussed, fishing was almost exclusively male work.

3. Because of the cultural commonalities linking the Pomor to Great Russians, most anthropologists classify the Pomor as a sub-ethnicity of the Russian people rather than as a full-fledged ethnicity or a mere tribe. See T. B. Bernshtam, *Pomory: formirovaniie gruppy i sistema khoziaistva* (Nauka, 1978); T. B. Bernshtam, *Russkaia narodnaia kul'tura Pomor'ia v XIX–nachale XX vv.* (Nauka, 1983).

4. The Russian historian Sergei Platonov first established that serfdom, the central institution of Russian society in the medieval and imperial periods, never played a part in the villages of the Russian Far North. See Platonov, *Proshloe russkogo severa* (Vremia, 1923).

5. V. V. Anufriev, "Sud'ba pomorov," *Barents zhurnal* 2 (2002): 50.

6. For thorough discussions of the Pomor dialect, see A. A. Krysanov et al., eds., *Ne vek zhit'—vek vspominat'* (Izd-vo Folium, 2006), 366–85; Kseniia Gemp, *Skaz o Belomor'e: slovar' pomorskikh rechenii* (Nauka, 2004).

7. The Russian word for this syncretism of paganism and Christianity is *dvoeverie* or "dual belief." See Stella Rock, "What's in a Word? A Historical Study of the Concept *Dvoeverie*," *Canadian-American Slavic Studies* 35, no. 1 (2001): 19–28; Eve Levin, "*Dvoeverie* and Popular

Religion," in *Seeking God: The Recovery of Religious Identity in Orthodox Russia, Ukraine and Georgia*, ed. S. K. Batalden (Northern Illinois University Press, 1993).

8. Although, strictly speaking, it is an anachronism to use the words "environmentalist" or "ecological" to describe people so far removed from the environmental movement of the twentieth century or the ecological studies that emerged in the nineteenth century, the Pomors' propensity to see themselves as part of a larger biological whole (as will be discussed presently) justifies the use of the terms.

9. Lynn White, "The Historical Roots of Our Ecologic Crisis," *Science* 155, no. 3767 (March 10, 1967): 1203–7.

10. Catherine Keller suggests that the continuing relevance of White's article can be explained by the fact that Christians felt sincerely stung by its observations; she talks about the debate that Christians, "either defensively or guiltily, cannot let go of." Catherine Keller, "The Lost Fragrance," in *Visions of a New Earth*, ed. Harold Coward and Daniel C Maguire (State University of New York Press, 2000), 82.

11. Baird J. Callicott, "Genesis Revisited: Murian Musings on the Lynn White, Jr. Debate," *Environmental History Review* 14, no. 1/2 (Spring–Summer 1990): 65.

12. White, "Historical Roots," 1206.

13. As Roderick Nash notes, White could have provided supporting arguments that he did not mention: the traditional Christian view of the wilderness as a cursed place or the Christian focus on the relative unimportance of the ephemeral, terrestrial world when compared with the hereafter. See Nash, *The Rights of Nature* (University of Wisconsin Press, 1989), 91.

14. An online bibliography of the "Lynn White Controversy," as the website's author calls it, demonstrating the number of publications challenging White's ideas, can be found online at http://ecoethics.net/bib/1997/.

15. René Dubos, *A God Within* (Charles Scribner's Sons, 1972), 158–61. Clarence Glacken also explores this theme in *Traces on the Rhodian Shore* (University of California Press, 1967).

16. Robin Attfield, *The Ethics of Environmental Concern* (University of Georgia Press, 1991), 23–24. Other authors who take White to task for a lack of historical nuance include Lewis Montcrief, "The Cultural Basis of Our Ecological Crisis," *Science* 170 (1970): 508–12; Manussos Marangudakis, "The Medieval Roots of Our Ecological Crisis," *Environmental Ethics* 23 (2001): 243–60; Elspeth Whitney, "Lynn White, Ecotheology, and History," *Environmental Ethics* 15 (1993): 151–69; John Passmore, *Man's Responsibility for Nature* (Druckworth, 1974); Keith Thomas, *Man and the Natural World: A History of the Modern Sensibility* (Pantheon Books, 1983).

17. Scholars who make such arguments include David Livingston, "The Historical Roots of Our Ecological Crisis: A Reassessment," *Fides et Historia* 26 (Winter–Spring 1994): 35–55; H. Paul Santmire, *The Travail of Nature* (Fortress Press, 1985).

18. Jayapaul Azariah and Darryl Macer, "Historical Roots of Our Ecological Crisis in East and West," *Eubios Journal of Asian and International Bioethics* 6 (1996): 125. Jeanne Kay adopts a similar approach; see Jeanne Kay, "Concepts of Nature in the Hebrew Bible," in *Judaism and Environmental Ethics: A Reader*, ed. Martin D. Yaffe (Lexington Books, 2001), 86–104.

19. The collected volume *Chrisitianity and Ecology*, ed. Dieter T. Hessel and Rosemary Radford Ruether (Harvard University Press, 2000), and the English journal *Ecotheology* are sources for the extensive scholarship of the stewardship school. See also Mark Bredin, "God the Carer: Revelation and the Environment," *Biblical Theology Bulletin* 38 (2008): 76–87; David G. Horrell, "A New Perspective on Paul? Rereading Paul in a Time of Ecological Crisis," *Journal for the Study of the New Testament* 33, no. 1 (2010): 4–31; Ben A. Minteer and Robert E. Manning, "An Appraisal of the Critique of Anthropocentrism and Three Lesser Known Themes in Lynn White's 'The Historical Roots of Our Ecological Crisis," *Organization and Environment* 18 (2005): 163–76. See also Rachel Havrelock, *River Jordan* (University of Chicago Press, 2011).

20. Heather Barnes Truelove and Jeff Joireman, "Understanding the Relationship between Christian Orthodoxy and Environmentalism: The Mediating Role of Perceived Environmental Consequences," *Environment and Behavior* 41 (2009): 806–21.

21. It is worth mentioning that the debate spurred by White has centered on Western Christianity rather than Eastern Orthodoxy, whose connections with environmentalism have scarcely received any scholarly attention at all.

22. See Glacken, *Rhodian Shore*, chapter 7. Other works that discuss the profound impact of medieval monasteries upon the landscape include Roland Bechmann, *Trees and Man: The Forest in the Middle Ages* (Paragon House, 1990); K. J. Kirby and C. Watkins, eds., *The Ecological History of European Forests* (CAB International, 1998); Ellen Arnold, "Engineering Miracles: Water Control, Conversion and the Creation of a Religious Landscape in the Medieval Ardennes," *Environment and History* 13 (2007): 477–502; Richard C. Hoffman, "Medieval Christendom in God's Creation: Environmental Continuities, Coevolutions, and Changes," in *Northern Europe: An Environmental History*, ed. Tamara White et al. (ABC-Clio, 2005), 45–72.

23. One might argue that the Amish pursue an environmentally sustainable lifestyle based upon religious precepts, but the relatively low environmental impact of their economic choices is epiphenomenal.

24. Roderick Nash, *The Rights of Nature* (University of Wisconsin Press, 1989), 89.

25. Although the Pomor specifically rejected pagan beliefs, deities, and territories as unclean, Terebikhin claims that the Pomor veneration of St. Nicholas as the god of the sea and the patron saint of most Pomor churches essentially continued an earlier tradition of recognizing Volos as the god of the old Russian North. See Nikolai Terebikhin, *Sakral'naia geografiia Russkogo Severa* (Izd-vo pomorskogo mezhdunarodnogo pedagogicheskogo universiteta imeni M. V. Lomonosova, 1993), chapter 1.

26. D. N. Zamiatin, "Geokul'turnoe prostranstvo i mifogeografiia evropeiskogo severa Rossii," *Pomorskie chteniia po semiotike kul'tury* (Pomorskii Universitet, 2006), 24.

27. V. V. Anufriev, *Russkie pomory: kul'turno-istoricheskaia identichnost', monografi͡ia* (Solti, 2008), 62–80.

28. Terebikhin, *Sakral'naia geografiia*, 145–83.

29. Terebikhin, *Sakral'naia geografiia*, 11–28.

30. Efimenko, *Materialy*, 186.

31. Although the names for these winds originated with the Pomor, they eventually found wider acceptance, especially in Siberia, where the names sometimes no longer made literal sense. For instance, the Pomor word for a southwesterly wind, the *shelonik*, was adopted by settlers elsewhere in Russia, despite the fact that it was named for the River Shelona, located to the southwest of the White Sea. See Valentin Rasputin, *Siberia, Siberia* (Northwestern University Press, 1996), 340.

32. Mikhail Prishvin, *Za volshebnym kolobkom* (Moskovskii rabochii, 1964), 179.

33. Valerii Lisnichenko, *Ekologiia pomora* (Pravda severa, 2007), 33.

34. Aleksandr Podvysotskii, *Slovar' oblastnogo arkhangel'skogo narechiia v ego bytovym i etnograficheskoi primenenii* (Tip. Imperatorskoi Akademii Nauk, 1885), 24.

35. B. A. Uspenskii, *Filologicheskie razyskaniia v oblasti slavianskikh drevnostei* (Izd-vo MGU, 1982), 36.

36. F. Selivanov, *Narodno-khristianskaia poeziia* (Sovietskaia Rossia, 1991), 20.

37. The identification of the easterly direction as the direction of Christ is not peculiar to the Pomor. Most Christian churches face east. The idea that the east is the direction of life, however, took on greater resonance for the Pomor, precisely because the sea that sustained them lay to the east. For the Pomor, to die at sea was not to perish but to rejoin nature and hence gain everlasting life. The Pomor as a rule did not say that someone died at sea but, rather, that "the sea took someone." See Terebikhin, *Sakral'naia geografiia*, 14–29.

38. A. K. Baiburin, *Kalendar' i trudovaia deiatel'nost' cheloveka* (Obshchestvo Znanie, 1989), 8.

39. It should not be thought that the conservationist attitudes of the Pomor were held by all inhabitants of the White Sea region; by no means did more recent immigrants to the area observe the same restrictions. Upon arriving at his new post as governor of Arkhangel'sk province, S. D. Bidikov toured the region and compiled his impressions into a published report. Whereas

the non-Pomor inhabitants of the Murmansk coast "caught as much as they could, paying no attention to the fact that their predatory actions were destroying their future prospects," the Pomor "halted their catch long before the fish had been exhausted." See S. D. Bidikov, *Arkhangel'skaia guberniia: eia bogatstva i nuzhdy* (Gubernskaia Tip., 1912), 51, 59.

40. "III-y Vserossiiskii s"ezd rybopromyshlennikov," *Izvestiia Arkhangel'skogo obshchestva izucheniia Russkogo severa* (hereafter cited as *Izvestiia AOIRS*) 4 (1910): 34.

41. I. Anufriev, "O granitsakh murmanskikh promyslovykh vod," *Severnoe khoziaistvo* 5 (1923): 48. Technically speaking, Russia had treaties in place to protect domestic fishing interests, but the tsarist government chose not to enforce these measures for fear of alienating Russia's allies.

42. A. A. Kamenev, "U pomorov sumskogo posada," *Izvestiia AOIRS* 17 (1910): 15.

43. A. A. Kamenev, "Iz zhizni pomor'ia," *Izvestiia AOIRS* 7 (1910): 18.

44. N. Zograf, "Nuzhdy nashego severnago rybolobstva," *Izvestiia AOIRS* 2 (1909): 23.

45. "Zhurnal obshchevo sobraniia chlenov Arkhangel'skogo obshchestva izucheniia Russkogo severa," *Izvestiia AOIRS*, addendum to no. 3 (1910): 3. Purse nets are large circular nets that are stretched out beneath one boat by another and then are gathered up toward the first boat like a purse, capturing all the fish in the water beneath. *Kil'not* nets, according to S. A. Mynikov, were "small standing nets designed for the catching of salmon near the coast." See S. A. Myznikov, *Russkie govory obonezh'ia* (Nauka, 2003), 399.

46. "Protokoly sektsii po rybmym promyslam," *Izvestiia AOIRS*, addendum to no. 3 (1910): 5.

47. "Pomorskii rybopromyshlennyi s"ezd," *Izvestiia AOIRS* 5 (1914): 129.

48. "Pomorskii rybopromyshlennyi s"ezd," 168.

49. V. V. Suslov, *Putevye zametki o severe Rossii i Norvegii* (Tip. A. F. Marksa, 1888), 25.

50. P. K., "Netronutye bogatstva," *Severnoe khoziaistvo* 1 (1924): 71–79.

51. A. L., "Kooperatsiia zveroboev," *Volna*, April 6, 1924, 3; E. Strogova, "Urozhai moria," *Volna*, August 24, 1926, 2.

52. The earliest rules applied by the Arkhangel'sk soviet allowed owners of sailboats and rowboats to keep 10 percent of their catch, while motorboat owners could keep only 5 percent. Gosudarstvennyi arkhiv Arkhangel'skoi oblasti (State Archive of Arkhangel'sk oblast; hereafter cited as GAAO) f. 150, op. 3, d. 10, l. 20.

53. A. V. Okorokov, "Promyslovaia deiatel'nost'," in *Kultura russkikh pomorov*, ed. E. L. Bazarov et al. (Nauchnyi mir, 2005).

54. Gosudarstvennyi arkhiv Arkhangel'skoi oblasti, Otdel sotsial'no-politicheskoi istorii (State Archive of Arkhangel'sk oblast', Social-Political History Section; hereafter GAAO-OSPI) f. 394, op. 3, d. 1, ll. 2–3.

55. For artel membership, see GAAO f. 266, op. 4, d. 46, l. 70. Another sign of the acceptance of the new labor collectives was the growing share of produce sold by collectives. By 1927, of the fish harvested in the White Sea, 48.1 percent came from cooperatives and 28.7 percent from private vendors; by 1928 these numbers were 69.8 percent and 14.1 percent, respectively. See V. L. Vyklovyskii, "Soiuznoe stroitel'stvo rybatskoi i okhotnich'ei kooperativatskii Severnogo kraia," *Severnoe khoziaistvo* 7–9 (1929): 71; Okorokov, "Promyslovaia deiatel'nost'," 35–44.

56. V. Griner, "Kooperirovat' li rybakov po mestu promysla ili po mestu zhitel'stva?" *Severnyi kooperator* 2 (1926): 14.

57. Av—v, "Gde prichina?" *Volna*, January 4, 1927, 6.

58. F. Dits, "Puti razvitiia murmanskikh treskovikh promyslov," *Severnoe khoziaistvo* 4 (1926): 43; N. Kuznetsov, "Ob obucheniia rybakov upravliat motorami," *Severnoe khoziaistvo* 3 (1928): 85. The number was still only ten in 1930; see "Treska mnogo, a treski malo," *Pravda severa*, April 4, 1931, 5.

59. N. Kuznetsov, "Nashi dostizheniia v rybnom khoziaisve k 10-y godovshchine Oktiabr'skoi revoliutsii," *Severnoe khoziaistvo* 9, no. 10 (1927): 27.

60. G. Krechkov, "Promyslovaia kooperatsiia Arkhangel'skoi gubernii," *Severnoi kooperator* 24 (1927): 28–43.

61. GAAO-OSPI f. 1, op. 1, d. 1469, l. 120.

62. A. Zhilinskii, "Promysel belukhi na Severe," *Severnoe khoziaistvo* 4 (1925): 16.

63. N. Kuznetsov, "Sel'danoi promysel v Kandalakshskom zalive," *Volna*, January 4, 1925, 2.

64. Fogel', "Severnyi morskoi zverinyi promysel v 1924 godu," *Severnoe khoziaistvo* 10 (1924): 56.

65. A. Sergeev, "Pod markoi arteli," *Volna*, November 18, 1928, 2.

66. E., "Nuzhen kultur'nyi rybak," *Volna*, June 30, 1928, 2.

67. N. Kuznetsov, "O strukture lovetskogo zveroboinogo promysla," *Severnoe khoziaistvo* 10, no. 12 (1928): 62–63.

68. Arkady Gaidar, "Rybaki," *Pravda severa*, November 30, 1929, 4. The term *kulak*, as many historians of the Soviet Union have observed, lacked any objective definition. In time, it came to connote anyone who lacked a pro-Soviet attitude.

69. Gaidar, "Rybaki," 4.

70. GAAO-OSPI f. 64, op. 1, d.19, l. 2.

71. P. Romanov, "Kulachestvo v Pomor'e," *Volna*, June 18, 1925, 4.

72. Kakhutov and Baklach, "O priznakakh kulatskikh khoziaistv na kotorye rasprostranaietsia kodeks zakonov o trude," *Pravda severa*, November 4, 1929, 4.

73. Krysanov et al., *Ne vek zhit'*, 17.

74. Exactly where the dekulakized families were to go was not indicated, although what they could take with them was specified: a suit of winter clothes, one of summer clothes, two pair of underwear, bedclothes, tools, and no more than five hundred rubles. See GAAO f. 1533, op. 2, d. 2, ll. 90–91.

75. A report from the local government authorities to the regional level complained about the fact that the kulaks were escaping before they could be removed: "Because of the strengthened pressure on the kulak, the kulak element is trying to get away from the village, to the timber mill, or to the Murman coast [the northernmost region of European Russia, and the least settled], where he has the possibility of being more successful" (GAAO f. 1533, op. 3, d. 25, l. 24). The local government asked for permission to recover the refugees, but subsequent reports show that all but a few got away (GAAO f. 1533, op. 2, d. 13, l. 1).

76. Krysanov et al., *Ne vek zhit'*, 242.

77. Krysanov et al., *Ne vek zhit'*, 272.

78. M. Kozhevnikov, "Na osennei putine my dolzhnyi pobedit'!" *Pravda severa*, September 19, 1930, 2.

79. In Onega district, for instance, to the west of Arkhangel'sk, the state ordered twenty-five boats—ten delivered by June 1, ten more by June 28, and the final five by July 25. None of these was delivered, leaving the 126 fishermen of the district waiting on the shore. Lobanov, Rych'evskii, and Baryshev, "Rybaksoiuz po-staromu likvidiruet proryv," *Pravda severa*, August 31, 1930, 2.

80. "Tresku mnogo, a treski malo," *Pravda severa*, April 4, 1931, 5.

81. GAAO f. 1533, op. 5, d. 1, l. 3.

82. GAAO f. 1533, op. 3, d. 2, l. 72.

83. GAAO f. 1533, op. 5, d. 1, l. 47 contains figures for the 1930 season; GAAO f. 1533, op. 3, d. 3, l. 10 contains the figures and description for 1931.

84. Krysanov et al., *Ne vek zhit'*, 152.

85. "Tresku mnogo," 5.

86. GAAO-OSPI f. 64, op. 1, d. 31, l. 200

87. The Christian aspect of Pomor resource exploitation, although for an earlier period, is underscored in Alexei Kraikovskii, "'The Sea on One Side, Trouble on the Other': Russian Marine Resource Use before Peter the Great," *Slavonic and East European Review* 93 (January 2015): 39–65.

Chapter 12: Experts on Unknown Waters

The chapter is part of a project titled "Natural Resources in History of Russia: Economic Institutes, Communities of Experts, and Infrastructures," funded by the Russian Science Foundation (N 16-18-10255).

1. For a general description of the area at the end of the nineteenth century, see Alexander Engelhardt, *A Russian Province of the North*, trans. Henry Cooke (Archibald Constable and Company, 1899), originally published in Russian as *Russkii Sever* (Izdanie A. S. Surovina, 1897).

2. The name Pomor descends from the Russian word *more* (the sea). The Pomors are the people who live near, travel to, and regularly work on the sea, and the term generally refers specifically to those residing near the White and Barents Seas. According to T. A. Bernshtam, in the sixteenth century the name Pomorye referred to the uninhabited areas of the western Barents Sea coast, where Russian fishermen came to fish. Gradually, the name Pomorye was extended to the inhabitants of the area, covering all people who took part in the Barents Sea cod fishery. Later, all populations of the White Sea coast were referred to as Pomors, although their local identity differs from place to place and is still under discussion. See T. A. Bernshtam, *Pomory. Formirovanie gruppy i sistema khoziaistva* (Nauka, 1978), 69–80.

3. See Kenneth Hewitt, *Regions of Risk: A Geographical Introduction to Disasters* (Pearson Education, 1997).

4. Hewitt, *Regions*, 8.

5. *Trudy Severnoi komissii 1897–1898 gg.* (Tip. Isidora Gol'dberga, 1898).

6. E. W. Clowes, S. D. Kassow, and J. L. West, eds., *Between Tsar and People: Educated Society and the Quest for Public Identity in Late Imperial Russia* (Princeton University Press, 1991). See also David Moon, "The Environmental History of the Russian Steppes: Vasilii Dokuchaev and the Harvest Failure of 1891," *Transactions of the Royal Historical Society* 15 (2005): 149–74; Bianka Pietrov-Ennker and Galina N. Ulianova, eds., *Grazhdanskaia identichnost i sfera grazhdanskoi deiatelnosti v Rossiskoi Imperii, vtoraia polovina XIX–nachalo XX veka* (ROSSPEN, 2007); Adele Lindenmeyer, *Poverty Is Not a Vice: Charity, Society and the State in Imperial Russia* (Princeton University Press, 1996); Joseph Bradley, *Voluntary Associations in Tsarist Russia: Science, Patriotism and Civil Society* (Harvard University Press, 2009); Joseph Bradley, "Subjects into Citizens: Societies, Civil Society, and the State in Tsarist Russia," *American Historical Review* 107, no. 4 (October 2002): 1094–123.

7. Carlo C. Jaeger et al. give the following definition of risk: "A situation or event in which something of human value (including humans themselves) has been put at stake and where the outcome is uncertain." See Carlo C. Jaeger et al., eds., *Risk, Uncertainty, and Rational Action* (Earthscan Publications, 2001), 17. On the religious world of the Pomor, see Stephen Brain (this volume).

8. For the history of major changes in understanding environmental risks in Europe, see Uwe Lübken and Christof Mauch, "Uncertain Environments: Natural Hazards, Risk, and Insurance in Historical Perspective," *Environment and History* 17, no. 1 (2011): 1–12; Frank Oberholzner, "From an Act of God to an Insurable Risk: The Change in the Perception of Hailstorms and Thunderstorms since the Early Modern Period," *Environment and History* 17, no. 1 (2011): 133–52. For the United States, see Arwen P. Mohun, *Risk: Negotiating Safety in American Society* (Johns Hopkins University Press, 2013), esp. chapter 6; Paul J. Best, "Insurance in Imperial Russia," *Journal of European Economic History* 18, no. 1 (1989): 139–69.

9. See Theodore Porter and William Gleason, "The Zemstvo and Public Initiative in Late Imperial Russia," *Russian History* 21 (1994): 419–37; Thomas E. Porter, *The Zemstvo and the Emergence of Civil Society in Late Imperial Russia, 1864–1917* (Carnegie Mellon University Press, 1991).

10. Ulrich Beck, *Risk Society: Towards a New Modernity* (SAGE, 2010), 27.

11. On the concept of a "knowledge society" in general, see Nico Stehr, *Knowledge Societies*

(SAGE, 1994); on the importance of studying the developments of knowledge societies and "knowledge regimes" for environmental history, see Sverker Sorlin and Paul Warde, "The Problem of the Problem of Environmental History: A Re-reading of the Field," *Environmental History* 12 (January 2007): 124–25; Frank Uekoetter, "The Knowledge Society," in *The Turning Points of Environmental History*, ed. Frank Uekoetter (University of Pittsburg Press, 2010), 132–45.

12. Gernot Böhme and Nico Stehr, eds., *The Knowledge Society: The Growing Impact of Scientific Knowledge on Social Relations* (Kluwer Academic Publishers, 1986), 18.

13. See Pavel Bogoslavskii, *O kupecheskom sudostroenii v Rossii, rechnom i pribrezhnom* (Morskoi uchenyi komitet, 1859–1863).

14. See Margarita Dadykina, Alexei Kraikovski, and Julia Lajus, "Mastering the Arctic Marine Environment: Organizational Practices of Pomor Hunting Expeditions to Svalbard (Spitsbergen) in the Eighteenth Century," *Acta Borealia* 34, no. 1 (2017): 61.

15. The real reasons for the cessation of hunting expeditions to Spitsbergen should be studied in more detail. There may have been a combination of backward technology, too much risk, and changes on the market that made such expeditions unprofitable. See T. S. Minaeva and V. Gorter, "Pomorskoe i norvezhskoe osvoenie Spitsbergena v pervoi polovine XIX veka," *Vestnik Severnogo (Arkticheskogo) Federal'nogo Universiteta. Seria: Gumanitarnye i sotsialnye nauki*, no. 4 (2013): 5–12.

16. For details on the use of Norway as a model for the economic development of the Russian North, see Julia Lajus, "In Search for Instructive Models: The Russian State at a Crossroads to Conquering the North," in *Northscapes: History, Technology, and the Making of Northern Environments*, ed. Dolly Jorgensen and Sverker Sorlin (University of British Columbia Press, 2013), 110–36.

17. See Charles C. Gillispie, "The Natural History of Industry," *Isis* 48, no. 4 (1957): 398–407.

18. Elisabeth A. Hachten, "In Service to Science and Society: Scientists and the Public in Late-Nineteenth-Century Russia," *Osiris* 17 (2002): 172.

19. Hachten, "In Service," 201–7; Moon, "Environmental History."

20. See *Trudy Severnoi komissii.*

21. The term *scientific-fisheries research* (*nauchno-promyslovye issledovaniia* in Russian) is a translation of the German term *Fishereiwissenschaft.*

22. This observation is in line with the statement made by Hachten that, in Russia in the late nineteenth century, natural science suddenly took on an aura of cultural authority, privileging it above many other forms of culture. See Hachten, "In Service," 184.

23. Knipovich also used the German spelling of his name, Knipowitsch, in his non-Russian publications.

24. For more details on the Murman expedition, see Julia A. Lajus, "Early Years of Biological Oceanography in the Russian North: Murman Scientific-Fishery Expedition, 1898–1908," in *Ocean Sciences Bridging the Millennia: A Spectrum of Historical Accounts. Proceedings of the VI International Congress on the History of Oceanography* (Intergovernmental Oceanographic Commission of UNESCO and China Ocean Press, 2004), 127–31; Julia A. Lajus, "'Foreign Science' in Russian Context: Murman Scientific-Fishery Expedition and Russian Participation in Early ICES Activity," *International Council for Exploration of the Sea [ICES] Marine Science Symposia* 215 (2002): 64–72.

25. Nikolai M. Knipovich, "Proekt nauchno-promyslovykh issedovanii u Murmanskogo berega," in *Trudy Severnoi komissii*, Supplement 1, 5–25.

26. For the discussion of the rhetoric of backwardness during the late tsarist era, see Yanni Kotsonis, *Making Peasants Backward: Agricultural Cooperatives and the Agrarian Question in Russia, 1861-1914* (St. Martin's Press, 1999).

27. For the full citation, see "Neobkhodimost' nauchnopromyslovykh morskikh issledovanii u beregov Murmana," in *Trudy Severnoi komissii*, Supplement 1, 1–3.

28. For a general description of Russian northern fisheries, see Julia Lajus, Alexei Kraikovski, and Alexei Yurchenko, "The Fisheries of the Russian North, c. 1300–1850," in *A History of*

the North Atlantic Fisheries, Vol. 1: From Early Times to the Mid-Nineteenth Century, ed. David J. Starkey, Jon Th. Thor, and Ingo Heidbrink (Verlag H. M. Hauschild, 2009), 41–64; also Alexei Kraikovski "'The Sea on One Side, Trouble on the Other': Russian Marine Resource Use before Peter the Great," *Slavonic and East European Review* 93, no. 1 (2015): 39–65. For the period after 1850, see Bjorn-Peter Finstad and Julia Lajus, "The Fisheries in Norwegian and Russian Waters, 1850–2010," in *A History of the North Atlantic Fisheries, Volume 2: From the 1850s to the Early Twentieth-First Century*, ed. David J. Starkey and Ingo Heidbrink (Verlag H. M. Hauschild, 2012), 226–37.

29. *Trudy Severnoi komissii*, 58, 51.

30. *Trudy Severnoi komissii*, 55.

31. *Trudy Severnoi komissii*, Supplement 1, 44.

32. See Tim D. Smith, *Scaling Fisheries: The Science of Measuring the Effects of Fishing, 1855–1955* (Cambridge University Press, 1994).

33. Smith, *Scaling*, 51.

34. Eric L. Mills, *Biological Oceanography: An Early History, 1870–1960* (Cornell University Press, 1989), 15–16.

35. Smith, *Scaling*, 51.

36. T. Solhaug and G. Saetersdal, "The Development of Fishery Research in Norway in the Nineteenth and Twentieth Centuries in the Light of the History of the Fisheries," *Proceedings of the Royal Society of Edinburgh (B)* 73 (1971/1972): 401.

37. Michael M. Sinclair and Tim D. Smith, "The Notion that Fish Species Form Stocks," *ICES Marine Science Symposia* 215 (2002): 297–304; Sarah Jansen, "Den Heringen einen Paß ausstellen: Formalisierung und Genauigkeit in den Anfängen der Populationsökologie um 1900," *Berichte zur Wissenschaftsgeschichte* 25, no. 3 (2002): 153–69.

38. Helen Rozwadowsky, *The Sea Knows No Boundaries: A Century of Marine Science under ICES* (ICES and University of Washington Press, 2002).

39. Jonathan D. Oldfield and Denis J. B. Shaw, *The Development of Russian Environmental Thought: Scientific and Geographical Perspectives on the Natural Environment* (Routledge, 2016).

40. For Dokuchaev, see Moon, "Environmental History." The development of soil sciences led by Dokuchaev was a reaction to the same complex of issues: risks for agriculture (droughts), attempts to understand the connections of plant distribution to the environmental factors, and a strong belief in science. See also David Moon, *The Plough that Broke the Steppes: Agriculture and Environment on Russia's Grasslands, 1700–1914* (Oxford University Press, 2013), and Mieka Erley (this volume). About Andrei N. Beketov's views on nature, see Daniel P. Todes, *Darwin without Malthus: The Struggle for Existence in Russian Evolutionary Thought* (Oxford University Press, 1989).

41. See Jens-Petter Nielsen, "The Murman Coast and Russian Northern Policies ca. 1855–1917," in *In the North My Nest Is Made: Studies in the History of the Murman Colonization, 1860–1940*, ed. Alexei Yurchenko and Jens Petter Nielsen (European University at St. Petersburg and University of Tromso Press, 2005); Julia Lajus, "Colonization of the Russian North: A Frozen Frontier," in *Cultivating the Colony: Colonial States and Their Environmental Legacies*, ed. Christina Folke Ax et al. (Ohio University Press, 2011), 164–90.

42. Dag Avango, Annika E. Nilsson, and Peder Roberts, "Assessing Arctic Futures: Voices, Resources and Governance," *The Polar Journal* 3, no. 2 (December 2013): 431–46.

43. For the need of resource units to be predictable for successful governance, see Elinor Ostrom, "A General Framework for Analyzing Sustainability of Social-Ecological Systems," *Science* 325 (2009): 419–22. "System dynamics need to be sufficiently predictable that users can estimate what would happen if they were to establish particular harvesting rules. . . . Some fishery systems approach mathematical chaos and are particularly challenging for users or government officials" (419).

44. *Trudy Severnoi komissii*, 53.

45. Nikolai M. Knipovich, *Polozhenie morskikh rybnykh i zverinykh promyslov Arkhan-*

gel'skoi gubernii (iz otchetvo Ministerstvu zemledeliia po komandirovkam 1893 i 1894 gg.) (Tip. V. Kirshbauma, 1895).

46. This practice of hiring unreliable and even exiled people to do research in the North or in Siberia had a long tradition in nineteenth-century Russia and even grew after the Revolution of 1905.

47. See Aleksei Karimov, *Dokuda topor i sokha khodili: Ocherki istorii zemelnogo i lesnogo kadastra v Rossii XVI–nachala XX veka* (Nauka, 2007), 182–84.

48. About the importance of internal colonization for understanding Russian history, see Alexander Etkind, *Internal Colonization: Russia's Imperial Experience* (Polity Press, 2011).

49. "Ulovy (nemtsev i anglichan) v Belom more," *Vestnik rybopromyshlennosti* 12 (1907): 570–71.

50. Zhaloby Pomorov, ["Complaints of Pomors,"] *Novoe vremia* 29 (July 1900): 4.

51. Leonid L. Breitfuss, "Kurzer Ueberblick ueber die Tatigkeit der wissenschaftlichen Murmanexpedition, 1898–1904," *Mitteilungen des Deutschen Seefisherei-Vereins* 7/8 (1905).

52. Leonid L. Breitfuss, *O Komitete dlia pomoschi pomoram Russkogo Severa* (Tipo-litografiia Pechatoe iskusstvo, 1910), 9–10.

53. Nikolai M. Knipovich, "Raboty ekspeditsii dlia nauchno-promyslovykh issledovanii Murmana s maia po noiabr 1900 g.," *Russkoe sudokhodstvo* 226/229 (1901): 52.

54. Knipovich, "Raboty," 54.

55. The discussion of beluga whales in Stephen Brain (this volume) is inaccurate. The Pomor did hunt these whales in quite large numbers, and statistics of catches as well as images of the hunting are well known. See, for instance, Yaroslava I. Alekseeva, "Istoriia promysla morskikh mlekopitaiuschikh v Belom i Barentsevom moriakh (XV v.–1915 g.)," in *Morskie mlekopitaiuschie Golarktiki: materialy Piatoi mezhdunarodnoi konf. Odessa, Ukraina, 14–18 oktiabria 2008 g.: sbornik nauchnykh trudov*, ed. A. Boltunov (Sovet po morskim mlekopitaiuschim, 2008), 35–37.

56. Leonid L. Breitfuss, *Ekspeditisiia dlia nauchno-promyslovykh issledovanii u beregov Murmana. Otchet o ee rabotakh v 1904 g.* (Komitet dlia Pomoshchi Pomoram Russkogo Severa), xxxiv.

57. Stephen Bocking, *Nature's Experts: Science, Politics, and the Environment* (Rutgers University Press, 2006), 88.

Chapter 13: Fishing, Settlement, and Conservation in the Russian Far East, 1860–1940

1. See Ryan Tucker Jones (this volume); Ryan Tucker Jones, "Running into Whales: The History of the North Pacific from below the Waves," *American Historical Review* 118, no. 2 (2013): 349–77.

2. For instance, Kendall E. Bailes, *Science and Russian Culture in an Age of Revolutions: Vol. I. Vernadsky and His Scientific School, 1863–1945* (Indiana University Press, 1990); Joan DeBardeleben, *The Environment and Marxism-Leninism: The Soviet and East German Experience* (Westview Press, 1985); Murray Feshbach, *Ecocide in the USSR: Health and Nature under Siege* (Basic Books, 1993); Paul R. Josephson, *Would Trotsky Wear a Bluetooth?* (Johns Hopkins University Press, 2010); D. J. Peterson, *Troubled Lands: The Legacy of Soviet Environmental Destruction* (Westview Press, 1993); Philip R. Pryde, *Environmental Management in the Soviet Union* (Cambridge University Press, 1991).

3. Micah Muscolino, "Fishing and Whaling," in *A Companion to Global Environmental History*, ed. J. R. McNeill and Erin Stewart Mauldin (Wiley, 2012), 629–30. See also Micah Muscolino, "Fisheries Build Up the Nation: Maritime Environmental Encounters between Japan and China," in *Japan at Nature's Edge: The Environmental Context of Global Power*, ed. Ian Jared Miller, Julia Adeney Thomas, and Brett L. Walker (University of Hawai'i Press, 2013), 56–70.

4. Tikhookeanskii institut geografii DVO RAN, *Atlas Primorskogo kraia* (Federal'noe gosudarstvennoe unitarnoe predpriiatie "PrimorAGP," 2007), 20; Erich Thiel, *The Soviet Far East: A Survey of Its Physical and Economic Geography* (Methuen, 1957), 109–10.

5. Robert H. G. Lee, *The Manchurian Frontier in Ch'ing History* (Harvard University Press, 1970), 48; Shira Sasaki, "Hunting Activities and Forest Management of the Udeghe People in Krasnyi Yar in the Russian Far East," in *Russia Country Report 2001*, ed. H. Kakizawa and M. Inoue (Institute for Global Environmental Strategies, 2002), 87.

6. Nikolay N. Kradin, Yuri G. Nikitin, and Nikolay A. Kliuev, "Hunting, Fishing and Early Agriculture in Northern Primor'e in the Russian Far East," in *Human-Nature Relations and the Historical Backgrounds of Hunter-Gatherer Cultures in Northeast Asian Forests: Russian Far East and Northeast Japan*, ed. Shira Sasaki (National Museum of Ethnology, 2009), 21.

7. Xanthippe Augerot, "An Environmental History of the Salmon Management Philosophies of the North Pacific: Japan, Russia, Canada, Alaska, and the Pacific Northwest United States" (PhD diss., University of Oregon, 2000), 54.

8. H. E. M. James, *The Long White Mountain, or A Journey in Manchuria* (Longmans, Green, 1888), 426–30; Ernest George Ravenstein, *The Russians on the Amur: Its Discovery, Conquest, and Colonization, with a Description of the Country, Its Inhabitants, Productions, and Commercial Capabilities* (Trübner, 1861), 369, 436.

9. N. M. Przheval'skii, *Puteshestvie v Ussuriiskom krae, 1867–1869 g.* (Gos. izd-vo geograficheskoi literatury, 1947), 66.

10. Przheval'skii, *Puteshestvie*, 90; N. Aliab'ev, *Dalekaia Rossiia: Ussuriiskii krai* (Obshchestvennaia pol'za, 1872), 61. See also A. V. Smoliak, *Ul'chi: Khoziaistvo, kul'tura i byt v proshlom i nastoiashchem* (Nauka, 1966), 29–42.

11. A. F. Budishchev, *Opisanie lesov Primorskoi oblasti*, vol. 5 of *Sbornik glavnykh offitsial'nykh dokladov po upravleniiu Vostochnoi Sibiri* (Tip. Kantseliarii Priamurskogo General-Gubernatora, 1898).

12. "Po naseleniia, promyshlennosti, i torgovle Priamurskogo kraia" (1869), Rossiiskii Gosudarstvennyi Istoricheskii Arkhiv (Russian State Historical Archive; hereafter cited as RGIA) f. 1315, op. 1, d. 3, ll. 201–202ob.

13. Przheval'skii, *Puteshestvie*, 83–85.

14. William M. Tsutsui, "The Pelagic Empire: Reconsidering Japanese Expansion," in *Japan at Nature's Edge: The Environmental Context of a Global Power*, ed. Ian Jared Miller, Julia Adeney Thomas, and Brett L. Walker (University of Hawai'i Press, 2013), 23–24; Eisuke Kaminaga, "Maritime History and Imperiology: Japan's 'Northern Fisheries' and the Priamur Governor-Generalship," in *Imperiology: From Empirical Knowledge to Discussing the Russian Empire*, ed. Kimitaka Matsuzato (Slavic Research Center, Hokkaido University, 2007), 263.

15. Quoted in François-Xavier Coquin, *La Sibérie: peuplement et immigration paysanne au XIXe siècle* (Institut d'études slaves, 1969), 638.

16. See for instance Przheval'skii, *Puteshestvie*, 70; "Imperatorskoe moskovskoe obshchestvo sel'skago khoziaistva," *Amur i Ussuriiskii krai* (Tip. I. D. Sytina, 1885), 125. Almost exactly the same phrase can be found in Aliab'ev, *Dalekaia Rossiia*, 88.

17. "Raport komandira Sibirskoi flotilii i portov vostochnogo okeana P. V. Kazakevich" (1866), Rossiiskii gosudarstvennyi arkhiv Voenno-Morskogo Flota (Russian State Naval Archive; hereafter cited as RGA VMF) f. 909, op. 1, d. 44, ll. 16–17, 28–28ob.

18. L. V. Aleksandrovskaia, *Opyt pervogo morskogo pereseleniia v Iuzhno-Ussuriiskii krai v 60-kh godakh XIX veka: dokumental'no-istoricheskoe povestvovanie* (OIAK, 2003), 9–20; L. V. Aleksandrovskaia, *Odisseia Fridol'fa Geka: dokumental'no-istoricheskoe povestvovanie* (OIAK, 1999); Anatolii Rekk-Lebedev, *Dal'nevostochnaia Lifliandiia: estontsy na Ussuriiskoi zemle* (Eesti raamat, 1989), 10.

19. "Zapiska o neobkhodimosti pereseleniia pomorov v Iuzhno-Ussuriiskii krai," 1896 (RGIA f. 391, op. 3, d. 1152, ll. 186–88). The Resettlement Administration used the term *Pomor* to refer to all seafaring peoples of European Russia, not just the Pomors of the White Sea and Barents Sea regions.

20. The value of the ruble fluctuated during the years under consideration here, but it was roughly equivalent to 50 cents in contemporary US currency. In 1885 the ruble was pegged to

the franc, and by 1897, when Russia adopted the gold standard, the ruble was equivalent of 2.75 francs, the equivalent of 0.77 grams of gold, or roughly 47 cents. In 1910 one ruble was worth about 51 cents. See Peter Lindert and Boris Mironov, "Russia_Ag_content_ruble_1535–1913," *Global Price and Income History Group*, February 2006, online at http://gpih.ucdavis.edu/files/Russia_Ag_content_ruble_1535–1913.xls/; National Mining Association, "Historical Gold Prices—1833 to Present," online at https://nma.org/wp-content/uploads/2016/09/historic_gold_prices_1833_pres.pdf; Carroll Davidson Wright, *The New Century Book of Facts: A Handbook of Ready Reference* (King-Richardson Company, 1910), 568.

21. "Otchet o pereselenii krest'ian morskim putem v Iuzhno-Ussuriiskii krai v 1891 godu" (1891), Gosudarstvennyi arkhiv Primorskogo kraia (State Archive of Primorskii krai; hereafter cited as GAPK) f. 1, op. 1, d. 32, ll. 54–64). Aleksandrovskaia suggests that the governor of Astrakhan was well aware that he was not sending actual fishermen and just wanted to rid himself of some poor workers, but I was unable to confirm this. See L. V. Aleksandrovskaia, "Osvoenie poberezh'ia Iuzhno-Ussuriiskogo kraia," *Zapiski Obshchestva izucheniia Amurskogo kraia* 30 (1996): 27.

22. RGIA f. 391, op. 3, d. 1152, l. 189; GAPK f. 1, op. 1, d. 32, l. 94.

23. "Delo ob organizatsii pereseleniia v Ussuriiskii krai," 1898–1899 (RGIA f. 1273, op. 1, d. 294, ll. 18–19ob.).

24. RGIA f. 1273, op. 1, d. 294, ll. 20–24.

25. Aleksandrovskaia, "Osvoenie," 30.

26. Rekk-Lebedev, *Lifliandiia*, 11–13.

27. A. T. Mandrik, *Istoriia rybnoi promyshlennosti rossiiskogo Dal'nego Vostoka (50-e gody XVII v.–20-e gody XX v.)* (Dal'nauka, 1994), 62.

28. "O polozhenii i nuzhdakh Iuzhno-Ussuriiskago kraia: Zapiska byvshago sekretaria Glavnago Nachal'nika russkikh morskikh sil v Tikhom Okeane," 1881 (RGA VMF f. 410, op. 2, d. 4046, ll. 237–39, 245–47).

29. A. A. Rittikh, *Pereselencheskoe i krest'ianskoe delo v Iuzhno-Ussuriiskom krae* (Pereselenchskoe upravlenie, 1898). Rittikh, it should be noted, advocated only controls on Japanese and Chinese fishing; he did not seek to limit the industry as a whole.

30. D. I. Shreider, *Nash Dal'nii Vostok: Tri goda v Ussuriiskom krae* (A. F. Devrien, 1897), 256.

31. Ferdinand Ossendowski, *Man and Mystery in Asia* (E. P. Dutton, 1924), 107.

32. "Ob okhrane morskikh promyslov v Tikhom Okeane," 1908 (RGA VMF f. 417, op. 1, d. 3795, ll. 62ob.–63ob.). On attitudes toward American whalers in Russian waters, see especially Ryan Jones (this volume).

33. P. F. Unterberger, *Priamurskii krai, 1906–1910 g.g.*, vol. 13 of *Zapiski imperatorskago russkago geograficheskago obshchestva po otdeleniiu statistiki* (Tip. V. O. Kirshbauma, 1912), 143–50.

34. "Pravila, utverzhdennye na 1894 g., po proizvodstvu morskikh promyslov . . . " (1894), RGIA Dal'nego Vostoka (Russian State Historical Archive of the Far East) f. 702, op. 2, d. 95, ll. 8–10).

35. Chancellery of the Priamur Governor-General, *O rybnom promysle v Primorskoi oblasti i na ostrove Sakhaline* (Tip. Kantseliarii Priamurskago General-Gubernatora, 1903), 8–9; "Zakony i instruktsii rybolovstva na Dal'nem Vostoke," April–May 1919, Gosudarstvennyi arkhiv Rossiiskoi Federatsii (State Archive of the Russian Federation; hereafter cited as GARF) f. 3773, op.1 d.1, ll. 1–5).

36. GARF f. 3773, op. 1 d. 1, ll. 20–22.

37. Aleksandrovskaia, "Osvoenie," 27.

38. Mandrik, *Istoriia (50-e gody XVII v.–20-e gody XX v.)*, 41.

39. P. Rusanov, "K zapreshcheniiu lova osetra i kalugi v basseine reki Amura," *Ekonomicheskaia Zhizn' DV* 9 (1927): 76–79.

40. Mandrik, *Istoriia (50-e gody XVII v.–20-e gody XX v.)*, 40.

41. Mandrik, *Istoriia (50-e gody XVII v.–20-e gody XX v.)*, 45–46, 51.

42. The Japanese had already gained fishing rights off Sakhalin Island through the Treaty of St. Petersburg (1875), which recognized Russian claims to the island itself. See G. Patrick March, *Eastern Destiny: Russia in Asia and the North Pacific* (Praeger, 1996), 90.

43. Tsutsui, "Pelagic Empire," 24.

44. Kaminaga, "Maritime History," 267–68.

45. "O posylke khodokov ot rybolovov nekotorykh gubernii Evropeiskoi Rossii v Primorskuiu oblast'," 1908 (RGIA f. 391, op. 3, d. 1129, ll. 171–72ob.).

46. RGIA f. 391, op. 3, d. 1152, ll. 82–83.

47. V. F. Romanov, *Nuzhdy Nikolaevskago raiona Primorskoi oblasti: Otchet o komandirovke v Nikolaevskii*, vol. 10 of *Trudy komandirovannoi po Vysochaishemu povleniiu Amurskoi Ekspeditsii* (Tip. V. O. Kirshbaum, 1911), 163–67.

48. Romanov, *Nuzhdy*, 79.

49. Quoted in Romanov, *Nuzhdy*, 83.

50. "O merakh bor'by protiv naplyva zheltoi rasy v Priamure," 1910 (RGIA f. 391, op. 4, d. 513, ll. 39–42ob.).

51. RGIA f. 391, op. 4, d. 513, ll. 77–78ob., 45–45ob.

52. John Stephan, *The Russian Far East: A History* (Stanford University Press, 1994), 146–47.

53. "O nezakonomernykh deistv iaponokomandov v voprose rasporiazheniia rybnymi promylsami," May 1921 (GARF f. 3722, op. 1, d. 9, ll. 254–45ob.).

54. "Dokumenty nauchno-issledovatel'skikh rabot V. K. Arsen'eva po izucheniiu morskikh promyslovykh zhivotnykh v vodakh Dal'nego Vostoka," 1922 (GAPK f. 633, op. 4, d. 100, l. 129).

55. Far Eastern Republic, *Fisheries of the Far Eastern Republic* (Special Delegation of the Far Eastern Republic in the United States of America, 1922), 18, 30.

56. S. Z—i, "Sovremennoe Polozhenie Russkoi Rybopromyshlennosti," *Russkoe Primor'e* 1 (July 1, 1922): 11–13.

57. "Kratkii otchet po rybnomu khoziaistvu DV kraia za 1923, 1924, 1925 gody," 1925 (GAPK f. 633, op. 4, d. 64, ll. 4–6).

58. V. O. Kolobov, "Perspektivy dal'nevostochnykh rybnykh promyslov," in *Rybnye i pushnye bogatstva Dal'nego Vostoka* (Izdanie nauchprombiuro Dal'rybokhoty, 1923), 343–57.

59. Rusanov, "K zapreshcheniiu," 76; "Perspektivnyi plan razvitiia rybnogo khoziaistva na 1928–33," 1927 (GAPK f. 633, op. 7, d. 43, ll. 9–10); "Tezisy o putiakh razvitiia rybnogo khoziaistva," 1926 (GAPK f. 633, op. 7, d. 25, ll. 1–11, 87–89).

60. Letter from Dal'ryba official Orel and Fishing Inspector Il'chenko, March 30, 1928 (GAPK f. 633, op. 7, d. 53, l. 3ob.); Letter from a senior fisheries inspector to the *nachal'nik* (head) of Dal'ryba, July 20, 1930 (GAPK f. 633, op. 7, d. 72, ll. 4–5); Barbara Wertheim, "The Russo-Japanese Fisheries Controversy," *Pacific Affairs* 8, no. 2 (June 1935): 193–94; "Dokumenty o khishchnichestva iapontsev v Dal'nevostochnykh vodakh za 1926–27," 1927 (GAPK f. 633, op. 7, d. 29, ll. 1–1ob., 19ob., 56).

61. GAPK f. 633, op. 7, d. 29, ll. 11–29ob., 51ob. These were narrow undecked vessels generally known as *kungasy* (singular: *kungas*) or simply "Korean boats."

62. Violet Conolly, *Soviet Trade from the Pacific to the Levant, with an Economic Study of the Soviet Far Eastern Region* (Oxford University Press, H. Milford, 1935), 9; A. S. Vashchuk et al., *Etnomigratsionnye protsessi v Primor'e v XX veke* (DVO RAN, 2002), 70.

63. Oddly, neither N. A. Bilim's study nor any Dal'ryba document provides exact figures for the 1930s, documenting only overall growth of kolkhoz and state fisheries in the 1930s, which derived from diverse population sources, including those displaced by collectivization, the reorganization of fisheries, and general migration. N. A. Bilim, "Pereselenie rybakov na sovetskii Dal'nii Vostok (1928–1941 gg.)," in *Iz istorii bor'by za sovetskuiu vlast' I sotsialisticheskoe stroitel'stvo na Dal'nem Vostoke*, ed. E. M. Shchagin (Khabarovsk gos. pedagogicheskii institut, 1965), 81–83, 87; "Dokumenty o pereselenii Volzhso-Kaspiiskikh rybakov v Primorskoi promyslovyi raion za 1928 g.," 1928 (GAPK f. 633, op. 7, d. 54, ll. 4–29).

64. Bilim, "Pereselenie," 76.

65. Wertheim, "Fisheries Controversy," 194–95.

66. "Otchet Dal'ryby po regulirovaniiu rybolovstva v Dal'nevostochnom raione za 1937 g.," 1937 (GAPK f. 633, op. 5, d. 95, ll. 1–3ob.); A.T. Mandrik, *Istoriia rybnoi promyshlennosti rossiiskogo Dal'nego Vostoka (1927–1940 gg.)* (Dal'nauka, 2000), 41.

67. Wertheim, "Fisheries Controversy," 191–92.

68. Primorskaia oblastnaia planovaia komissia, *Primorskaia oblast' v tsifrakh* (Dal'giz, 1933), 40–47; GAPK f. 633, op. 5, d. 95, l. 8.

69. Wertheim, "Fisheries Controversy," 195–96.

70. "Spravka o sostoianii zapasov nekotorykh promyslovykh ryb I o meropriiatiiakh po ikh okhrane u uvelicheniiu v vodakh Primorskogo kraia," 1946 (GAPK f. 633, op. 5, d. 75, ll. 12–20).

71. Augerot, "Environmental History," 71–72.

72. Joseph E. Taylor, *Making Salmon: An Environmental History of the Northwest Fisheries Crisis* (University of Washington Press, 1999); Richard White, *The Organic Machine* (Hill and Wang, 1995), 47–48.

Chapter 14: The Tragedy of Captain Ligov

1. See Rex Weyler, *Greenpeace: How a Group of Ecologists, Journalists, and Visionaries Changed the World* (Raincoast Books, 2004).

2. See, most comprehensively, Yulia Ivashchenko, "Soviet Whaling: Past History and Present Impacts" (PhD diss., Southern Cross University, 2013), 19.

3. A. A. Vakhov, *Shtorm ne utikhaet: roman* (Magadanskoe Knizhnoe Izdatel'stvo, 1957; A. A. Vakhov, *Fontany na gorizonte* (Magadanskoe Knizhnoe Izd-vo, 1951).

4. "Zabytie Genii," *Khabarkovskii Ekspress*, June 19, 2006, online at http://www.debri-dv.com/article/1641/.

5. Several excellent chapters by Sokolsky, Lajus, and Brain (all in this volume) suggest new attention to Russia's role in the environmental history of the world's oceans. Also see David Moon, "The Curious Case of the Marginalisation or Distortion of Russian and Soviet Environmental History in Global Environmental Histories," *International Review of Environmental History* 3, no. 2 (2017), 31–50.

6. Other sources, such as whalers' onboard newspapers, reflect similar understandings of whales and whaling. See Ryan Tucker Jones, "Cetus Soveticus: Exotic and Collective Cultures within Soviet Whaling in the 1960s," in *Whaling and History IV: Papers Presented at a Symposium in Sandefjord on the 20th and 21st of June 2013*, ed. Jan Erik Ringstad (Kommandør Chr. Christensens Hvalfangstmuseum, 2014).

7. Nick Breyfogle discusses the ramifications for Russia's relatively quick but still post-Western adoption of industrial technology in his introduction (this volume).

8. Ernst C. Webermann, *The Whale Fishery in Russia. Part I: A History of the Whale Fishery* (n.p., 1914), 21, 22; Igor Krupnik, *Arctic Adaptations: Native Whalers and Reindeer Herders of Northern Eurasia*, trans. Marcia Levenson (University Press of New England, 1993), 208; Iu. N. Bespiatyh, *Archangelsk nakanune i v gody Severnoi Voiny, 1700–1721* (Izd-vo BLITs, 2010), 131. On Pomor resistance to modern whaling, see Stephen Brain (this volume).

9. Alexey Kraikovsky, "The Governmental Projects for the Development of Russian Whaling in the 18th Century," in *Whaling and History III: Papers Presented at a Symposium in Sandefjord on the 18th and 19th of June, 2009*, ed. Jan Erik Ringstad (Kommandør Chr. Christensens Hvalfangst Museum, 2009); Alexey Kraykovskiy and Zoya Dmitrieva, "Sources on the History of Usage of Marine Resources in the Russian North in the 17th and 18th Centuries," in *Regionalnye tendentsii vzaimodeistviia cheloveka i prirody v processe perehoda ot agrarnogo k industrialnomu obshestvu* (Zolotaia bukva, 2003).

10. J. N. Tonnessen and A. O. Johnsen, *The History of Modern Whaling* (University of California Press, 1982), chapter 8.

11. S. E. Kleinenberg, *Kitoboinyi promysel Sovetskogo Soyuza* (Rybnoe Khoziaistvo, 1955), 22, 23.

12. See Irina Gan, "'The First Practical Soviet Steps towards Getting a Foothold in the Antarctic': The Soviet Antarctic Whaling Flotilla *Slava*," *Polar Record* 47, no. 240 (2011): 21–28.

13. On blue whales, see International Union for Conservation of Nature, "Balaenoptera musculus," *Red List of Threatened Species*, online at http://www.iucnredlist.org/details/2477/0/. On the aftereffects of industrial whaling, see (somewhat controversially) A. M. Springer et al., "Sequential Megafaunal Collapse in the North Pacific Ocean: An Ongoing Legacy of Industrial Whaling?" *Proceedings of the National Academy of Sciences of the United States of America* 100, no. 21 (October 14, 2003): 12223–28.

14. This is the well-argued viewpoint in Yulia Ivashchenko, Philip Clapham, and Robert Brownell, "Soviet Illegal Whaling: The Devil and the Details," *Marine Fisheries Review* 73, no. 3 (2011): 1–19. See also Philip Clapham and Yulia Ivashchenko, "A Whale of a Deception," *Marine Fisheries Review* 71, no. 1 (2009): 44–52; Alfred Berzin, "The Truth about Soviet Whaling: A Memoir," *Marine Fisheries Review* 70, no. 2 (2008): 1–3; A. V. Iablokov and V. A. Zemskii, eds., *Soviet Whaling Data (1949–1979)* (Center for Russian Environmental Policy, 2000).

15. Berzin, "Soviet Whaling," 6.

16. Charles Homans, "The Most Senseless Environmental Crime of the 20th Century," *Pacific Standard*, November 12, 2013, online at https://psmag.com/the-most-senseless-environmental-crime-of-the-20th-century-9594972483d1#.hawa57tve/.

17. See, for example, "Pis'ma Finliandskago matrosa s kitolovnogo sudna 'Suomi,'" *Morskoi sbornik* 20, no. 2 (February 1856); *Morskoi sbornik* 39, no. 2 (February 1854), "Bibliographia," 33.

18. Otto Lindholm, "Kitoboinoi promysel," *Otdel'nye Otpiski iz Zhurnala "Russkoe Sudokhodstvo"* (Litografiia R. Golike, 1888), 10.

19. Otto Lihndholm, "Whales and How Tides and Currents in the Okhotsk Sea Affect Them," trans. Lydia A. Hutchinson, *Scripps Institution of Oceanography Technical Report* (February 1963): 1–12, online at http://escholarship.org/uc/item/8j93s3pv/.

20. A. Ia. Maksimov, "Kitoboi," in *Na Dalekom Vostoke: Polnoe sobranie sochinenii* (M. K. Maksimov, 1898), 2.

21. Maksimov, "Kitoboi," 14.

22. Maksimov, "Kitoboi," 19.

23. K. N. Nosilov, "Na kitoboinom parokhode iz puteshestviia po Severu," in *Chitalnaia narodnoi shkoli* (Tipo-Litografiia K. L. Pentkovskago, 1911), 21, 23.

24. Nosilov, "Na kitoboinom parakohde," 25.

25. William E. Lenz, "Narratives of Exploration, Sea Fiction, Mariners' Chronicles, and the Rise of American Nationalism: 'To Cast Anchor on that Point Where all Meridians Terminate,'" *American Studies* 32, no. 2 (Fall 1991): 41–61; Philip Armstrong, "'Leviathan Is a Skein of Networks': Translations of Nature and Culture in Moby-Dick," *English Literary History* 71, no. 4 (Winter 2004): 1039–62.

26. B. A. Zenkovich, *Vokrug sveta za kitami* (Molodaya Gvardiya, 1936), chapter 4.

27. Zenkovich, *Vokrug sveta*, 11.

28. A. N. Solyanik, *Cruising in the Antarctic: An Account of the Seventh Cruise of the* Slava *Whaling Flotilla*, trans. O. Gorchakov (Foreign Languages Publishing House, 1956).

29. Solyanik, *Cruising*, 13.

30. About Solyanik's honesty, see for example, the report that the New Zealand commissioner to the IWC made to the 1958 IWC meeting: "Brief to Secretary of External Affairs, Wellington, July 12, 1958," Archives New Zealand M 2/7.

31. V. A. Arsen'ev and V. A. Zemskii, *V strane kitov i pengvinov* (Izd-vo Moskovskogo obshchestva ispytatelei prirody, 1951).

32. Yuri Mikhalev, *Kity yuzhnogo polushariia: biologiia, promysel, perspektivy vosstanovleniia populiatsii* (Soviet po morskim mlekopitaiushchim, 2008), 33.

33. Arsen'ev and Zemskii, *V strane kitov*, 69, 63.

34. Arsen'ev and Zemskii, *V strane kitov*, 235.

35. Mikhalev, *Kity*, 33.

36. Rossiiskii gosudarstvennyi arkhiv ekonomiki (Russian State Archive of the Economy) f. 9242, op. 1, d. 324, ll. 66, 67.

37. J. R. McNeill and Corinna R. Unger, "Introduction: The Big Picture," *Environmental Histories of the Cold War*, ed. J. R. McNeill and Corinna R. Unger (Cambridge University Press, 2010), 4.

38. "Piatnadsatyi kit na linii," *Ogonek*, no. 21 (May, 1954), 22.

39. A. A. Vakhov, *Garpuner (Ocherk o luchshem garpunere kitoboinoi flotilii* "Aleut" *F. D. Prokopenko* (Primizdat, 1949).

40. Marc Slonim, *Soviet Russian Literature: Writers and Problems, 1917–1977* (Oxford University Press, 1977), 268.

41. Edith Rogovin Frankel, "Literary Policy in Stalin's Last Year," *Soviet Studies* 28, no. 3 (July 1976): 394.

42. Deming Brown, *Soviet Russian Literature since Stalin* (Cambridge University Press, 1978), 253.

43. Frankel, "Literary Policy," 395.

44. Lindholm, "Kitoboinii promysel," 13.

45. L. V. Aleksandrovskaia and Birgitta Ingemanson, *Udivetel'naya zhizn' Otto Lindgol'ma: Dokumental'noe-istoricheskoe povestvovanie* (OIAK, 2003).

46. Vakhov, *Shtorm*.

47. Vakhov, *Tragedia kapitana Ligova: Roman* (Khabarovskoe regional'noe otdelenie Soiuza pisatelei Rossii, 2008), 17.

48. Avrahm Yarmolinsky, *Literature under Communism: The Literary Policy of the Communist Party of the Soviet Union from the End of World War II to the Death of Stalin* (Indiana University Press, 1960), 101.

49. Vakhov, *Tragedia*, 47. See also Alain Corbin, *The Lure of the Sea: The Discovery of the Seaside in the Western World, 1750–1840* (Penguin Books, 1995).

50. See Lajus and Brain (both in this volume).

51. Vakhov, *Tragedia*, 36.

52. See especially Krupnik, *Arctic Adaptations*, chapter 2.

53. Vakhov, *Tragedia*, 33.

54. Vakhov, *Fontany na gorizonte*, 296. There exists also an excellent historical novel by the Chukchi Soviet writer Yuri Ritkheu, *A Dream in Polar Fog* (Archipelago Books, 2005), first published in the Soviet Union as *Son v nachale tumana: Roman* (Sovietskaia Rossiia, 1970). This book only tangentially touches on whaling itself, however, and its provenance and themes are sufficiently unique as to merit historical treatment on its own.

55. "Minutes of the Fourth Session, Nov. 25, 1946," in IWC Archives, *Proceedings of the 1946 Conference*, 1:26.

56. "Spravka po rabote kitiboinoi flotilii 'Slava,'" Rossiiskii gosudarstvennyi arkhiv ekonomiki f. 9242, op. 1, d. 409, l. 69. The idea of "total use" certainly was not exclusive to the Soviets, but other governments thought it less crucial as a conservation measure. See for example, Birger Birgerson letter to Remington Kellogg, October 28, 1946, in box 1, Remington Kellogg Papers Record Unit 074343, Smithsonian Archives.

57. Vakhov, *Tragedia*, 61.

58. See Krupnik, *Arctic Adaptations*, 76–81.

59. See, among many, the analysis of the whale in New Zealand fiction in Phillip Armstrong, "The Whale Road," in *A New Zealand Book of Beasts: Animals in Our Culture, History, and Everyday Life*, ed. Phillip Armstrong and Annie Potts (University of Auckland Press, 2014).

60. For Japanese officials, see Jakobina Arch, "From Meat to Machine Oil: The Nineteenth-Century Development of Whaling in Wakayama," in *Japan at Nature's Edge: The Environmental Context of a Global Power*, ed. Ian Jared Miller, Julia Adeney Thomas, and Brett L. Walker (Hawaii University Press, 2013), 40, 41.

61. See Hiroyuki Watanabe, *Japan's Whaling: The Politics of Culture in Historical Perspective* (Trans Pacific Press, 2009).

62. For an early example of agonizing about the sufferings of whales, see Frank Bullen, "Humpbacking in the Friendly Islands," *Colonial Good Words* 36 (1897): 628.

63. "Pis'ma Finliandskago matrosa," 83, 86.

64. See Nosilov, "Na kitoboinom parakhode," 27.

65. Nosilov, "Na kitoboinom parakhode," 20.

66. Maksimov, "Kitoboi," 7.

67. Otto Lindholm, "Kitoboinoi promysel," 29.

68. See Lajus (this volume).

69. Zenkovich, *Vokrug sveta*, chapter 7.

70. Vakhov, *Fontany na Gorizonte*, 276.

71. Vakhov, *Tragedia*, 56, 60.

72. Solyanik, *Cruising*, 32. This essay cannot deal with professional Soviet science, which was directed at a different audience but also probed with unusual interest into whale behavior, especially compared with Western cetology of the 1950s to 1970s, which had mostly abandoned such questions for being insufficient in scientific rigor. For trends in Western cetology, see Amy Samuels and Peter Tyack, "Flukeprints: A History of Studying Cetacean Societies," in Janet Mann et al., eds., *Cetacean Societies: Field Studies of Dolphins and Whales* (University of Chicago Press, 2000).

73. For example, Zenkovich, *Vokrug sveta*, chapter 10.

74. Arsen'ev and Zemskii, *V strane kitov*, 98.

75. For Western whaling fiction, see Mary McCallum's explanation for early twentieth-century New Zealand whalers; Mary McCallum, *The Blue* (Penguin Books, 2007). For the Japanese, see Arch, "Meat to Machine Oil," 45.

76. For more on Russian hunting culture, see Ian Helifant, "The Savage Gaze: The Contested Portrayal of Wolves in Nineteenth-Century Russia," in Jane Costlow and Amy Nelson, eds., *Other Animals: Beyond the Human in Russian Culture and History* (Pittsburgh University Press, 2010).

77. Arsen'ev and Zemskii, *V strane kitov*, 80.

78. Arsen'ev and Zemskii, *V strane kitov*, 105.

79. Arsen'ev and Zemskii, *V strane kitov*, 102.

80. P. A. Zarva, "'Aleut'—'Slava,'" in *Slava: Zapiski sovetskikh kitoboev* (Odesskoe oblastnoe izd-vo, 1952), 88.

81. Zenkovich, *Vokrug sveta*, chapter 1.

82. Arsen'ev and Zemskii, *V strane kitov*, 81.

83. Vakhov, *Tragedia*, 27.

84. William Husband, "'Correcting Nature's Mistakes': Transforming the Environment and Soviet Children's Literature, 1928–1941," *Environmental History* 11 (April 2006): 303, 312.

85. Zenkovich, *Vokrug sveta*, chapter 10.

86. Vakhov, *Shtorm*, 10.

87. Berzin, "Soviet Whaling," 4.

Chapter 15: Strengthening the Tsarist Empire's Immune System

1. The society was founded in 1877 in St. Petersburg with the objective of safeguarding public health and promoting the improvement of public health and sanitation. It was among the first Russian scientific societies with a hygienic orientation. See F. T. Shtangeev, *O lecheniie i rezhime pri chakhotke* (Yalta, 1898).

2. E. Elenin, *Ialta i Okrestnosti* (Tip. Vakhtina, 1901–1902), 17. See also F. T. Shtangeev, *O lecheniie i rezhime pri chakhotke* (Yalta, 1898).

3. Dym, "Berega Zdorov'ia," *Krymskii kurortnyi listok*, May 8, 1911.

4. From antiquity through to the nineteenth century, medical theorists proposed and navigated between two underlying theories of etiology. On the one hand, physicians had long recognized many diseases as contagious, having recognized transmission from person to person through touch, the air, objects, or animals. This represents the contagionist viewpoint. On

the other hand, since Hippocrates and Galen wrote basing their thought on humoral theories, disease had also been understood as the product of local circumstances, individual predisposition, and particular environmental factors. This represents the localist/environmentalist stance. Considerations grew increasingly sophisticated as the nineteenth century progressed, with most physicians falling somewhere between strict contagionists and strict localists. For a nuanced and sophisticated discussion of broader European discussions of etiology and policies derived from those discussions, see Peter Baldwin, *Contagion and the State in Europe 1830–1930* (Cambridge University Press, 1999), 2–4.

5. For an extended introduction to medical geography, see Converey Bolton Valencius, "Histories of Medical Geography," in *Medical Geography in Historical Perspective*, ed. Nicholaas A. Rupke (The Wellcome Trust Centre for the History of Medicine at UCL, 2000), 19.

6. Baldwin, *Contagion and the State*, 43–44; John Davis, *Russia in the Time of Cholera* (I. B. Tauris, 2018); Charlotte Henze, *Disease, Health Care and Government in Late Imperial Russia: Life and Death on the Volga, 1823–1914* (Routledge, 2011).

7. See, for one example, Richard Harrison Shyrock, *The Development of Modern Medicine: An Interpretation of the Social and Scientific Factors Involved* (Victor Gollanz, 1948).

8. Linda Nash, *Inescapable Ecologies: A History of Environment, Disease, and Knowledge* (University of California Press, 2006).

9. On the literature and the exotic in the Caucasus, see Susan Layton, *Russian Literature and Empire: The Conquest of the Caucasus from Pushkin to Tolstoy* (Cambridge University Press, 1994). On the Black Sea, see Charles King, *The Black Sea: A History* (Oxford University Press, 2004).

10. Unlike cholera, tuberculosis has been understudied in English language literature on the tsarist era, despite some reports naming it the cause of 25 percent of Russian deaths at the time. On the history of tuberculosis in western Europe and the United States, see Sheila M. Rothman, *Living in the Shadow of Death: Tuberculosis and the Social Experience of Illness in American History* (HarperCollins, 1994); F. B. Smith, *The Retreat of Tuberculosis, 1850–1950* (Croom Helm, 1988); David S. Barnes, *The Making of a Social Disease: Tuberculosis in Nineteenth Century France* (University of California Press, 1995).

11. N. P. Ivanov, *Russkii al'manakh po otechestvennym vodam, morskim kupan'iam, sanitarnym stantsiiam i dr. lechebnym mestam Rossii* (F. V. Shchepanskago, 1894), 4.

12. N. Golovkinskii, *Putevoditel' po Krymu* (Tip. Spiro, 1894), 46.

13. K. A. Belilovskii, *G. Yalta: sanitarnyi ocherk* (Tavricheskaia Gubernskaia Tip., 1912), 3.

14. *Prakticheskii spravochnik "V kurortakh kryma" izdatel'stvo Tavrida 1909–1910g* (Tip. S. M. Brun, 1909), 4–5.

15. V. I. Kosarev, *Ialta: Gorno-primorskaia stantsiia: Kratkii ocherk sostoianiia sanit. Dela v ialte* (Tip. I. I. Rogovenko, 1913), 3.

16. N. Pravdin, *Alupka: Prakticheskii putevoditel'-spravochnik i flora Vorontsovskago parka* (Tip. N. P. Lupandinnoi, 1913), 4.

17. K. Iu. Bumber et al., *Krym: putevoditel'* (Krymskoe ob-vo estestvoispytatelei i liubitelei prirody, 1914), 319–20.

18. Ivanov, *Al'manakh*, 93–98.

19. Ivanov, *Al'manakh*, 103. Dry winds from the northwest were also known as *krymskii sprokko*.

20. Bumber et al., *Krym*, 327–32.

21. M. P. Ogranovich, *Proekt ustroistvo sanitarnoi stantsii barachnoi systeme v Krymu* (Tip. ObshchRasprPolezKnig, 1884), 1.

22. *Prakticheskii spravochnik*, 1.

23. Ivanov, *Al'manakh*, 1.

24. Ivanov, *Al'manakh*, 2.

25. N. P. Dmitriev, *Morskoe kupan'e v Ialte i voobshche na iuzhnom beregu Kryma* (Tip. Tsentr, 1895).

26. Dmitriev, *Morskoe kupan'e*, 36–40.
27. Dmitriev, *Morskoe kupan'e*, 28.
28. Dmitriev, *Morskoe kupan'e*, 38.
29. Dmitriev, *Morskoe kupan'e*, 40.
30. Dmitriev, *Morskoe kupan'e*, 28.
31. *Sputnik po gorodu Evpatoriia* (Izd. Gorodskoi Upravy, 1916), 3.
32. Dmitriev, *Morskoe kupan'e*, 47.
33. "Kurortnyia Nuzhdy: Solnechnoe lechenie v Krymu," *Russkaia Riv'era*, July 17, 1909.
34. Golovkinskii, *Putevoditel'*, 59.
35. Ivanov, *Al'manakh*, 130. See also V. K. Vinberg, *Prakticheskoe rukovodstvo vinogradstva i vinodeliia* (A. F. Debriena, 1896).
36. Iu. V. Vasil'chikov, *Ialta i blizhaishiiā okresnosti: putevoditel'-spravochnik 1911–1912 g.* (Tip. Lupandinnoi, 1911). Medical reports on grape therapy were often republished, word for word, in local newspapers. On grape treatments, see "Kurortnye Nuzhdy," *Russkaia Riv'era*, August 22, 1909.
37. V. K. Vinberg, *Prakticheskoe rukovodstvo vinogradstva i vinodeliia* (A. F. Debriena, 1896). See also Pravdin, *Alupka*; Vasil'chikov, *Ialta*.
38. Ivanov, *Al'manakh*, 121–22.
39. Golovkinskii, *Putevoditel'*, 60–61.
40. Ivanov, *Al'manakh*, 121–28.
41. Golovkinskii, *Putevoditel'*, 60–61.
42. Ivanov, *Al'manakh*, 129.
43. Z. B. Kaganovich, *Noveishii putevodil' po Krymu na 1913 god* (Tip. Kh. Vulikhmana, 1913), 81–82; "Evpatoriia kurort: Moinakskaia griazelechebnitsa," *Evpatoriiskii vestnik*, March 18, 1909.
44. Andrei Malgin, *Russkaia riv'era: Kurorty, turizm, i otdykh v Krymu v epokhu Imperii* (SONAT, 2006), 285.
45. "O sostave i svoistve tselebnikh griaz i limanov," *Evpatoriiskii vestnik*, July 11, 1909.
46. Yu. A. Morgulis, *Sakskye griazy: Putevoditel' dlia iedushchikh lechit'sia v saki* (Tip. Spiro, 1889), 23–24.
47. Bariatinskaia's sanatoria speaks to the ways in which charity operated in tsarist Russia as the responsibility of both local governments (particularly the zemstvo) and private individuals. Even if the sanatorium was perceived as being funded by a private individual, it is clear that public institutions on multiple levels contributed as well. This was not an unusual situation. See Adele Lyndenmeyr, *Poverty Is Not a Vice: Charity, Society, and the State in Imperial Russia* (Princeton University Press, 1996).
48. *Putevoditel'-spravochnik: Ialta i blizhaishiia okrestnosti, II izd.* (Tip. N. P. Lupandinnoi, 1913), 84–85.
49. *Putevoditel'-spravochnik: Ialta*, 68–72.
50. *Almanakh-putevoditel' po otechestvennym kurortam i lechebnym mestam* (V. M. Strunskago, 1915–1916), 103.
51. Ronald L. Numbers, "The Significance of Regions in American Medical History," in *Disease and Medical Care in the Mountain West: Essays on Region, History, and Practice*, ed. Martha L Hildreth and Bruce T. Moran (University of Nevada Press, 1998).
52. Rothman, *Shadow of Death*, 132.
53. Numbers, "Regions," 8–9.
54. Rothman, *Shadow of Death*, 131–60.

Chapter 16: Reshaping the Land, Chasing the Mosquito

The author wishes to thank Alla Mikhailovna Baranova, Yusuf Nuralievich Nuraliev, Viktor Dubovitsky, and Hamroh Fozylov. Thanks as well to the staff of Tajikistan's central Ministry of Health library and the Central State Historical Archive (Boygonii Markazii Dawlatii Jumhurii Tojikiston); to Khudoyar, archivist of Tajikistan's Academy of Sciences archive; and to the librarian at the Moscow Tropical Institute, today the Martsinovskii Institute of Parasitology and Tropical Medicine. Support for this research came from the National Council for Eurasian and East European Research. Saifudin S. Karimov in the Republic of Tajikistan's Ministry of Health and Vladimir P. Sergiev at Moscow's Martsinovskii Institute were very forthcoming with institutional support and permissions, for which I am grateful.

1. Geographical names in this paper are generally reproduced as in the original source (often the Russian equivalent), sometimes with the current or original Tajik version of the name in parentheses. Names familiar in the West such as Amu Darya and Bukhara are given only in their widely used forms.

2. According to a 1934 report, malaria incidence reached 75 percent among Red Army units serving in Eastern Bukhara in 1922, and mortality from malaria in the army ranged from 10 to 25 percent. See A. N. Guterts, "Lechenie plazmotsidom sredneaziatskoi maliarii," *Za sotsialisticheskoe zdravookhranenie Uzbekistana* 7 (1934): 11–14. One historian of public health cites the following rates of malaria infection in 1925–1926 among the general population in several specific locations in Tajikistan: Kulob, 88 percent; Muminabad, 96 percent; Khovaling, 86 percent; Bol'dzhuan, 90 percent; Gissar (Hisor), 94 percent. He claims that 80 percent of the cases were the tropical (*falciparum*) form, which could have a mortality rate of up to 30 percent. See Ia. T. Tadzhiev, *Zdravookhranenie Tadzhikistana* (Tip. Gosmedinst., 1974).

3. The notion of the revolution "betrayed" originated with Leon Trotsky, but coming from a very different political perspective Nicholas Timasheff came to a similar verdict with his *Great Retreat: The Growth and Decline of Communism in Russia* (E. P. Dutton, 1946). See also Moshe Lewin, *The Making of the Soviet System: Essays in the Social History of Interwar Russia* (Methuen, 1985). For a recent work that argues credibly for continuity and that points to a common cultural ecosystem from which both early Bolshevik and Stalinist culture emerged, see David L. Hoffmann, *Stalinist Values: The Cultural Norms of Soviet Modernity* (Cornell University Press, 2003).

4. On the return of malaria to Tajikistan beginning in the 1980s and extending into the troubled period following the collapse of the USSR, see V. P. Sergiev et al., "Importation of Malaria into the USSR from Afghanistan, 1981–89," *Bulletin of the World Health Organization* 71, no. 3–4 (1993): 385–88; A. Ia. Lysenko, A. V. Kondrashin, and M. N. Ezhov, *Maliariologiia*, 2nd ed. (Otkrytye sistemy, 2003); S. S. Karimov, T. A. Sherkhonov, and D. S. Saiburkhonov, "*Maliariia v Tadjikistane: ot bor'by k eliminatsii* [Malaria in Tajikistan: From Control to Elimination]," *Meditsinskaia parazitologiia i parazitarnye bolezni* (2009): 23–26; B. Matthys, T. Sherkanov, S. S. Karimov, et al., "History of Malaria Control in Tajikistan and Rapid Malaria Appraisal in an Agro-ecological Setting," *Malaria Journal* 7 (2008): 217.

5. Treatments of the history of malaria in other geographical contexts and more globally have applied this approach with success, including James L. A. Webb Jr., *Humanity's Burden: A Global History of Malaria* (Cambridge University Press, 2008).

6. I have in mind here an approach that stems from Donald Worster's observations in "Doing Environmental History," in Donald Worster and Alfred W. Crosby, eds., *The Ends of the Earth: Perspectives on Modern Environmental History* (Cambridge University Press, 1988). Worster reminds us that we must not overlook plant matter (and the human agricultural activity related to it) when we address environmental history, noting that we have more before us than just the land and the human and animal life upon it. "Where people and vegetation come together more issues in environmental history cluster than anywhere else," Worster notes. I would argue that in the case of malaria control we have a similar situation submerged in the water, where if we take away plant matter, our perspective on the environmental underpinnings of malaria lose their anchor and their coherence.

7. Cited in E. S. Kalmykov, "Landshaftno-maliariologicheskoe raionirovanie Tadzhikskoi SSR v tseliakh ratsionalizatsii protivomaliariinykh meropriiatii" (PhD diss., Dushanbe Institute of Epidemiology and Hygiene, Ministry of Health, Tajik SSR, 1966), 247–48.

8. A. A. Makar'in, "Postanovka bor'by s maliarii v b. Vostochnoi Bukhare (Tadzhikistan) v

napravlenii protivomaliariinykh meropriiatii v 1924–27gg.," in *Zhivotnye parazity i nekotorye parazitarnye bolezni cheloveka v Tadzhikistane*, ed. E. N. Pavlovskii (Tip. AN SSSR, 1929), 15.

9. For an overview of malaria in its changing environmental context over time, see Webb, *Humanity's Burden*. Randall M. Packard, *The Making of a Tropical Disease: A Short History of Malaria* (Johns Hopkins University Press, 2007), contains a useful survey of malaria control methods in the modern period. Sonia Shah presents new ways of viewing humans' relationship to malaria over the *longue durée* in *The Fever: How Malaria Has Ruled Humankind for 500,000 Years* (Farrar, Straus, and Giroux, 2010).

10. It does bear noting that Russian medical thought, even into the early Soviet period, involved a continued reliance on environmental etiological thinking. This was in line with the thinking highlighted by George Lywood (this volume), where he shows that Russian medicine in the late imperial period was characterized by a strong environmentalist bent, even after the bacteriological revolution broadly took hold. I would not argue that this marks Russian medical thought as exceptional, however. Adherence to a contagionist or an environmentalist viewpoint about the causes of infectious disease has never been an all or nothing affair. Even in contemporary medicine, we see the remnants of this in our ability to recognize microorganisms as the root cause of infectious disease, even while we also acknowledge that a range of other factors can be at play in facilitating or exacerbating infection. For an interesting analysis of how environment and disease were linked in medical thought and writing about California from the late nineteenth into the late twentieth century, see Linda Nash, *Inescapable Ecologies: A History of Environment, Disease, and Knowledge* (University of California Press, 2006).

11. On the limits that geography, impassibility, and *bezdorozh'e* (lack of roads) placed on Soviet officials in Tajikistan, see Botakoz Kassymbekova, "Humans as Territory: Forced Resettlement and the Making of Soviet Tajikistan, 1920–38," *Central Asian Survey* 30, no. 3–4 (2011): 349–70.

12. Cited in Kalmykov, "Landshaftno-maliariologicheskoe raionirovanie," 247–48.

13. Willi Rickmer Rickmers, "Travels in Bokhara," *Geographical Journal* 14, no. 6 (December 1899): 596–617.

14. See especially the essays in David L. Hoffmann and Yanni Kotsonis, eds., *Russian Modernity: Politics, Knowledge, Practices* (Palgrave Macmillan, 2000); Paul R. Josephson, *Industrialized Nature: Brute Force Technology and the Transformation of the Natural World* (Island Press, 2002); James C. Scott, *Seeing like a State: How Certain Schemes to Improve the Human Condition Have Failed* (Yale University Press, 1998). For a view of efforts to radically transform the land and waterscape of Germany over the course of the modern period and the ideas that underpinned those efforts, see David Blackbourn, *The Conquest of Nature: Water, Landscape, and the Making of Modern Germany* (W. W. Norton, 2006).

15. Two good examples of such programs with much in common with Soviet malaria control in Tajikistan, especially in its Red Army origins, are the US military-led campaigns to control mosquito-borne yellow fever in Cuba and the campaign to control yellow fever and malaria in the Panama Canal Zone, both led by William Gorgas. See Packard, *Making of a Tropical Disease*, 111–49; James L. A. Webb Jr., "*Aedes aegypti* Suppression in the Americas: Historical Perspectives," *The Lancet* 388, no. 10044 (2016): 556–57. The campaigns were surveyed by one of Gorgas's key lieutenants in Joseph A. Le Prince, *Mosquito Control in Panama: The Eradication of Malaria and Yellow Fever in Cuba and Panama* (G. P. Putnam's Sons, 1916).

16. L. F. Paradoksov, "Nekotorye dannye ob istorii zdravookhraneniia i zabolevanii naseleniia Tadzhikistana," *Zdravookhranenie Tadzhikistana* 1, no. 1 (1933): 19.

17. Hélène Carrere d'Encausse, "Civil War and New Governments," in *Central Asia: 130 Years of Russian Dominance, a Historical Overview*, ed. Edward Allworth, 3rd ed., Central Asia Book Series (Duke University Press, 1994).

18. Makar'in, "Postanovka," 13–27.

19. Kalmykov, "Landshaftno-maliariologicheskoe raionirovanie," 250.

20. For the early plans to form the Tropical Institute in 1921 in Moscow and gear up to

combat malaria in various locations in Russia and other Soviet territories, see "Perepiska s gubernskimi otdelami Nar obrazovaniia ob otpuske khinina, o vyrabotke plana meropriiatiia po bor'be s maliariei," January–December 1921, Gosudarstvennyi arkhiv Rossiiskoi Federatsii (State Archive of the Russian Federation) f. A-482, op. 4, d. 354, ll. 2, 51), inter alia.

21. Mark Popovskii, "Tot, kto sporil. Povest' o Leonide Isaeve," *Vivos Voco*, online at http://vivovoco.astronet.ru/VV/PAPERS/BIO/ISAEV/CONTENT.HTM/.This story is based on M. Popovskii, *Liudi sredi liudei: Povesti* (Detskaia literatura, 1972).

22. Makar'in, "Postanovka," 15.

23. Makar'in, "Postanovka," 14.

24. Makar'in, "Postanovka," 14.

25. Paradoksov, "Nekotorye dannye," 23.

26. Tadzhiev, *Zdravookhranenie*.

27. Tadzhiev, *Zdravookhranenie*; E. N. Pavlovskii, ed., *Zhivotnye parazity i nekotorye parazitarnye bolezni cheloveka v Tadzhikistane* (Tip. AN SSSR, 1929).

28. The approach had much in common with Colonel Gorgas's campaigns in Havana and Panama, and it is likely that these represented a model for the Red Army, although no direct reference to these examples was made in the sources.

29. Paris green, or copper acetoarsenite, is a chemical compound that was used widely as an insecticide at the turn of the twentieth century. It is highly toxic and takes the form of a bright green powder, and for this reason it also served as a pigment popular among artists in the nineteenth century. Paris green was commonly used in efforts to combat the mosquito vector of malaria and other insects and pests, from rats to potato bugs, in the early twentieth century, before the development and widespread adoption of DDT.

30. *Anopheles superpictus*, *An. claviger*, *An. artemievi*, and *An. pulcherrimus* appear to have been the most prevalent species of malaria-transmitting mosquito in Tajikistan. Zamonidin Habirov et al., "Malaria and the Anopheles Mosquitoes of Tajikistan," *Journal of Vector Ecology* 37, no. 2 (2012): 419–27.

31. Makar'in, "Postanovka"; I. I. Masaitis, "Protivomaliariinye raboty v g. Kuliabe (Tadzhikistan)," in *Zhivotnye parazity i nekotorye parazitarnye bolezni cheloveka v Tadzhikistane*, ed. E. N. Pavlovskii (Tip. AN SSSR, 1929), 28–39.

32. For more on the origins of the debate between a so-called medical approach and a technological approach, and how the debate evolved in the international sphere and Europe in particular, see Hughes Evans, "European Malaria Policy in the 1920s and 1930s: The Epidemiology of Minutiae," *Isis* 80, no. 1 (March 1989): 40–59.

33. A. Shchurenkova, "Problema nomer odin. (O bor'be s maliariei v tadzhikistane v 30e gody: Vospominanii. Zapisala G. Vanina.)," *Pamir* 1 (1987): 124–25.

34. For the ideas behind such revolutionary reform and the activity included under its umbrella in various spheres, see inter alia Richard Stites, *Revolutionary Dreams: Utopian Vision and Experimental Life in the Russian Revolution* (Oxford University Press, 1989); Elizabeth A. Wood, *The Baba and the Comrade: Gender and Politics in Revolutionary Russia* (Indiana University Press, 2000); Glennys Young, *Power and the Sacred in Revolutionary Russia: Religious Activists in the Village* (Pennsylvania State University Press, 2008). On the forms this took in public health in particular, see Tricia Starks, *The Body Soviet: Propaganda, Hygiene, and the Revolutionary State* (University of Wisconsin Press, 2009); Frances Lee Bernstein, *The Dictatorship of Sex: Lifestyle Advice for the Soviet Masses* (Northern Illinois University Press, 2007). For the forms this took in Central Asia, see Douglas Taylor Northrop, *Veiled Empire: Gender and Power in Stalinist Central Asia* (Cornell University Press, 2004); Paula A. Michaels, *Curative Powers: Medicine and Empire in Stalin's Central Asia* (University of Pittsburgh, 2003).

35. Makar'in, "Postanovka," 19.

36. L. F. Paradoksov, "10 let zdravookhraneniia v iuzhnom Tadzhikistane (iz vospominanii vracha)," *Zdravookhranenie Tadzhikistana* 2, no. 2 (1936): 5–9.

37. Boygonii Markazii Dawlatii Jumhurii Tojikiston (Central State Archive of the Republic of Tajikistan; hereafter cited as BMDJT) f. R172, op. 1, d. 28, ll. 12–13.

38. BMDJT f. R172, op. 1, d. 28, l. 3.

39. Tadzhiev, *Zdravookhranenie*; M. N. Keshish'ian, "Organizatsiia bor'by s maliariei v TadzhSSR," *Zdravookhranenie Tadzhikistana* 1, no. 1 (1933): 11–17.

40. See many correspondence and reporting documents in BMDJT f. R172, op. 1 (Narkomzdrav TASSR), and f. 279, op. 6 (Narkomzdrav TSSR).

41. Makar'in, "Postanovka," 25.

42. Keshish'ian, "Organizatsiia bor'by," 11.

43. BMDJT f. R172, op. 1, d. 28, l. 29.

44. U. M. Muminov, "Maliariia v Tadzhikistane," *Zdravookhranenie Tadzhikistana* 3, no. 1 (1937): 3.

45. See L. J. Bruce-Chwatt, "Malaria Research and Eradication in the USSR: A Review of Soviet Achievements in the Field of Malariology," *Bulletin of the World Health Organization* 21 (1959): 737–72.

46. P. G. Sergiev and N. N. Dukhanina, *Malaria in the Union of Socialist Soviet Republics and Planning Measures of Malaria Prophylaxis* (World Health Organization, 1966).

Chapter 17: Conclusions

1. Kenneth Pomeranz, "Introduction: World History and Environmental History," in Edmund Burke III and Kenneth Pomeranz, eds., *The Environment and World History* (University of California Press, 2009), 3–32.

2. John F. Richards, *The Unending Frontier: An Environmental History of the Early Modern World* (University of California Press, 2003); Joseph Fletcher, "Integrative History: Parallels and Interconnections in the Early Modern Period, 1500–1800," *Journal of Turkish Studies* 9 (1985): 37–57; Victor Lieberman, ed., *Beyond Binary Histories: Re-imagining Eurasia to c. 1830* (University of Michigan Press, 1999); Victor Lieberman, *Strange Parallels: Southeast Asia in Global Context, c. 800–1300, Vol. 1: Integration on the Mainland*, and *Vol. 2: Mainland Mirrors: Europe, Japan, China, South Asia and the Islands* (Cambridge University Press, 2003, 2009).

3. H. R. Fisher, *The Russian Fur Trade, 1550–1700* (University of California Press, 1943); Janet Martin, *Treasure of the Land of Darkness: The Fur Trade and Its Significance for Medieval Russia* (Cambridge University Press, 1986); Douglas R. Weiner, "The Predatory Tribute-Taking State: A Framework for Understanding Russian Environmental History," in Burke and Pomeranz, *Environment and World History*, 276–315; Alexander Etkind, "Barrels of Fur: Natural Resources and the State in the Long History of Russia," *Journal of Eurasian Studies* 2 (2011): 164–71, esp. 167.

4. Joseph Stalin, "Speech to Industrial Managers, February 1931," *Pravda*, February 5, 1931, in J. V. Stalin, *Problems of Leninism* (Foreign Languages Publishing House, 1953), 454–58.

5. As classically developed in Jürgen Habermas, *The Structural Transformation of the Public Sphere: An Inquiry into a Category of Bourgeois Society*, trans. Thomas Burger (MIT Press, 1989).

6. Alexander Etkind, *Internal Colonization: Russia's Imperial Experience* (Polity, 2011).

7. Alfred Crosby, *Ecological Imperialism: The Biological Expansion of Europe, 900–1900* (Cambridge University Press, 1986).

8. Literature on the nature and process of Russian expansion and settlement is extensive. For a study of the hopes pinned on development of lands in the Russian Far East, see Mark Bassin, *Imperial Visions: Nationalist Imagination and Geographical Expansion in the Russian Far East, 1840–1865* (Cambridge University Press, 1999). On the Pontic Steppe of southern Ukraine, see Willard Sunderland, *Taming the Wild Field: Colonization and Empire on the Russian Steppe* (Cornell University Press, 2004) and David Moon, *The Plough that Broke the Steppes: Agriculture and Environment on Russia's Grasslands, 1700–1914* (Oxford University Press, 2013). For the Caucasus, see Nicholas Breyfogle, *Heretics and Colonizers: Forging Russia's Empire in the South Caucasus* (Cornell University Press, 2005); Thomas Barrett, *At the Edge of Empire:*

The Terek Cossacks and the North Caucasus Frontier, 1700–1860 (Westview Press, 1999); Austin Jersild, *Orientalism and Empire: North Caucasus Mountain Peoples and the Georgian Frontier, 1845–1917* (McGill-Queen's University Press, 2002).

9. On the eastern steppes and Central Asia, see Michael Khodarkovsky, *Russia's Steppe Frontier: The Making of a Colonial Empire* (Indiana University Press, 2002); Jeff Sahadeo, *Russian Colonial Society in Tashkent, 1865–1923* (Indiana University Press, 2007); Alexander Morrison, *Russian Rule in Samarkand, 1868–1910: A Comparison with British India* (Oxford University Press, 2008); Ihor Stebelsky, "The Frontier in Central Asia," in *Studies in Russian Historical Geography*, ed. James H. Bater and R. A. French (Academic Press, 1983), 1:143–74. Finally, a key discussion of these issues is contained in the anthology edited by Nicholas Breyfogle, Abby Schrader, and Willard Sunderland, *Peopling the Russian Periphery: Borderland Colonization in Eurasian History* (Routledge, 2007) and the forum on "Colonialism and Technocracy at the End of the Tsarist Era," especially Willard Sunderland, "The Ministry of Asiatic Russia: The Colonial Office that Never Was but Might Have Been," *Slavic Review* 69, no. 1 (2010): 120–50, and Peter Holquist, "'In Accord with State Interests and the People's Wishes': The Technocratic Ideology of Imperial Russia's Resettlement Administration," *Slavic Review* 69, no. 1 (2010): 151–79.

10. William Cronon, *Changes in the Land: Indians, Colonists, and the Ecology of New England* (Hill and Wang, 1983); Eric R. Wolf, *Europe and the People without History* (University of California Press, 1982).

11. Charles Halperin, *Russia and the Golden Horde: The Mongol Impact on Medieval Russian History* (Indiana University Press, 1985); Donald Ostrowski, *Muscovy and the Mongols: Cross-cultural Influences on the Steppe Frontier, 1304–1589* (Cambridge University Press, 1998); Janet Martin, *Medieval Russia, 980–1584* (Cambridge University Press, 1993); Martin, *Treasure of the Land*; George Vernadsky, *A History of Russia, Vol. 3: The Mongols and Russia* (Yale University Press, 1943); Lev Nikolaevich Gumilev, *Drevniaia Rus' i Velikaia step'* (Mysl,' 1989).

12. See Albert Seaton, *The Horsemen of the Steppes: The Story of the Cossacks* (Bodley Head, 1985).

13. For a discussion of this, see Mikail Mamedov, "'Going Native' in the Caucasus: Problems of Russian Identity, 1801–1864," *Russian Review* 67 (April 2008): 275–95; Willard Sunderland, "Russians into Iakuts? 'Going Native' and Problems of Russian National Identity in the Siberian North, 1870s–1914," *Slavic Review* 55 (Winter 1996): 806–25; Thomas Barrett, *At the Edge of Empire: The Terek Cossacks and the North Caucasus Frontiers, 1700–1860* (Westview Press, 1999).

14. Such a wide-ranging landscape history has yet to be written. S. V. Kirikov, in his *Izmeneniia zhivotnogo mira v prirodnykh zonakh SSSR (XIII–XIX vv.)* (Izd-vo AN SSSR, 1959), analyzes changes in distribution and numbers of fauna from the fourteenth century through the nineteenth century, and V. K. Rakhilin has written a broad-brush overview in his *Obshchestvo i zhivaia priroda: Kratkii ocherk istorii vzaimodeistviia* (Nauka, 1989).

15. William Cronon, *Nature's Metropolis: Chicago and the Great West* (W. W. Norton, 1991); John R. McNeill, *The Mountains of the Mediterranean World: An Environmental History* (Cambridge University Press, 1992); Simon Schama, *Landscape and Memory* (Vintage, 1995).

16. David Schimmelpenninck van der Oye, *Russian Orientalism: Asia in the Russian Mind from Peter the Great to the Emigration* (Yale University Press, 2010); Sunderland, "Ministry of Asiatic Russia"; Adeeb Khalid, "Russian History and the Debate over Orientalism," *Kritika* 1, no. 4 (2000): 691–99; Alexander Morrison, "Metropole, Colony, and Imperial Citizenship in the Russian Empire," *Kritika* 13, no. 2 (Spring 2012): 327–64.

17. On Russia and the history of firearms, see Kenneth Warren Chase, *Firearms: A Global History to 1700* (Cambridge University Press, 2003).

18. See Mamedov, "Going Native"; Alexander Morrison, "Introduction: Killing the Cotton Canard and Getting Rid of the Great Game: Rewriting the Russian Conquest of Central Asia, 1814–1895," *Central Asian Survey* 33, no. 2 (June 2014): 131–42.

19. Howard Lamar and Leonard Thompson, *The Frontier in History: North America and South Africa Compared* (Yale University Press, 1981); John C. Weaver, *The Great Land Rush and the Making of the Modern World, 1650–1900* (McGill/Queens University Press, 2003);William Beinart and Lotte Hughes, *Environment and Empire* (Oxford University Press, 2007); Thomas Dunlap, *Nature and the English Diaspora: Environment and History in the United States, Canada, Australia, and New Zealand* (Cambridge University Press, 1999); Richards, *Unending Frontier*. It is important to note that even among educated professionals serving the tsarist regime (the Soviet case is less well researched), there were attempts to respect and preserve local lifeways and autonomy despite the regime's drive for a takeover. See Ian W. Campbell, "Settlement Promoted, Settlement Contested: The Shcherbina Expedition of 1896–1903," *Central Asian Survey* 30, no. 3–4 (2011): 423–36.

20. Virginia Martin, "Kazakh Chingissids, Land and Political Power in the Nineteenth Century," *Central Asian Survey* 20, no. 1 (2010): 79–102; Paul W. Werth, "From Resistance to Subversion: Imperial Power, Indigenous Opposition, and Their Entanglement," *Kritika* 1, no. 1 (Winter 2000), 21–43.

21. John LeDonne, *The Grand Strategy of the Russian Empire, 1650–1831* (Oxford University Press, 2004); Stebelsky, "Frontier." For the US experience, see Andrew Isenberg, *The Destruction of the Bison: An Environmental History, 1750–1920* (Cambridge University Press, 2000).

22. Stephen C. Brain, "The Great Stalin Plan for the Transformation of Nature," *Environmental History* 15, no. 4 (October 2010): 1–31.

23. See especially Francine Hirsch, *Empire of Nations: Ethnographic Knowledge and the Making of the Soviet Union* (Cornell University Press, 2005); Robert Argenbright, "Vanguard of 'Socialist Colonization'? The Krasnyi Vostok Expedition of 1920," *Central Asian Survey* 30, no. 3–4 (2011): 437–54.

24. Certainly the feeling was in good measure mutual, as Muslim elites would be reticent to remain in situ in Central Asia and convert to the Christian faith. Yet, some Muslims converted to Christianity closer to Moscow, among the Volga Tatars. On these Kräshens, see Agnès Nilüfer Kefeli, *Becoming Muslim in Imperial Russia: Conversion, Apostasy and Literacy* (Cornell University Press, 2014).

25. Of course, the Great Game existed more in the overheated imagination of regimes than in actual strategic planning and probably played a much less important role than previously ascribed. On this, see Alexander Morrison, "'Nechto eroticheskoe,' 'courir après l'ombre': Logistical Imperatives and the Fall of Tashkent," *Central Asian Survey* 33, no. 2 (2014): 153–69; David Schimmelpennick van der Oye, "Paul's Great Game: Russia's Plan to Invade British India," *Central Asian Survey* 33, no. 2 (2014): 143–52.

26. Isabelle Ohayon, *La sédentarisation des Kazakhs dans le URSS de Stalin: collectivisation et changement social (1928–1945)* (Maisonneuve et Larose, 2006).

27. John Opie, *Ogallala: Water for a Dry Land* (University of Nebraska Press, 1993).

28. Ulrike Herzschuh, "Palaeo-moisture Evolution in Monsoonal Central Asia during the Last 50,000 years," *Quaternary Science Reviews* 25 (2006): 163–75; Fa-Hu Chen et al., "Moisture Changes over the Last Millennium in Arid Central Asia: A Review, Synthesis and Comparison with Monsoon Region," *Quaternary Science Reviews* 29 (2010): 1044–68; Ian Boomer et al., "Advances in Understanding the Late Holocene History of the Aral Sea Region," *Quaternary International* 194 (2009): 79–90; Philippe Sorrel, "Climate Variability in the Aral Sea Basin (Central Asia) during the Late Holocene Based on Vegetation Changes," *Quaternary Research* 67 (2007): 357–70; Jinsong Wang et al., "Characteristics of the Dry/Wet Trend over Arid Central Asia over the Past 100 Years," *Climate Research* 41 (2010): 51–59.

29. Vladimir Lenin, *Polnoe sobranie sochinenii* (Izd-vo politicheskoi literarury, 1963), 43:200, quoted in Teichmann (this volume).

30. James C. Scott, *Seeing like a State: How Certain Schemes to Improve the Human Condition Have Failed* (Yale University Press, 1998).

31. Teichmann (this volume). Philip Hanson, in *The Rise and Fall of the Soviet Economy: An Economic History of the USSR, 1945–1991* (Longman, 2003), makes the point that the Soviet economy could function efficiently only through the application of terror.

32. Julia Obertreis (this volume). Viktor Dukhovnyi's memoirs were published as *Zov vody* (Izd-vo Rudomino, 2006). Additional memoirs are contained in the collection *Ozhivshaia legenda. Vospominaniia veteranov osvoeniia Golodnoi Stepi* (Gulistan, 2002).

33. Jeff Sahadeo, "Cultures of Cotton and Colonialism: Politics, Society and the Environment in Central Asia, 1865–1923" (paper delivered at the AAASS [now ASEEES] Annual Convention, Toronto, Ontario, 2003). Muriel Joffe, "Autocracy, Capitalism, and Empire: The Politics of Irrigation," *Russian Review* 54, no. 3 (July 1995): 365–88, makes the point that the late tsarist regime also sought self-sufficiency in cotton.

34. David Kinkela, *DDT and the American Century: Global Health, Environmental Politics, and the Pesticide that Changed the World* (University of North Carolina Press, 2011).

35. Chris Otter, *The Victorian Eye: A Political History of Light and Vision in Britain, 1800–1910* (University of Chicago Press, 2008), 137.

36. For early Soviet environmental management, see Douglas R. Weiner, *Models of Nature. Conservation, Ecology and Cultural Revolution in Soviet Russia* (Indiana University Press 1988; 2nd ed., Pittsburgh University Press, 2000).

37. James Van Horn Melton, *The Rise of the Public in Enlightenment Europe* (Cambridge University Press, 2001); Margaret Jacob, "The Mental Landscape of the Public Sphere: A European Perspective," *Eighteenth-Century Studies* 28 (1994): 95–113; Edith W. Clowes, Samuel D. Kassow, and James L. West, eds., *Between Tsar and People: Educated Society and the Quest for Public Identity in Late Imperial Russia* (Princeton University Press, 1991); Joseph Bradley, *Voluntary Associations in Tsarist Russia: Science, Patriotism and Civil Society* (Harvard University Press, 2009); David Wartenweiler, *Civil Society and Academic Debate in Russia, 1905–1914* (Clarendon Press; Oxford University Press, 1999); Ilya V. Gerasimov, *Modernism and Public Reform in Late Imperial Russia: Rural Professionals and Self-Organization, 1905–30* (Palgrave Macmillan 2009); Vera Romanovna Leikina-Svirskaia, *Intelligentsia v Rossii vo vtoroi polovine XIX v.* (Mysl,' 1971); Vera Romanovna Leikina-Svirskaia, *Russkaia intelligentsia v 1900–1917 godakh* (Mysl,' 1981); Harley D. Balzer, ed., *Russia's Missing Middle Class: The Professions in Russian History* (M. E. Sharpe, 1996); Alfred B. Evans, Laura A. Henry, and Lisa Sundstrom, eds., *Russian Civil Society: A Critical Assessment* (M. E. Sharpe, 2006); Douglas R. Weiner, *A Little Corner of Freedom: Russian Nature Protection from Stalin to Gorbachev* (University of California Press, 1999).

38. Carl Bridenbaugh, "Baths and Watering Places of Colonial America," *William and Mary Quarterly*, 3rd ser., 2 (1946): 151–81; Tracy J. Revels, *Sunshine Paradise: A History of Florida Tourism* (University Press of Florida, 2011).

39. This literature had its origins in Crosby's *Columbian Exchange*; for one of the best of the recent work, see Marcy Norton, *Sacred Gifts, Profane Pleasures: A History of Tobacco and Chocolate in the Atlantic World* (Cornell University Press, 2008).

40. See especially George Konrád and Ivan Szelenyi, *The Intellectuals on the Road to Class Power*, trans. Andrew Arato and Richard E. Allen (Harcourt, Brace, Jovanovich, 1979).

41. Iurii Nikiforovich Gladkii, *Rossiia v labirintakh geograficheskoi sud'by* (R. Aslanov and Iuridicheskii tsentr Press, 2006).

42. Mike Davis, *Late Victorian Holocausts: El Niño Famines and the Making of the Third World* (Verso Books, 2002).

LIST OF CONTRIBUTORS

STEPHEN BRAIN is associate professor of history at Mississippi State University. He is the author of *Song of the Forest: Russian Forestry and Stalin's Environmentalism, 1905–1953* (University of Pittsburgh Press, 2011) and multiple articles on Russian/Soviet environmental history.

NICHOLAS BREYFOGLE is associate professor of history at The Ohio State University. The author or editor of five volumes, he specializes in environmental history, especially the human history of water. He is currently completing the book "Baikal: The Great Lake and Its People." For more than a decade he has served as founding editor of *Origins: Current Events in Historical Perspective* (http://origins.osu.edu/). He was a coinvestigator on the Leverhulme Trust International Network "Exploring Russia's Environmental History and Natural Resources" (2013–2016).

JOHN L. BROOKE is Warner Woodring Chair and Arts and Sciences Distinguished Professor of History and professor of anthropology at The Ohio State University (OSU). The director of the OSU Center for Historical Research (CHR), he co-chaired the 2011–2013 OSU-CHR Program on Health, Disease, and Environment in World History. His most recent book, *Climate Change and the Course of Global History: A Rough Journey*, was published by Cambridge University Press in 2014.

ANDY BRUNO is associate professor of history and environmental studies at Northern Illinois University. He is the author of *The Nature of Soviet Power: An Arctic Environmental History* (Cambridge University Press, 2016). With the support of a fellowship in aerospace history from NASA and the History of Science Society, he is currently conducting research on the history of the 1908 Tunguska explosion.

SARAH CAMERON is assistant professor of history at the University of Maryland–College Park. Her book, *The Hungry Steppe: Famine, Violence and the Making of Soviet Kazakhstan,* is forthcoming with Cornell University Press.

IAN CAMPBELL is associate professor of history at UC-Davis. In his first book, *Knowledge and the Ends of Empire* (Cornell University Press, 2017), he discusses the role of Kazakh intermediaries alongside tsarist scholars and bureaucrats in producing environmental and ethnographic knowledge about the steppe.

PEY-YI CHU received her PhD from Princeton University in 2011 and is currently assistant professor of history at Pomona College. She has held fellowships at the Davis Center for Russian and Eurasian Studies and the Rachel Carson Center for Environment and Society. Her work has appeared in *Environmental History* and *Environment and History*.

MARC ELIE is a researcher at the Centre d'étude des mondes russe, caucasien et centre-européen (Paris). He has published a book with Juliette Cadiot, *Histoire du Goulag* (La Découverte, 2017). He writes on the environmental history of the eastern steppes of western Siberia and northern Kazakhstan.

MIEKA ERLEY is assistant professor of Russian and Eurasian studies at Colgate University. She studies the cultural history of Russia and Central Asia in the imperial and Soviet periods, with a focus on the relationship between cultural production and knowledge production in the social and natural sciences. She is currently working on a book entitled "Black Soil, Red Earth: A Cultural History of Russian Soil." Her work has been published in *Slavic Review* and *Novoe literaturnoe obozrenie*.

RYAN TUCKER JONES is associate professor in the Department of History at the University of Oregon. His book, *Empire of Extinction: Russians and the North Pacific's Strange Beasts of the Sea, 1741–1867*, was published by Oxford University Press in 2014.

JULIA LAJUS is an associate professor in the Department of History, St. Petersburg School of Social Sciences and Humanities at the Higher School of Economics National Research University. She is a head of the Laboratory for Environmental and Technological History and academic director of the International Master's Program in Applied and Interdisciplinary History, "Usable Pasts." She has participated in several international projects, such as "History of Marine Animal Populations," ESF EUROCORES "Boreas: Histories from the North," "Assessing Arctic Futures: Voices, Resources and Governance," and "Exploring Russia's Environmental History," sponsored by the Leverhulme Trust. Currently she leads the project "Natural Resources in the History of Russia: Economic Institutes, Communities of Experts and Infrastructures," funded by the Russian Science Foundation. She has served as vice-president of the European Society of

Environmental History (2011–2015), and has coauthored and edited four books and more than twenty articles on environmental history and other subjects.

GEORGE LYWOOD received his PhD from The Ohio State University in 2012, completing the dissertation "Our Riviera, Coast of Health: Environment, Medicine, and Resort Life in Fin-de-Siècle Crimea." He is the author of "Stories from Crimea" (http://origins.osu.edu/connecting-history/51314-top-ten-origins-stories-crimea).

DAVID MOON is Anniversary Professor in History at the University of York, in England. He is the author of *The Plough that Broke the Steppes: Agriculture and Environment on Russia's Grasslands, 1700–1914* (Oxford University Press, 2013). In 2013–2016 he was lead investigator on the Leverhulme Trust International Network "Exploring Russia's Environmental History and Natural Resources" and in 2015–2017 he held a Leverhulme Trust Major Research Fellowship to investigate connections between the steppes and the Great Plains of North America.

PROF. DR. JULIA OBERTREIS holds the Chair of Modern and East European History at the University of Erlangen-Nürnberg in Germany. She has published on the history of everyday life, oral history, and environmental and infrastructure history of the Soviet Union and eastern Europe, in particular on Leningrad and Central Asia. She received a fellowship of the Freiburg Institute of Advanced Studies, School of History, in 2008–2009. Her most recent book is *Imperial Desert Dreams: Cotton Growing and Irrigation in Central Asia, 1860–1991* (V&R Unipress, 2017).

MARK SOKOLSKY received his PhD from The Ohio State University in 2016. His doctoral dissertation, "Taming Tiger Country: Colonization and Environment in the Russian Far East, 1860–1940," was completed with support from the Social Sciences and Humanities Research Council of Canada and The Ohio State University.

CHRISTIAN TEICHMANN received his PhD from Humboldt University, Berlin, where he worked as a lecturer until 2016. His book, *Macht der Unordnung: Stalins Herrschaft in Zentralasien, 1920–1950*, is a study of water management, the cotton economy, and Stalinist governance in Soviet Central Asia.

LISA K. WALKER received a PhD in European history from UC Berkeley in 2003. She has divided her time since then between research on the Soviet Union's experience of infectious disease and work in international development, with a focus on health and infectious and emerging diseases in Russia and Central and Southeast Asia. She currently works for the US Agency for International Development (USAID) in Phnom Penh, Cambodia.

DOUG WEINER is professor of history at the University of Arizona. He is the author of *Models of Nature: Ecology, Conservation and Cultural Revolution in Soviet Russia* (Indiana University Press, 1988) and *A Little Corner of Freedom: Russian Nature Protection from Stalin to Gorbachev* (University of California Press, 1999). He has most recently written "Communism and the Environment" for the *Cambridge Encyclopedia of World Communism* (Cambridge University Press, 2017), vol. 3.

INDEX